Averting Disaster Before It Strikes

Dmitry Chernov · Ali Ayoub ·
Giovanni Sansavini · Didier Sornette

Averting Disaster Before It Strikes

How to Make Sure Your Subordinates Warn You While There is Still Time to Act

Dmitry Chernov
Reliability and Risk Engineering
ETH Zürich
Zürich, Switzerland

Giovanni Sansavini
Chair of Reliability and Risk Engineering
ETH Zürich
Zürich, Switzerland

Ali Ayoub
Department of Nuclear Science
and Engineering
Massachusetts Institute of Technology
Cambridge, MA, USA

Didier Sornette
Southern University of Science
and Technology
Shenzhen, Guangdong, China

ISBN 978-3-031-30771-3 ISBN 978-3-031-30772-0 (eBook)
https://doi.org/10.1007/978-3-031-30772-0

This Springer imprint is published by the registered company Springer Nature Switzerland AG
The registered company address is: Gewerbestrasse 11, 6330 Cham, Switzerland

ACKNOWLEDGEMENTS

The authors of the handbook are grateful to the Foundation for Science of SCOR, one of the world's largest reinsurers, and ETH Zürich Foundation which provided the funding to conduct 100 interviews with the executives of critical infrastructure companies around the world and for the preparation of this handbook.

The authors are grateful to Pavel Zakharov, Euan Mearns, Igor Bevzenko, Sergey Titov, and Mikhail Guleykov for their help in the project behind this handbook.

The authors are also grateful to the senior management of the industrial company, the world's largest in its sector, for their desire to test the results of the study in the work of a real critical infrastructure company. The senior management helped to implement a pilot project at four of the company's production sites, involving more than 400 managers and employees, in order to prevent potential disasters.

The authors would like to thank Nick Onley and Tom Barber for patiently refining the drafts of the handbook, and Alexey Iorsh for the amusing cartoons that make the text a pleasure to read.

And finally, a special thank you to Dr. Hisako Niko, a senior editor at Springer, who followed the development of this project.

EXECUTIVE SUMMARY

This handbook is about how to transform the way large critical infrastructure companies communicate about safety and technological risks. It aims to support senior managers to get the information they need from their subordinates concerning the risks they are facing, in order to prevent accidents before it is too late.

The handbook is written for the owners, senior managers, and industrial safety directors of critical infrastructure companies. It is also relevant to consultants in the field of labor protection and industrial safety, specialists in the field of industrial risk insurance, and regulators of critical infrastructure facilities.

This summary will provide leaders with a brief overview of the book's main ideas. Furthermore, a detailed summary is offered for the relevant chapters of the book.

This handbook has several goals:

- to show that the problem with the prompt and accurate reporting of risks exists in many critical infrastructure companies and has been the cause, or one of the causes, of several major disasters at critical infrastructure facilities worldwide (Chapter 1);
- to elaborate the reasons why information about risks is concealed within large industrial companies — why subordinates hide the risks they can see in their area of competence from management, and why managers do not want to hear about the problems and risks their subordinates face (Chapter 2);
- to make practical recommendations to the owners and senior managers of critical infrastructure companies on how they can significantly improve intra-organizational risk communication (Chapter 3);
- to give a practical example of a pilot project to radically improve the quality and speed of risk information transmission in a world-leading industrial company (Chapter 4);
- finally, to discuss (I) the prospects for automating the collection of risk-related information when it comes to the operation of equipment in critical infrastructures, and the potential role of artificial intelligence in this endeavor, (II) the potential benefits and drawbacks of disclosing information about the critical risks of large industrial companies to insurance companies in exchange for lower insurance

premiums, and (III) the discernible variations in the way risk-related information is communicated in companies across different countries, cultures, and regions (Discussion section).

The recommendations in this handbook are based on interviews with 100 executives at various levels, working in 65 critical infrastructure companies around the world, in power, oil and gas, metals, chemicals and petrochemicals, mining and other industries.

The recommendations of these leaders were also tested in the pilot project, in an industrial company which is the world leader in its sector. More than 400 managers at various hierarchical levels and employees at several of the company's industrial plants took part in the pilot project.

Most importantly, the handbook explains what senior managers can do for improving the quality and speed of reporting about safety and technological risks within companies that operate critical infrastructure facilities. The handbook is intended to provide leaders of these companies with simple and practical solutions to overcome the problems of intra-organizational transmission of information about risks.

THE PROBLEM

After a major disaster, when investigators are piecing together the story of what happened, a striking fact often emerges: before disaster struck, some people in the organization involved were aware of dangerous conditions that had the potential to escalate to a critical level. But for a variety of reasons, this crucial information did not reach decision-makers. Therefore, the organization kept moving ever closer to catastrophe, effectively unaware of the possible threats. In the event of an accident, losses and costs for dealing with the consequences are often hundreds — or even thousands — of times greater than the finances that would have been required to deal with the risks when they were first recognized, and before they led to a major accident. Due to the asymmetry of risk information at different levels of the corporate hierarchy of critical infrastructure companies, preventive decisions were not taken in a timely manner. Ultimately, this led to the organizations facing catastrophic events. This observation has been documented in the following major technological accidents: Challenger space shuttle explosion (USA, 1986); Chernobyl nuclear power plant disaster (USSR, 1986); Sayano-Shushenskaya hydropower plant accident (Russia, 2009); Deepwater Horizon oil spill (USA, 2010); Fukushima-1 nuclear power plant disaster (Japan, 2011); methane explosions at American and Russian coal mines in the 2010s, and in several other disasters. Detailed information on the importance of this issue for many critical infrastructure companies worldwide is presented in Chapter 1.

WHY THE PROBLEM EXISTS

Chapter 2 takes a detailed look at the reasons why there is a problem with transmission of objective information about safety and technological risks in large critical infrastructure companies.

WHO CREATES AN INTERNAL CLIMATE WITHIN AN ORGANIZATION WHERE IT IS NOT ACCEPTABLE TO TALK ABOUT PROBLEMS?

97% of interviewees (97 out of 100 respondents) answered that most of the blame lies with managers. 2% of respondents argued that the responsibility is equally shared by managers and subordinates. 1% of respondents believed that the reasons for such an internal corporate atmosphere lay mostly in the personal qualities of individuals and their relationship with colleagues, and not in their organizational roles, whether manager or subordinate. None of the respondents placed the main responsibility on employees.

TOP 10 REASONS WHY LEADERS DO NOT WANT TO HEAR ABOUT PROBLEMS FROM THEIR SUBORDINATES

1. **Tackling reported problems will be costly, and owners and shareholders are imposing strict financial and production targets (58%: 58 out of 100 respondents).** Senior managers do not want to hear about problems from their subordinates because the costs of addressing any serious issue in a critical infrastructure company will be very high. In addition, owners and shareholders are often imposing strict financial and production targets on their senior managers already. Reports from employees about any serious safety and technological problem may threaten the implementation of these plans, as well as negatively affecting the career and the earnings of senior managers.
2. **Managers are afraid of being seen as incompetent if they take responsibility for previous bad decisions that have created current problems (38%).** When employees inform managers about any serious problem and risk, they are indirectly hinting towards the bad decisions and mistakes made by managers in the past that have led to the problem developing in the first place. Rather than admitting that they may have made a mistake, managers try not to hear about or respond to current problems.
3. **Senior management assume that once they have been told about a problem, they will need to solve it (36%).** Managers are afraid that, if an employee informs them about a problem, the responsibility to solve it is automatically transferred on their shoulders.
4. **Senior managers expect employees to solve problems independently in their area of responsibility (28%).** Some managers prefer not to pay attention to warnings from employees, because they believe that employees are paid well enough and should be able to deal with problems that arise independently, without involving them.
5. **Senior management prefer not to know about risks, in order to avoid being held responsible (including legal responsibility) if things go wrong (27%).**

Some managers do not want to hear about existing risks from their employees because they do not want to be held legally responsible for an accident or emergency. Irrationally but perhaps understandably, they believe that, if risk information does not reach management, the responsibility for the onset of an emergency remains entirely with their subordinates who are managing the facility involved. This has some basis in experience. During investigations following major accidents in critical industries worldwide, some senior executives were able to avoid criminal liability because they claimed they had not been aware of the problems that ultimately led to the accidents — while their subordinates, unable to plead ignorance, were punished.

6. **Leaders do not want to step out of their comfort zone to solve complex questions (26%).** Some leaders do not want to step out of their comfort zone, change their routine, and take on extra work to react to problems their employees have warned them about. This may as well sometimes imply that managers have to rush to a production site in a remote region to deal with the problem on the spot.
7. **Leaders are people too — like anyone, they would rather hear good news than bad ones (24%).** One should remember that leaders are just humans underneath, and it is just human nature to prefer good news rather than bad.
8. **Managers see issues reported by employees as unimportant (23%).** From the perspective of some managers, most of the problems employees bring to senior management are insignificant. As a result, some executives are reluctant to hear about the concerns of rank-and-file employees and do not want to have to respond, as for them these are minor issues. But with this approach, there is the chance that vital information about critical risks may be overlooked.
9. **Short-term contracts for managers (19%).** The reluctance of some managers to hear about serious problems is influenced by their own short-term contracts, as part of a company's short-term corporate goals. The short-term contracts of senior managers (up to 3 years) are detrimental to creating a favorable environment for the reporting of information about risks. Leaders feel under pressure to show shareholders a quick positive result, therefore they are unwilling to receive bad news about production issues that will require time and money to rectify. Solving serious problems in critical infrastructure companies generally requires sustained effort over many years.
10. **A common corporate leadership culture pervades the entire company and industry (15%).** In some large companies, a corporate culture of "*no bad news*" accumulates over decades. Entire generations of leaders have grown up in a culture in which only good news can be brought to the authorities.

TOP 10 REASONS WHY EMPLOYEES ARE RELUCTANT TO DISCLOSE RISKS TO THEIR SUPERVISORS

1. **Fear of blame and punishment from executives: subordinates assume that they will be held responsible for the occurrence of any problem they report to their managers (63%).** Employees are afraid that, if they raise the alarm about a problem, management will accuse them of having caused or exacerbated

it through their mistakes. In most cases, problems do not arise from nothing. They usually develop partly because of poor decision-making by managers (for example a refusal to approve adequate resources to keep facilities running safely) and partly through the actions of employees who may have been forced to violate operational safety to meet production targets dictated by managers.

2. **Employees are afraid of losing income and ruining their career prospects by looking incompetent in the eyes of their bosses (48%).** The fear of losing earnings and damaging their own career prospects stops many employees from reporting serious problems and risks within their area of competence.
3. **Inertia of corporate culture (43%).** Most employees will go along with the corporate culture that exists in their company. If that culture dictates secrecy about problems, demands only good news, and punishes staff for the presence of problems on their watch, then most employees will simply not inform the authorities. If managers are unwilling to listen when employees raise concerns, this will eventually lead to an ingrained culture of lies at every level of an organization.
4. **Fear of destroying relationships with colleagues or line managers (32%).** Many employees do not disclose risks to their bosses because they think this will ruin their relationship with their colleagues, or with their immediate supervisor.
5. **Fear that employees will be expected to solve any problem they report (27%).** Employees are afraid that, if they report a problem to senior management, they will be left to handle it themselves with no extra resources to do so.
6. **Employees do not fully understand the risks they are running, and lack the training or experience to assess their criticality (22%).** Sometimes employees do not realize the risks they encounter in their day-to-day work, so they do not inform their superiors about them. If people are not aware they are taking risks, they are unlikely to think that they are doing anything wrong, and will see no reason to inform anyone. It will generally be employees who are unqualified or inexperienced, who fail to recognize or assess risks. Sometimes employees only care about their own area of work and do not want to look at the risk picture across the entire production process. In this case, they are unlikely to report problems outside their own limited area of competence.
7. **Employees feel it is pointless to report risk information because managers failed to respond to similar messages in the past (21%).** Some employees see little point in informing their superiors about problems or risks, because there has been no response to previous warnings and the problems have remained unsolved. Frustrated by this lack of action, some employees simply stop telling their superiors about problems, assuming that their efforts will be futile.
8. **Fear of being seen as disloyal to a company, as a rebel who wants to "*rock the boat*" or as a "*bad news guy*" (20%).** When employees start to "*ring the alarm bells*" and draw attention to problems, they will be perceived by their superiors as rebels, *"black sheep"*, troublemakers who want to *"rock the boat"* or "*go on the rampage*". Most managers are afraid of such potential disruption, which could lead to earlier management mistakes coming to light. Consequently,

they will often berate would-be whistle-blowers: "*All your colleagues are quite happy, but you always seem to have a problem with something. You think you're special and you want to wash our dirty linen in public*".

9. **Industrial safety performance indicators and reward systems encourage concealment (15%).** Corporations use many key performance indicators to manage their productivity. Some of these metrics can incentivize employees to downgrade an incident, and under-report equipment problems or anything else that could stand in the way of hitting ambitious corporate targets.
10. **Some employees are confident that they can solve the problem on their own (13%).** Employees can sometimes be overconfident in their own capabilities to solve a problem. If subordinates have a strong sense of ownership, they will be tempted to solve problems by themselves and then report their success, rather than reporting the problem to the manager and waiting for them to come up with a solution. In doing so, they may overestimate their capabilities and convince themselves that there is no immediate need to inform their superiors about it.

HOW THE PROBLEM CAN BE SOLVED

RECOMMENDATIONS FOR OWNERS AND SENIOR MANAGEMENT: TEN PRACTICAL WAYS TO IMPROVE THE QUALITY AND SPEED OF RISK INFORMATION TRANSMISSION WITHIN CRITICAL INFRASTRUCTURE ORGANIZATIONS

The handbook draws on information received from 100 practitioners in industry, and the results of a decade of research on the reasons for concealing risks before and after major technological accidents. Together they inform some clear practical recommendations for owners and managers of large industrial companies, who want to fundamentally improve the transmission of risk information within their organization, in order to prevent serious industrial accidents. A detailed account of the recommendations is presented in Chapter 3.

1. **Owners and senior management should be willing to give up short-term profits in exchange for the long-term stability of critical infrastructure.** Fundamental improvements in the quality and speed of reporting critical risks within a critical infrastructure company are possible only when the owners and senior management are willing to focus on the long-term ownership of the company. This involves accepting that the significant costs required to manage existing serious and critical risks may impact short-term profits, but are essential to protect the long-term reliability, sustainability, and value of the company. The owners will make more money on the long-run and with much less risk in case of investing in reliability of critical infrastructure, which increases the long-term return and decreases the short-term risks and volatility. If owners are willing to allocate resources to prevent critical problems, then their managers will follow suit and begin to pay attention to these safety issues. The changed view regarding

safety in the minds of senior management will lead over time to a change in attitudes and working practices throughout the company. To operate sustainably in the long term, a critical infrastructure company must find a balance between safety, finance, and production. The task of top management is to create a system that allows managers at every level to freely analyze and discuss risks and to find the acceptable balance.

2. **Senior management should be approachable about problems, and have the desire and resources to control and mitigate identified risks.** Everything comes from leadership. Employees will report problems if managers want to hear them. Managers should want as much information as possible about potential risks. If the management support is not there, all other interventions are doomed to fail. The only way to improve the situation regarding feedback in an organization is if leaders have a genuine desire to hear about risks from their subordinates — and communicate this to them — and then take decisions and allocate resources to stop risk escalation. Senior managers should have the necessary support — moral and practical — from owners and shareholders to implement risk reduction measures. Having secured this, they should then take the initiative to implement cultural change, dismantling any system of penalties for reporting risks or incidents, and making it clear that they actively want to hear about problems. Only then will employees, inspired by the evident commitment of their leaders to a safer workplace, be willing to report the risks they have encountered.
3. **Risks must be prioritized, as it is impossible to manage every risk within an organization simultaneously.** It is impossible to effectively manage all risks — prioritization is essential. Resources are always limited and will never be sufficient to mitigate every possible risk. Without establishing clear priorities, managers have so much information to handle that they cannot distinguish what is important from what is not. A gradation of risks immediately makes the situation clearer — what further information is required, which risks need monitoring, and which demand urgent action so that *"major negative events"* can be prevented. It is vital that critical risks and problems that may threaten the work of an entire enterprise come swiftly to the attention of senior managers so that they can immediately inform the highest level of the hierarchy, while less serious risks can be delegated to appropriate lower levels of management for further action. For effective decision-making, you need to have a system in place to deliver an integrated risk assessment of production processes, where all the key risks of an industrial facility are assessed and then ranked by severity. This will enable an organization to prioritize the allocation of risk management resources. Not all employees in the organization are dealing with critical risks — only a limited circle of managers and employees is responsible for this. Senior managers should start by working on the control and reduction of critical risks with those managers and employees who manage them.
4. **Senior managers must be leaders in safety.** It is imperative that any initiative to prioritize safe operation of critical infrastructure comes from senior management. In highly hierarchical companies, the example set by the leader

is paramount. Most critical infrastructure companies have several management levels and are quite bureaucratic. If subordinates see that safety is extremely important to the CEO, and the entire corporate system makes it a top priority, then most employees will imitate senior management and follow the principles they are espousing. If safety is made the top priority by the CEO, then production site workers have no grounds for relegating it down the list of their own priorities, and will instead be willing to place it first, above production and profitability indicators.

5. **Senior management should build an atmosphere of trust and security, so that employees feel safe to disclose risk-related information.** Without trust in the leadership, there can be no high-quality feedback from employees on the problems of an organization. Often, employees evaluate the possible consequences of disclosing risk information based on rumors about how senior management reacted in a previous situation with colleagues. Employees project both the positive and negative experiences of their colleagues onto themselves. Employees need to have security guarantees, both for their careers and for their colleagues. Managers must guarantee the security of their sources and take responsibility for solving any significant problem they are informed about. If an environment can be established where employees do not feel under threat, they will begin to give candid feedback. To increase employee confidence, it is essential to reduce their uncertainty about the actions of managers. Managers need to demonstrate exactly how employees are treated when they give honest feedback. Only through repeated positive responses from managers will it be possible to dispel the common perception that an organization can be dangerous to employees who speak out. The first step is for senior management to make a declaration that feedback is encouraged at all levels of a company. Nevertheless, this is not enough in itself: employees must see the truth of the statement applied in practice, with employees receiving praise and not punishment for offering honest feedback. The message that senior management actively want to hear about problems, and that it is safe for employees to tell them, should come right from the top of the hierarchy. It is important that the CEO and senior executives give employees specific examples of their colleagues' positive experiences of communicating problems to their seniors. It is also vital that managers demonstrate respect for their subordinates, including a sincere interest in their well-being, safety, and progress. If these principles are applied reliably across the board, then even the most cautious employees will gradually come round to the idea that a company is a safe environment, where they can confidently reveal their concerns about the situation on the ground without any negative consequences.
6. **Middle management are allies of senior management in building an organization where active dialogue between superiors and subordinates is welcomed.** Senior management can only build an effective system to obtain accurate information about risks, and change the safety culture in a company, by working with the middle managerial level. Therefore, the best strategy is to make middle managers allies and not enemies. The middle managers in charge

of the production facilities know more about the situation at an organization than shop floor employees and lower-level managers. If senior managers only ask for the opinion of shop floor employees about the critical risks of an organization, they may not always get an accurate assessment of the situation. Getting information about critical risks only from lower-level employees may just lead to an increase in information noise, making it more difficult for senior management to understand the true picture of safety at a site. Once honest dialogue has been established between senior management and owners about critical risks and how to handle them, the next step is to establish the same honest dialogue between the leadership and the middle management level. Senior management should emphasize that they trust middle management. They must ensure that middle managers disclosing risks and problems are not penalized or dismissed. They must show that they want to work together with middle managers to solve problems, and not leave them to tackle issues alone. They must appreciate and reward subordinates who provide accurate information. It is also important that middle managers have the opportunity to adjust the production plans set by headquarters, so that they have the authority to stop suspicious pieces of asset for repair and the resources to carry out these repairs.

7. **Use different upward risk transmission channels.** In addition to receiving information through the traditional management hierarchy, senior managers are encouraged to regularly visit industrial sites to hear directly from managers and employees regarding the critical risks they are facing. It is also recommended to use other alternative channels for obtaining information about risks, such as: fault logs or risk registers/databases; safety training observation program cards; smartphone apps to allow shop floor employees or lower-level managers to timely report risks to senior managers directly; independent production monitoring systems; process improvement proposals; problem-solving boards; and anonymous mailboxes and helplines.
8. **The words of leaders should be supported by their actions: problems once identified need to be solved.** Leaders should never say one thing and then do another — their words must be matched by their deeds. This is especially relevant if senior managers call for risk disclosure, and then consistently address the issues that their subordinates bring to their attention. When employees report risks and problems, they do so in the belief that managers will make the right decisions to solve the problem or at least reduce the risk. A critical infrastructure company may well not have enough resources to solve all the problems identified at any given time. If this is the case, then managers must be sure to feed back to the employee who reported an issue, and assure them that they will tackle the problem when they can. If identified problems are not satisfactorily solved, then employees will inevitably lose faith, and will not bother to disclose risks to their superiors anymore.
9. **Do not penalize specific employees: look for systemic defects within the organization.** Executives should not penalize individual employees for incidents, but instead look for the systemic shortcomings in a company's operations that forced the employees to commit safety breaches.

10. **Reward employees for disclosure of safety and technological risks.** The best way to reward employees is to recognize their important contribution to an organization, as everyone derives fulfillment from having their work appreciated and praised. Management should deliver this not just through a private conversation, but in front of the whole workforce. Expressing gratitude publicly in this fashion provides an opportunity for senior management to highlight the kind of behavior and performance they wish to see from all their employees. Public recognition will motivate the employee to even greater efforts and encourage colleagues to communicate new risks and problems up through the hierarchy. According to most respondents, non-financial motivation is more effective than material incentives, which have many disadvantages. There are many effective ways to motivate employees for disclosing information about risks, which do not involve financial reward.

PILOT PROJECT EXPERIENCE OF INTRODUCING A SYSTEM FOR TRANSMITTING INFORMATION ON SAFETY AND TECHNOLOGICAL PROBLEMS WITHIN A CRITICAL INFRASTRUCTURE COMPANY

Chapter 4 presents detailed information about the pilot project, which tested various methods for significantly improving the quality and speed of reporting information about safety and technological problems within the critical infrastructure company. The project involved more than 400 employees (from senior management to shop floor employees) of an industrial company that is a world leader in its field. Within the just first few months of the introduction of the project, shop floor employees and line managers disclosed seven critical risks to senior management that they believed had the potential to lead to accidents resulting in either the death of personnel, long-term decommissioning of production facilities, and significant environmental issues. All these risks were quickly addressed by senior management and production site leaders. In several cases, these prompt disclosures and interventions prevented serious incidents from developing. Employees also disclosed to senior management 104 other problems that were compromising the industrial safety of four of the company's production sites. Most of these issues have also now been resolved.

The success of the project indicates that, with suitable information transmission systems in place, shop floor employees and line managers are willing to disclose to senior management serious safety and technological problems in their area of responsibility, in order to prevent emergencies.

The authors of the handbook aim to create a proven mechanism — a universal standard — over the next 10 years to fundamentally improve the quality and speed of reporting about critical risks in companies operating critical infrastructure by implementing similar projects in different countries worldwide. The overarching goal is clear: to prevent industrial accidents and disasters from occurring in the first place, to save people's lives, reduce environmental damage, and increase the resilience of critical infrastructure facilities.

DISCUSSION

The handbook ends with a discussion on the following three topical issues.

AUTOMATING THE COLLECTION OF INFORMATION ABOUT EQUIPMENT OPERATION IN CRITICAL INFRASTRUCTURES, AND THE PROSPECTS FOR ARTIFICIAL INTELLIGENCE (AI)

Most of the interviewees are positive about developing automated systems to collect complex information about the functioning of critical equipment, which continuously transmit feedback on their condition and operation to headquarters.

Most of the respondents stressed that the degree of automation of information collection depends primarily on economic feasibility. The main criterion for assessing the feasibility of introducing an automated system should be the level of risk that it can remove.

The obvious advantage of such automation is its ability to reduce the influence of the subjective human factor: once it is set up and running reliably, there is no further need for the manual collection, processing, and transmission of information about critical risks through the traditional management hierarchy. It is very important that, in such automated systems, there is no manual data entry to exclude the possibility of any manipulation of data by employees or managers at different levels. However, assessing feedback from the system and informing a management decision is hardly possible without human involvement. Therefore, at this stage automation is only possible up to a certain extent.

It is important to note that the influence of the human factor will never drop to zero in the coming decades. Automated systems cannot replace highly professional employees, who can diagnose the operation of complex but outdated equipment — basing their assessments not only on data from sensors, but also on the intuition they have developed over many years of experience. This experientially grounded intuition is particularly important when analyzing the work of complex interdependent technical systems.

Cyber risks should also be considered when implementing automated systems: with the growth of automation, the risk to companies from network failures and unauthorized access will grow.

Respondents expressed skepticism about extending AI to making decisions in the operation of critical infrastructure. Many risks can be introduced if AI is allowed to independently decide on serious operational issues. Therefore, the final decision must still be left to professional operators, supported by analytical information from the AI system. AI can be allowed to make secondary decisions, where the scale of any possible damage is limited. AI is best used to analyze large amounts of data, creating broader analytics to inform smarter leadership decisions. Additionally, AI is helpful for generating various scenarios of future situations in the company.

DISCLOSURE OF CRITICAL RISKS TO INSURANCE COMPANIES IN EXCHANGE FOR REDUCED PREMIUMS

As part of the in-depth interviews with the executives of critical infrastructure companies worldwide, the authors wanted to know their views on whether it is worth disclosing the critical risks of their businesses to insurance companies in exchange for lower insurance premiums.

In the study, 93 respondents answered this question. 57 of them (61%) reacted positively to the idea that a critical infrastructure company should fully disclose to the insurer all information it knows about its own critical risks in exchange for a reduction in insurance premium. 24 respondents (26%) expressed skepticism or were against such an exchange. 12 respondents (13%) found it difficult to answer this question.

This section portrays the divergent views on the pros and cons of a proposal to disclose information about risks in exchange for a reduction in insurance premiums. Successful practical examples of interaction between critical infrastructure companies and insurance companies are also presented.

IMPACT OF NATIONAL CULTURE ON RISK INFORMATION TRANSMISSION WITHIN CRITICAL INFRASTRUCTURE COMPANIES

Some of the respondents have worked in several countries and continents. They were asked if they noticed an effect of national and cultural differences on the reporting and discussion of risk. All the leaders interviewed, who have international work experience, agreed that communication about risks within organizations is significantly influenced by the peculiarities of national culture, religion, and worldview. The interviewers asked the respondents to compile their subjective ratings of the quality of internal risk communication in the countries where they have worked. First, the respondents gave examples of countries and cultures where they felt that risk information from subordinates to superiors was significantly distorted in reports. Then, they described countries where risk information was transmitted without significant distortion. They explained why they thought some countries and cultures have problems with objective feedback, while in other countries this problem does not seem to be so pronounced.

Many respondents expressed the view that, in all cultures on the planet, people want to present themselves to others in the best possible light. In any society, any group of people, nobody likes to receive bad news — so nobody wants to be the bearer of bad news. The only question is how it is customary in different societies to react to it. There are hierarchies in every society, but the management style — the way managers manage their subordinates — differs.

All the interviewees in their own way conveyed the idea that the key factors affecting the quality of information sent up a company hierarchy are the power distance between managers and employees, and related to that, the traditions of authoritarian (monologue) or democratic (dialogue) governance in the country.

CONCLUSION

The goal of writing this handbook was to provide executives that operate critical infrastructure with practical tools and solutions, so that they can improve the quality and speed of risk communication in their companies. Better information makes for better decisions, and these in turn have an impact on reducing the likelihood of severe accidents at industrial facilities. The authors hope that this handbook will help prevent major emergencies and save many lives.

Contents

The Problem

1 Importance of Risk Information Transmission in Critical Infrastructure Organizations 3

Why the Problem Exists

2 Factors That Obstruct the Reporting of Information About Risks in Critical Infrastructure Companies 15

2.1 Causes of Risk Concealment Based on the Analysis of Past Disasters 15

2.2 Main Factors of Intra-organizational Risk Concealment That Discourage Subordinates from Reporting Risk-Related Information Internally, or Encourage Managers to Ignore Early Warnings When They Are Reported (Based on Analysis of 20 Major Historical Accidents and Disasters) 20

2.3 Views of Practitioners Managing Critical Infrastructure About Why Managers Are Reluctant to Receive Risk-Related Information, and Why Employees Are Reluctant to Disclose Risks 27

2.3.1 Who Creates an Internal Climate Within an Organization Where It Is not Acceptable to Talk About Problems? 28

2.3.2 Reasons Why Leaders Do not Want to Hear About Problems from Their Subordinates 34

2.3.3 Reasons Why Employees Are Reluctant to Disclose Risks to Their Managers 68

How the Problem Can Be Solved

3 Recommendations for Owners and Senior Management: Ten Practical Ways to Improve the Quality of Risk Information Transmission Within Critical Infrastructure Organizations 95
3.1 Recommendation No. 1: Owners and Senior Management Should Be Willing to Give Up Short-Term Profits in Exchange for the Long-Term Stability of Critical Infrastructure 96
3.2 Recommendation No. 2: Senior Management Should Be Approachable About Problems, and Have the Desire and Resources to Control and Mitigate Identified Risks 113
3.3 Recommendation No. 3: Risks Must Be Prioritized, as It Is Impossible to Manage Every Risk Within an Organization Simultaneously 122
3.4 Recommendation No. 4: Senior Managers Must Be Leaders in Safety 131
3.5 Recommendation No. 5: Senior Management Should Build an Atmosphere of Trust and Security, so that Employees Feel Safe to Disclose Risk-Related Information 141
3.6 Recommendation No. 6: Middle Management Are Allies of Senior Management in Building an Organization Where Active Dialogue Between Superiors and Subordinates Is Welcomed 159
3.7 Recommendation No. 7: Use Different Upward Risk Transmission Channels 187
3.8 Recommendation No. 8: The Words of Leaders Should Be Supported by Their Actions: Problems Once Identified Need to Be Solved 219
3.9 Recommendation No. 9: Do not Penalize Specific Employees: Look for Systemic Defects Within the Organization 227
3.10 Recommendation No. 10: Reward Employees for Disclosure of Safety and Technological Risks 238
3.11 Other Recommendations for Improving the Quality of Risk Communication in Critical Infrastructure Companies 257

4 A Pilot Project—Introducing a System for Transmitting Information on Safety and Technological Problems Within a Critical Infrastructure Company 263

Conclusion 337

Discussion: Automating the Collection of Information About Equipment Operation, and the Prospects for Artificial Intelligence in the Operation of Critical Infrastructure 339

Discussion: Disclosure of Critical Risks to Insurance Companies in Exchange for Reduced Premiums 353

Discussion: Impact of National Culture on Risk Information Transmission Within Critical Infrastructure Companies 363

ABOUT THE AUTHORS

Dr. Dmitry Chernov is a senior researcher at the Chair of Reliability and Risk Engineering at the Swiss Federal Institute of Technology in Zurich (ETH Zurich). He has more than 15 years of experience as a crisis communication and disaster management consultant for large corporate clients working in oil and gas, electric power, metals and mining, chemical, telecommunication, transport, utilities, retail manufacturing, etc. He first recognized the importance of intra-organizational risk concealment in 2007 during one of his seminars for a critical infrastructure company. Since then he has focused on researching solutions to improve intra-organizational risk communication, in order to enable timely decision-making before and during industrial disasters. Additional information: www.riskcommunication.info.

Dr. Ali Ayoub is a postdoctoral researcher at the Department of Nuclear Science and Engineering at the Massachusetts Institute of Technology. He received his Ph.D. and M.Sc. in nuclear engineering from the ETH Zurich after finishing his undergraduate training at the American University of Beirut. Dr. Ayoub is a member of the European Commission ESReDA "*Risk, knowledge, and management*" project group. Besides his interests in tackling the problems of risk communication and risk information transmission in critical industries, his research interests include: risk analysis and nuclear safety, uncertainty quantification and model-data integration, atmospheric dispersion, resilience engineering, and timely decision-making.

Prof. Giovanni Sansavini is an associate professor of Reliability and Risk Engineering at the Institute of Energy and Process Engineering, ETH Zurich. Currently, he is the chairperson of the ETH Risk Center and of the Technical Committee on Critical Infrastructures of the European Safety and Reliability Association (ESRA). His research focuses on the development of hybrid analytical and computational tools suitable for analyzing and simulating failure behaviors of engineered complex systems, with focus on physically networked critical infrastructures and sustainable energy systems. He aims to quantitatively define reliability, vulnerability, resilience and risk within these systems using a computational approach based on physical system modeling, advanced Monte Carlo simulation, soft computing techniques, and optimization.

Prof. Didier Sornette is Chair Professor and dean of the Institute of Risk Analysis, Prediction and Management (Risks-X) at the Southern University of Science and Technology (SUSTech), Shenzhen, China. He is professor emeritus of ETH Zurich since August 2022, and research professor emeritus of the National Center for Scientific Research of France since 2020. He is a member of the Swiss Academy of Engineering Sciences and of the Academia Europaea, and Fellow of the American Association for the Advancement of Science. He uses rigorous data-driven mathematical statistical analyses combined with nonlinear multi-variable dynamical models with positive and negative feedbacks to study the predictability and control of crises and extreme events in complex systems, with applications to all domains of science and practice.

The Problem

Chapter 1
Importance of Risk Information Transmission in Critical Infrastructure Organizations

Compelling evidence from industrial working practice demonstrates that in many critical infrastructure companies, decision-makers find it difficult to get objective information about safety and technological risks.

Management theory postulates that executives manage their subordinates through information: they receive information from various sources, process it, take decisions and convey these to their subordinates. The quality of information received by executives about the real situation inside and outside an organization affects the quality of their decisions, and ultimately the adequacy of an organization's response to any changes in the internal and external environment. Getting feedback from subordinates about the real situation at the very bottom of the corporate hierarchy is crucial for the survival of an organization in the long-term, as it allows executives to detect risks in time and take measures to mitigate them. But in reality, for a number of reasons, feedback from subordinates to executives is often distorted. As a result, executives often receive unrealistically reassuring reports from subordinates—assurances that in general, everything at the bottom of the hierarchy is fine and all risks are under proper control.

The seriousness of the problem of "*embellished*" feedback about technological risks first became clear to the first author of this handbook during a 2007 seminar for an industrial company, one of the three largest in the world in its field. The seminar was devoted to management decisions and communication in emergencies. More than 120 senior managers from headquarters and directors of the company's production sites attended this seminar. Crisis response and communication solutions are a well-established theme in modern risk management: there are many practical manuals on this topic, with numerous clear examples of positive and negative responses by companies during and after various incidents. One of the postulates on how to respond effectively to major accidents is that managers should make sure they have reliable information about the preliminary causes of an accident, the extent of possible damage and the resources available to them, in order to determine how to

D. Chernov et al., *Averting Disaster Before It Strikes*,
https://doi.org/10.1007/978-3-031-30772-0_1

tackle emergencies. If this inside information is honestly and immediately brought to the attention of all interested parties—authorities, victims, employees, the general public, investors, etc.—then the social crisis caused by an accident can be quickly resolved.

After the seminar, one of the top managers of this company told the seminar facilitator (the first author of this handbook) about the unfortunate situation with reporting technological risks and incidents within the company. According to him, the directors of production sites (middle management) did not generally disclose anything negative to senior management at headquarters about what was happening at their enterprises. They preferred to send reports reassuring headquarters that "*everything is fine*", "*everything is under control*". Unable to rely on the official channels for a true picture of what was happening at the company's industrial sites, senior management were forced to establish parallel channels to gather alternative information about operations at the sites. They asked the company's internal security department to be responsible for creating this alternative flow of risk information from the bottom up. With this in mind, the manager asked the seminar facilitator a very serious question: "*What should managers do in the event of an emergency if they cannot get reliable information from their subordinates in the first hours or even days after the incident itself?*". The facilitator found it difficult to offer any recommendations for improving risk communication from this standpoint: most of the existing solutions in the field of crisis response assume the presence of reliable information in the hands of managers immediately after an incident. However, if the site managers have initially misinformed headquarters about the situation there, then it makes no sense to recommend that the company's senior management should promptly and comprehensively inform the public: how can executives, when they do not have reliable inside information from which to make a statement? The question of how, in practice, to improve intra-organizational risk communication was never raised again during subsequent crisis response seminars for this company. The participants agreed that the first step for managers at various levels in responding to emergences is to request, receive and properly transmit (without distortion) all the available information about the details of incidents.

A year after the last crisis response seminar, the company experienced the largest accident in its history. It became a national emergency, and the biggest disaster in the global industry in decades. The accident killed dozens of workers, the cost to the company of recovery exceeded several billion dollars, and the full reconstruction of the affected production facility took about ten years. However, in terms of informing affected families, emergency services and the wider public about the accident, the company and the national and regional authorities worked quite effectively. Immediately afterwards, they were able to organize the quick dissemination of information about the details of the accident among the residents of nearby settlements, and almost entirely allay any panic among the population. During the investigation that followed, it turned out that one of the causes of the accident was a very rare failure in the operation of a critical piece of machinery. The facility's technicians were unaware of this specific risk, even though the same failure had caused a similar incident at another production site decades earlier. However, the earlier incident was local in nature, caused only limited damage to the site, and none of the workers were injured. As a result, this first incident was dismissed as relatively insignificant, and the risk of similar equipment failure at other sites was not communicated across the company. Especially in retrospect, it is clear that, with essentially the same piece of equipment installed at other sites and being operated in a similar way, there was a real risk of the same fault recurring. However, with no warning after that first incident, no special measures were taken to control this risk at other sites of the company. Moreover, in the months preceding the later more serious accident, the site management observed unexpected deviations in the operation of the equipment, but did not inform senior management at headquarters about it. Even the alternative channels set up by the company's internal security department failed to warn headquarters about the unsafe operation of the equipment, or to identify the increased likelihood of the risk occurring. There was a potential conflict of interest at the level of the site managers in hiring only a limited number of contractors to repair the equipment operated there. This conflict of interest was spotted by the internal security department several months before the accident, but this did not prompt an emergency audit to assess the performance of the contractor involved or the quality of the equipment repairs they were making. The ingrained practice in the company of sending inaccurately reassuring reports about technological risks, highlighted so clearly by the investigation, reminded the author about his conversation with the executive after the seminar more than a year before the accident. It would appear from the account above that staff at the site were aware of a possible critical escalation of risks, but did not communicate this to executives at headquarters—if they had, a quick intervention would probably have prevented the eventual catastrophe.

A few years later, the first author was involved in the investigation of a major accident at a critical infrastructure site that was unprepared for abnormal weather conditions. This unreadiness came as a surprise to headquarters. To make matters worse, the first reports to headquarters from the scene of the emergency assured senior management that the facility could withstand severe bad weather, and its functionality would be restored in a very short time. However, the reality on the ground turned out to be far worse than the optimistic reports of subordinates. As part of the investigation of this emergency, in-depth interviews were conducted with workers involved from all levels of the corporate hierarchy. In one such interview, a lower-level manager shared his vision of how the leadership of the company generally comprehend what is actually happening at production facilities. He maintained that senior management at headquarters understand only 30% of what is really happening on the production sites regarding the management of critical matters. In other words, 70% of the major problems faced by people at the bottom of the corporate hierarchy remain unknown to senior management. He argued that the prevailing management system in this company discourages employees in the field from disclosing the risks they encounter to higher authorities or internal auditors. Very severe punishments for any shortcomings and mistakes are common in the company. The interviewer gave an example of how an internal infrastructure status report is generated and sent to headquarters. On the spot, lower-level managers can create a report that honestly outlines the difficulties they are facing with the equipment failure rate, and requests funding to mitigate the risks. However, as the report makes its way through the traditional hierarchy of the corporate bureaucracy, it may change significantly. There is a significant possibility that such a critical risk report will be blocked and not even reach headquarters—the immediate supervisors of the lower manager who sent it will just say: "*What are you doing? Why are you rocking the boat?*". After all, if senior managers get to know the truth about the unflattering situation on the spot, they may punish the whole line of managers down to the lower-level manager who highlighted the problem. Even if such a report does reach senior management, it will have been retouched: information about the situation on the ground will be significantly embellished and assurances will be given that in reality everything is not so bad, and the risks identified are under control. With such embellished information coming up through the company hierarchy, senior managers' sense of what is happening on the ground will be distorted by the reassurances of their subordinates, and may be a long way from the real picture. This is a very serious corporate problem for many critical infrastructure companies. The practice of reporting mainly good news upwards means that in general, the leadership do not understand the critical risks encountered by employees on the shop floor. This may lead to management decisions that do not address the real situation in the running of critical infrastructure.

The investigation into the Fukushima Daiichi nuclear accident in 2011 highlighted the inability of the internal hierarchy of the nuclear power plant operator to pass on warnings from various experts to senior management that the plant was not prepared for a possible large tsunami. A few years before the accident, a group of young specialists had proposed installing special protective structures to defend the plant and the emergency power supply in case of a beyond-design tsunami, and thus reduce the likelihood of a nuclear fuel meltdown. The cost of the proposed protective structures was approximately US $50 million. However, some of the top managers of the company operating the plant never saw these proposals, and those who did rejected them—partly because they were unwilling to consider such high costs, and partly because they simply could not believe that a beyond-design tsunami at the plant would ever happen. According to some estimates, the consequences of the accident could eventually cost more than $200 billion and take two decades for the cleanup. By comparison, the preventive measures to prepare the plant for such a tsunami would have cost the operator four thousand times less.[1]

Many of organizational problems in communicating risks that led to the Fukushima Daiichi accident were reminiscent of failures in the transmission of risk information that occurred before and during the Chernobyl accident. The developers of the RBMK reactors operated in Chernobyl did not inform the Politburo or the plant operators about some shortcomings in the reactor design, which had led to various incidents at Soviet nuclear power plants in the decade preceding the accident. Despite

[1] Dmitry Chernov, Didier Sornette, Giovanni Sansavini, Ali Ayoub, Don't Tell the Boss! How poor communication on risks within organizations causes major catastrophes, Springer, 2022, subchapter 1.16. Fukushima-Daiichi nuclear disaster (Japan, 2011), https://link.springer.com/book/10.1007/978-3-031-05206-4.

these incidents, the necessary changes to the RBMK reactor design were only implemented after the Chernobyl accident. Beforehand, none of the staff at the Chernobyl plant knew about the design shortcomings. The plant contractor made several errors in drawing up a test program which involved a controlled rundown of the turbine at reactor No. 4. Additionally, the plant personnel implemented the test incorrectly—violating reactor operating regulations. The test regime put the reactor into extreme (beyond design) operation, at which point the design defects of the RBMK became critical—causing the largest accident in the history of civil nuclear power. Moreover, a few hours after the accident, the director of the plant reported to Moscow, reassuring them: "*The reactor is intact... The radiation situation is within normal range*". This could hardly have been further from the truth: already in the first hours after the accident, he had clear evidence that the reactor was damaged, and the radiation at the site was at least a million times beyond environmental background radiation. With only these false reassurances to go on, the Kremlin initially assumed the Chernobyl accident was a relatively insignificant event. This led to a belated evacuation of the population in the area around the plant, and to delays and inaccuracies in informing the Soviet public and the international community in the first days and weeks after the accident. The USSR spent about US $27 billion dollars to deal with the consequences of the catastrophe alone. According to estimates by the Soviet academician Valery Legasov, the total damage caused by the Chernobyl accident amounted to approximately 300 billion rubles[2] (US $450 billion in 1990 prices or more than $1 trillion in 2022 prices).

Ultimately, the constant recurrence of similar intra-organizational risk communication failures across countries and industries over decades has led two of the authors of this handbook to suggest that this problem has not been given due attention worldwide, and has not been addressed at a practical level in industries that operate critical infrastructure. This fact prompted the authors to initiate a detailed cross-sectoral study of the reasons for concealing risk information before and during various major disasters.

Between 2013 and 2015, two of the present authors studied hundreds of major incidents across multiple industries. They identified tens of risk information concealment cases—both internally between members of an organization at different levels, and externally between an organization and external audiences—that helped to cause the onset or the aggravation of an emergency. In addition to the Fukushima Daiichi and Chernobyl nuclear accidents, there have been many disastrous accidents in the critical infrastructure sector where delayed, misleading or withheld communication about risks played an important part. Here are some notorious examples: the collapse of the Vajont dam in the Piave river valley, where at least 1,921 people died (Italy, 1963); a toxic leak at a pesticide factory in Bhopal that killed several thousand people and damaged the health of more than half a million (India, 1984); the Exxon Valdez oil spill, when the slow and inadequate response resulted in extensive pollution of over 2,000 km of Alaska's coastline (USA, 1989); the explosion of a natural gas liquids

[2] Valery Legasov, Problems of Safe Development of the Technosphere, Communist Journal, #8, 1987, pp. 92–101.

pipeline near Ufa, which killed 575 people who were on passing passenger trains (USSR, 1989); a turbine failure at the Sayano-Shushenskaya HPP that killed 75 plant workers (Russia, 2009); the explosion on the Deepwater Horizon platform, which led to the largest offshore oil spill in world history (USA, 2010); methane explosions at the Raspadskaya mine, where 91 people died (Russia, 2010); and many more.

Studying the available accounts of these disasters, two of the authors focused on situations where some of the managers, employees or contractors involved chose not to inform various audiences about risks they had encountered or been told about. This enabled them to establish 30 constantly recurring factors that foster the formation of an atmosphere that favors the concealment of risks within an organization, or encourages those involved to delay informing internal or external audiences about them (these reasons are discussed in more detail in Sect. 2.1 of this handbook). The results of this study were published in 2016 by Springer Switzerland in the book "*Man-made Catastrophes and Risk Information Concealment: Case Studies of Major Disasters and Human Fallibility*".[3] It was translated into Japanese in 2017.[4]

In 2022, Springer Nature Switzerland published another book by the present authors: "*Don't Tell the Boss! How poor communication on risks within organizations causes major catastrophes*".[5] In this book, 20 different accidents and disasters were examined with a focus on intra-organizational concealment of information about risks: when employees failed to inform managers on time and in full about existing critical problems, or when managers ignored warnings from subordinates about the dangerous development of events occurring in their organizations. In addition to the disasters mentioned above, the book analyzed the following catastrophes: the great famine in China (1958–1962), which claimed the lives of more than 20 million people; the collapse of two reservoir dams in China (1975); problems with the rear cargo door of McDonnell Douglas DC-10 aircraft (USA, 1970s); staphylococcus food poisoning from Snow Brand Milk Products Co. (Japan, 2000); the train derailment at Amagasaki (Japan, 2005); the methane explosion at the Upper Big Branch coal mine in the United States (2010); the collapse of the Fundão tailings facility in Brazil (2015); the methane explosion at the Severnaya coal mine (Russia, 2016); the African swine fever epidemic in China (2018–2018) and the cover-up of the novel coronavirus outbreak in Wuhan (China, 2019–2020). The authors identified factors that led employees or contractors of organizations involved to delay or withhold reporting risk information up, down, or across their corporate hierarchies. They also investigated why managers preferred to ignore existing risks, despite warnings from their subordinates. The main factors encouraging the intra-organizational

[3] Dmitry Chernov, Didier Sornette, Man-made Catastrophes and Risk Information Concealment: Case Studies of Major Disasters and Human Fallibility, Springer, 2016, https://www.springer.com/gp/book/9783319242996.

[4] Dmitry Chernov, Didier Sornette, Man-made Catastrophes and Risk Information Concealment, Soshisha Publishing Co., 2017, https://www.amazon.co.jp/大惨事と情報隠蔽-原発事故-大規模リコールから金融崩壊まで-ドミトリ-チェルノフ/dp/4794222955.

[5] Dmitry Chernov, Didier Sornette, Giovanni Sansavini, Ali Ayoub, Don't Tell the Boss! How poor communication on risks within organizations causes major catastrophes, Springer, 2022, https://link.springer.com/book/10.1007/978-3-031-05206-4.

concealment of information about risks in these emergencies were systematized and ranked by frequency of occurrence (these reasons are described in more detail in Sect. 2.2 of this handbook).

These studies have led to some answers to the question of why critical risks were not reported in organizations before and during various emergencies. However, they have not helped answer the question of what needs to be done in practice to improve the speed and quality of risk reporting from shop floor workers to decision-makers. The main directions for organizational change were clear—these followed from the analysis of the reasons that encourage employees to hide the true situation in their area of competence. However, based only on this analysis it was not possible to give specific practical recommendations for companies that operate critical infrastructure. A thorough review of the scientific and business literature published on the topic of "*organizational silence*"[6] also failed to yield an adequate answer in this regard. Therefore a new study was initiated in 2018, aiming to develop practical recommendations for the leaders of critical infrastructure companies on how to improve the quality of information reported from employees to senior management about the risks and problems of a critical infrastructure company. From October 2018 until June 2021, the present authors conducted in-depth interviews with 100 top managers, regulators, technical managers, middle and lower managers, and occupational health and safety managers of leading industrial companies in Western Europe (41% of all respondents), Russia (32%), North America (10%), the Middle East (9%), Africa (5%) and Australia (3%). The respondents were drawn from the following sectors of critical infrastructure: power (40% of all respondents representing nuclear, thermal, wind and hydro generation, as well as power transmission), oil and gas (35%), chemicals and petrochemicals (9%), mining (6%), metallurgy (6%) and other industries (3%). The choice of these industries was dictated by the potentially huge damage caused by emergencies, and the importance of critical infrastructure facilities within the economy of any country. The practical recommendations for better risk communication established during this study are discussed in Chap. 3. The authors place great value on the opinion of practitioners-managers in industry, as they are unlikely to propose academic solutions that are difficult to deploy in the practice of a large industrial company. The interviewers sought to answer these questions: why managers are reluctant to receive information about critical safety and technological risks; why employees are reluctant to disclose such risks to their leaders; and whether it is primarily managers or employees who are responsible for the creation of a climate within organizations which discourages the reporting or discussion of problems. Responses were categorized to clearly understand what practitioners see as the reasons for the concealment of information about risks in organizations (these reasons are set out in Sect. 2.3 of this handbook—and should have some parallels

---

[6] Dmitry Chernov, Didier Sornette, Giovanni Sansavini, Ali Ayoub, Don't Tell the Boss! How poor communication on risks within organizations causes major catastrophes, Springer, 2022; subchapter 2.2 Results of other research on the challenges of voice and silence in an organization, https://link.springer.com/book/10.1007/978-3-031-05206-4.

with the factors listed in Sect. 2.2, which the authors deduced from their study of previous disaster accounts).

In planning a study focused on developing practical recommendations for improving the quality and speed of risk reporting within critical infrastructure companies, the authors envisaged that, after its completion, they would be able to test the solutions developed. The last of the 100 interviews was conducted in June 2021. The following month, the first author of the handbook was invited to conduct a seminar on management decisions and communications in emergencies for an industrial company that is a world leader in its field. The company has experienced a string of incidents over recent years; so its leaders wanted to improve the quality of the management team's response, and develop more effective communication with external audiences in emergency situations.

The studies outlined above clearly suggest that the first step to effective emergency management is for senior management to get accurate information as quickly as possible, from managers at the industrial site where an accident occurred. Decision-makers need to know the scale of the emergency, as much as can be established about what happened, and an estimate of how long it will take to contain the emergency and deal with the aftermath. Therefore, a significant part of the two-day seminar—which took place in October 2021 and brought together 104 executives of this company—was devoted to the problems of communicating information about risks before emergencies, and ensuring that accurate information gets to senior managers quickly in the first minutes and hours after the onset of an accident. The seminar cited some of the accidents mentioned above as examples. It was noted that underplaying the scale of an emergency or otherwise hiding information about the real situation on the ground in reports to company headquarters leads to: (I) a slow and inadequate response from senior management and the rest of a company to a developing crisis; (II) the absence of a company's senior leadership at the scene of an accident; (III) insufficient or delayed allocation of emergency resources to deal with the situation; (IV) an information vacuum around the accident, which generates rumors and panic among various audiences.

As part of the seminar, an anonymous survey was conducted about the current status quo in the company regarding internal communication about safety-related issues and technological risks. The results indicated that the company had serious problems in reporting objective risk-related information internally.

These results were presented to senior management of the company with a proposal to launch a pilot project at their most critical industrial sites, aiming to avoid emergencies by fostering better reporting of critical safety problems from ordinary employees to senior management. Part of the purpose of this pilot project was to test in the work of a real industrial company the recommendations received from 100 leaders around the world in 2018–2021, as well as the solutions that logically followed from the analysis of dozens of disasters.

In November 2021, the company's senior management gave the green light to the implementation of the pilot project. In December 2021, the first author of the handbook visited the four selected production sites. Over the ten months of the project implementation, 15 seminars were held for 422 people: top managers at company headquarters, directors and employees at the four pilot sites, and specialists from the Health, Safety, and Environment (HSE) department. During the project, the employees and managers at the selected sites willingly started to share information about the critical risks there with senior management. The seminars catalyzed the process of information sharing within the company. In other words, employees and lower managers became less timid and started to share information to senior management that they would probably not have done otherwise. Within the framework of the seminars, seven critical (high) risks were revealed that could have caused accidents involving the death of personnel, the long-term decommissioning of production facilities or serious environmental damage. All these risks were taken under immediate control by senior management and the directors of the sites concerned. As a result, potential accidents were prevented. The seminars also identified 104 problems at the four pilot production sites that, while less dangerous than the seven critical risks, still had a negative impact on risk management and the industrial safety of the company. The majority of these problems were taken under the control of the special working team of the project. Detailed information about the pilot project is presented in Chap. 4.

Why the Problem Exists

Chapter 2
Factors That Obstruct the Reporting of Information About Risks in Critical Infrastructure Companies

2.1 Causes of Risk Concealment Based on the Analysis of Past Disasters[1]

In 2013–2015 two of the present authors conducted the study of tens of major accidents and incidents[2] sought to identify the reasons that prevent the transmission of relevant, clear and accurate information about risks, both within an organization (intra-organizational) and between different organizations (inter-organizational). More than 30 recurring factors were identified that appeared repeatedly in major disasters around the world and within different historical periods. The study concluded that, when employees distort information about risks before or during a disaster, they do it not because of their own individual characteristics or personal motives, but because the internal environment of the organization motivates people working there to hide risks from both internal and external audiences.

These 30 intra-organizational and inter-organizational factors preventing adequate risk transmission were divided into 5 groups: (1) the nature of the external environment surrounding an organization and the incentives that it creates; (2) the corporate objectives and strategy of an organization, and internal managerial practices; (3) the conditions of the internal system for communicating and gathering information about risks within an organization (formal and informal channels); (4) internal practices for managing risk assessment; (5) the psychological characteristics of employees within an organization.

[1] This section includes previously published materials [©Springer, All rights reserved, Man-made Catastrophes and Risk Information Concealment, 2016], permission to reproduce this had been gained from the respective copyright holder.

[2] Dmitry Chernov, Didier Sornette, Man-made Catastrophes and Risk Information Concealment, Springer, 2016, https://www.springer.com/gp/book/9783319242996.

D. Chernov et al., *Averting Disaster Before It Strikes*,
https://doi.org/10.1007/978-3-031-30772-0_2

(1) **EXTERNAL ENVIRONMENT OF AN ORGANIZATION**

- The short-term focus of global political and business philosophy
- Deregulation
- Mutually beneficial relationships between government representatives and private industries which do not serve the public interest
- Low status and entry criteria, and unattractive wages, for employment with government regulators
- Weak control over complex systems and fragmentary perception of the whole risk picture
- Political instability and struggle between political camps
- National arrogance
- Fear of widespread public panic
- National security secrecy

(2) **INTERNAL ECOLOGY OF AN ORGANIZATION**

- Short-term financial & managerial objectives and unrealistic forecasts for future development
- Permanent "*rush work*" culture
- "Success at any price" and "no bad news" culture
- "*Ivory tower syndrome*" or the fragmentary perception of the whole picture of risks among top managers
- Lack of specific knowledge and experience among members of boards of directors
- Weak internal control within an organization
- Frequent labor turnover
- Habituation (problems and risks seem inconceivable because nothing has gone wrong in the past)
- Wishful thinking/Self-suggestion/Self-deception among decision-makers
- The remoteness of units/facilities from headquarters

(3) **RISK COMMUNICATION CHANNELS**

- Long chains of communication for risk information. Absence of a direct, urgent 24-7-365 channel between field staff and executives. Field staff who do not have authority to immediately stop a process if they suspect evidence of risk
- No internal or external incentives for whistleblowers
- Poor inter-organizational risk communication
- Absence of direct horizontal communication between departments of an organization (communication between units only occurs through superiors)

(4) **RISK ASSESSMENT AND RISK KNOWLEDGE MANAGEMENT**

- Absence of a prompt industry-wide risk assessment system
- Unwillingness to investigate in detail the causes of an accident, and absence of established risk assessment systems within organizations (recording, evaluating and ranking risks over decades)
- High frequency of unconfirmed alerts
- Ignorance among critical personnel and managers of other accidents or near miss cases within an organization, the industry and abroad. Absence of a system to manage risk knowledge (accumulation, systemization and transmission)

(5) **PERSONAL FEATURES OF MANAGERS AND EMPLOYEES**

- Desire to "*look good in the eyes of superiors*" and fear of being seen as incompetent, leading to reluctance to admit personal mistakes
- Unrealistic projections of personal performance
- Fear of criminal prosecution after a serious incident

Detailed analysis of each of these factors can be found in Chap. 3 of the book devoted to the results of this study.[3] In the present subchapter, several factors are highlighted that are most significant in discouraging employees of critical infrastructure companies from reporting risks to their superiors, and making managers reluctant to receive such reports.

SHORT-TERM FINANCIAL & MANAGERIAL OBJECTIVES, UNREALISTIC PROJECTIONS OF FUTURE GROWTH, AND ANNUAL BONUS SYSTEMS

Across organizations worldwide, there is a prevalence of short-term development strategies, due to widespread pressure from shareholders on management to achieve ambitious financial results as quickly as possible. Even organizations that operate critical infrastructure are often required to set profitability goals that can be to the detriment of production safety, as meeting them usually involves reducing capital investment, delaying the modernization of equipment and undermining the long-term interests of organizational development. Setting such ambitious goals creates an unhealthy psychological climate within an organization. Employees from the shop floor to the headquarters feel they must hide the true internal situation from shareholders, regulators and other audiences, to create the appearance of a successful company that is achieving, or will soon achieve, phenomenal short-term results. The

[3] Dmitry Chernov, Didier Sornette, Man-made Catastrophes and Risk Information Concealment, Springer, 2016, https://www.springer.com/gp/book/9783319242996.

situation is exacerbated by the widespread prevalence of annual bonuses for executives, which tempts managers to focus on short-term profitability and to "*embellish*" an organization's annual results in order to make the grade.

PERMANENT "*RUSH WORK*" CULTURE

Short-term business development goals, pressure from competitors to develop and launch new products, scientific and technological progress—these and other factors can lead to an intra-corporate culture that promotes constant haste in all aspects of an organization's activities. Employees are always in a hurry. With no time to test solutions, they inevitably make mistakes, but under relentless pressure to deliver, they prefer not to report their problems. Instead of honestly acknowledging the real situation, they will send placatory reports to their superiors and colleagues. Employees will assure them that everything is going to plan, when in fact the quality of decision-making is often poor, risks are ignored and alternative solutions are not pursued due to lack of time. As a result, an organization generates sub-standard products/solutions and hides the risks and internal shortcomings caused by haste.

"*SUCCESS AT ANY PRICE*" AND "*NO BAD NEWS*" CULTURE

Ambitious business development goals impose high pressure on all employees to demonstrate personal achievements and improvement. Companies often create a climate in which goals must be achieved at all costs. Managers will not tolerate subordinates bringing them bad news, demanding to see only successful results. In some organizations, a state of total fear of the management develops: it becomes almost impossible to admit any professional mistake without risking sanctions or punishment, including dismissal. Some employees cannot function well under this pressure and, to preserve the impression of success, they falsify their achievements and embellish reality. Hearing nothing but good news from their cowed employees, executives are under the illusion that everything is going well. When a crisis finally comes, it turns out that the real situation was being concealed to fit in with the impossible standards promoted in an organization and the demand to achieve success at any price.

LONG CHAINS OF COMMUNICATION FOR RISK INFORMATION, AND ABSENCE OF AUTHORITY TO ACT

In some cases, disaster investigation shows that prior to an accident, operators observed a potentially dangerous deviation in the operation of equipment, but they lacked the authority to turn it off or take any extraordinary action to prevent the situation worsening: there was no established "*stop-the-job*" system. Operators also lacked an emergency communication channel with senior management to request exceptional authority to stop suspiciously functioning facilities. Existing regulations made the approval process very time-consuming and bureaucratic, which made it impossible to get a prompt decision from the top.

ABSENCE OF A SYSTEM FOR ENCOURAGING AND SUPPORTING EMPLOYEES WHO HAVE VALID CONCERNS (WHISTLEBLOWERS)

In some cases, an employee may be concerned about the existence of risks in their area of competence, which their colleagues do not wish to acknowledge or even want to actively suppress. Often a risk exists because of managerial misjudgments by immediate superiors, who are unwilling to admit their mistakes by informing top management. Corporate culture frowns on perceived insubordination, and will not allow employees to take the initiative and pass their concern directly to senior management.

IGNORANCE AMONG CRITICAL PERSONNEL AND MANAGERS OF OTHER ACCIDENTS OR NEAR MISS CASES WITHIN AN ORGANIZATION, AN INDUSTRY AND ABROAD. ABSENCE OF A RISK KNOWLEDGE MANAGEMENT SYSTEM (ACCUMULATION, SYSTEMATIZATION, AND TRANSMISSION)

Many managers believe that their problems and risks are unique, leading them to try to find their own solutions. Often though, they are simply unaware of potentially relevant experience in other departments, companies, industries or countries because there is no accurate, systematized, detailed knowledge bank of previous accidents. Unfortunately many organizations, including government ministries and think-tanks for a given industry, do not systematize sector risks, or collect information about near miss cases on an ongoing basis. In other words, no one describes and studies in detail the causes of accidents elsewhere in an industry or abroad. And in internal corporate journals, there are few articles sharing the experience of other departments of the same organization, let alone that of competitors or foreign enterprises.

RELUCTANCE TO FULLY INVESTIGATE ACCIDENTS OR ISSUE DETAILED REPORTS ON THE CAUSES OF ACCIDENTS AND THE SHORTCOMINGS OF AN ORGANIZATION

Many organizations that have encountered an emergency have been reluctant to assist in subsequent investigation of what happened, or to produce detailed reports on the causes of accidents and their own organizational shortcomings. While this reluctance is understandable, the downside is that no one can then fully establish the mistakes that led to the accident and no sector-wide learning can take place. Further accidents can thus occur under the same scenario, which might have been avoided had an honest and thorough corporate investigative report been issued.

PRESSURE TO LOOK GOOD IN THE EYES OF SUPERIORS AND RELUCTANCE TO ADMIT PERSONAL MISTAKES FOR FEAR OF BEING SEEN AS INCOMPETENT AND/OR BEING FIRED

There is a universal aspect of human nature: people wish to present themselves to others in a good light in order to receive approval and, in a professional context, career promotion. When this is played out in an organizational environment like the ones described above, human nature and corporate culture work together to exacerbate people's tendency to say what they think others want to hear, rather than speaking unpalatable truths. Many in a subordinate position will distort information about the real situation at ground level when communicating with executives, because they want to look good in the eyes of their superiors. They are unwilling to admit their own mistakes, fearing that they will be perceived as incompetent, and ultimately that they could be fired. In order to demonstrate their competence to managers and colleagues, some employees set themselves unrealistic targets for work progress and achievement. Then, unable to cope with the workload but lacking the courage to recognize that they have overestimated their real strengths, some will feel forced to start embellishing their real achievements. This can lead to inaccurate and misleading information being sent up the chain of command. Such distortion affects the quality of information received by executives, and thus the quality of the decisions they make.

2.2 Main Factors of Intra-organizational Risk Concealment That Discourage Subordinates from Reporting Risk-Related Information Internally, or Encourage Managers to Ignore Early Warnings When They Are Reported (Based on Analysis of 20 Major Historical Accidents and Disasters)[4]

Between 2015–2022, the present authors gathered information on accidents and disasters where there was evidence of internal concealment of risks by employees and contractors, which subsequently led to emergencies. They also looked for incidents in which employees, contractors and lower/middle managers had warned senior management about the risks long before the accident, but where, for various reasons, the managers ignored these warnings, and accidents then occurred in line with the concerns that the subordinates had reported. From this initial research, 20 such major accidents and catastrophes were identified that have occurred across different industries and different countries over the past 80 years.[5]

---

[4] This section includes previously published materials [©Dmitry Chernov, Ali Ayoub, Giovanni Sansavini, Didier Sornette, All rights reserved, Don't Tell the Boss!, 2022], permission to reproduce this had been gained from the respective copyright holder.

[5] List of 20 accidents and disasters where there was internal concealment of information about risks: unpreparedness of the Red Army for the Nazi invasion (USSR, 1941); Great Chinese Famine (China, 1958–1962); collapse of the Banqiao and Shimantan reservoir dams (China, 1975); problems with the rear cargo door in McDonnell Douglas DC-10 aircraft (USA, 1970s); Challenger space shuttle explosion (USA, 1986); Chernobyl nuclear power plant disaster (USSR, 1986); collapse of Barings Bank (Singapore-Great Britain, 1995); food poisoning caused by staphylococcus in Snow Brand dairy products (Japan, 2000); SARS outbreak (China, 2002–2003); Amagasaki train crash (Japan, 2005); Sayano-Shushenskaya hydropower plant accident (Russia, 2009); methane explosion at the Upper Big Branch coal mine (USA, 2010); Deepwater Horizon oil spill (USA, 2010); methane explosions at the Raspadskaya mine (Russia, 2010); forest fires in the European part of Russia

The analysis of these accidents revealed several main factors that motivate subordinates not to disclose risks to their supervisors, or supervisors to ignore the warnings of their subordinates. A detailed presentation of each of the factors can be found in Chap. 2 of the book, where the results of this study were published.[6] Below is a brief summary of the main factors identified, listed in decreasing order of prevalence.

PRIORITY OF SHORT-TERM SOCIO-ECONOMIC, FINANCIAL AND OPERATIONAL GOALS OVER THE LONG-TERM SAFETY AND WELL-BEING OF CITIZENS, CUSTOMERS AND EMPLOYEES

(this factor was identified in 90% of the accidents studied)

Most of the accidents analyzed involved situations dominated by short-term development strategies, arising from pressure on management to meet the demands of owners/shareholders/politicians to achieve specific financial/production/socio-political results in a short period of time. For critical infrastructure companies, such short-term profitability and production targets are often detrimental to the safety and long-term stability of production. Achieving these goals is usually associated with increased load on obsolete equipment, reduced capital investments, delays in equipment upgrades, and so on.

Prioritizing short-term profitability creates an unhealthy psychological climate within organizations, and puts pressure on senior managers to achieve goals at any cost. In many cases, managers are well aware of the serious problems that these short-term production goals might create. However, they are afraid to challenge the decisions of owners and shareholders because they fear accusations of incompetence and disloyalty, and ultimately the loss of their positions due to the dissatisfaction of owners and shareholders. To replace them, owners can always find new managers who are willing to accept higher risks to meet their targets.

Field staff, lower and middle managers are often aware of the negative consequences of pursuing such short-term goals, and in some cases attempt to warn senior management of the likely problems. However, these messages are often ignored as they threaten the fulfillment of the goals. If the warnings are acted on and appropriate actions taken to manage the risks, this will lead to an increase in costs and reduce the profitability of the organization—and as a result, targets are unlikely to be achieved. In order to avoid questioning the competence of the owners, some senior managers dismiss the warnings of their subordinates, insisting that their subordinates independently find ways to safely control the risks in their area of responsibility and do not

(Russia, 2010); Fukushima-1 nuclear power plant disaster (Japan, 2011); Volkswagen diesel emissions scandal (Germany-USA, 2000-2010s); collapse of the Fundão tailing dam at Samarco iron ore mining site (Brazil, 2015); Severnaya coalmine blowouts (Russia, 2016); African swine fever epidemic (China, since 2018).

[6] Dmitry Chernov, Didier Sornette, Giovanni Sansavini, Ali Ayoub, Don't Tell the Boss! How poor communication on risks within organizations causes major catastrophes, Springer, 2022, https://link.springer.com/book/10.1007/978-3-031-05206-4.

bother headquarters with their fears. Those who do not comply with this demand or fail to solve emerging problems on their own are liable to be punished, up to and including dismissal.

In the face of this demand from senior management, employees in the field will in future do their utmost to deal with the risk issue on their own. Only good news will be sent upstairs, and subordinates will avoid talking about any problems they observe. Gradually, the entire management hierarchy enters a state of near euphoria from the continual positive news communicated up through the corporate body, all indicating the unimpeded growth of production and profitability. Meanwhile, risks and issues are accumulating—but it is only at the grassroots level that this is recognized, with senior management not knowing, or wanting to know, about the real state of affairs. Such an organization is heading for disaster.

Frequently when accidents do occur in the operation of critical infrastructure, they come as a surprise to managers, owners and shareholders. However, a detailed independent investigation often reveals a causal connection between the accidents and the existence of short-term goals imposed by superiors, in conjunction with tacit pressure on subordinates within the corporate hierarchy to achieve these goals at any cost—even by violating safety rules.

The situation is exacerbated by the ubiquity of the annual bonus system, which encourages senior managers to focus on achieving short-term profitability to secure their bonus, despite the increase in risks this may create in the longer-term operation of a critical infrastructure company.

In many cases, short-term financial goal setting is a false, and indeed dangerous, economy. In the event of an accident, losses and costs for dealing with the consequences are often hundreds—or even thousands—of times greater than the finances that would have been required to deal with the risks when they were first recognized, and before they led to a major accident.

OVER-AMBITIOUS ORGANIZATIONAL GOALS

(this factor was identified in 75% of cases)

Short-term goals are frequently associated with the achievement of ambitious results. Owners/shareholders/politicians set very ambitious goals for senior managers, but in many cases do not provide them with the necessary additional resources—money, time, materials, and equipment—to achieve these goals. At the same time, senior managers warn their subordinates that they will be punished or fail to achieve promotion if they do not attain these ambitious goals, even when the necessary resources are absent.

Ambitious goals set at headquarters are often impossible to achieve without major safety breaches. Senior managers look to identify ambitious and loyal middle and lower managers who are ready to take responsibility for achieving results—even if this means taking risks and tacitly violating the vital safety rules that cover the operation of critical infrastructure facilities.

Some subordinates may point out that the established production and financial indicators are unrealistic or unsafe within the current state of the equipment and allocated resources, and might even criticize the senior management for trying to impose impossible goals. However, most subordinates cannot challenge the decisions of their superiors without consequences for their own careers, so there is pressure on them to fall into line and comply with instructions from above. In order not to jeopardize their career, most lower and middle managers will avoid mentioning to executives the risks that might arise when implementing such plans. Instead, they and their work teams on the production sites will try to achieve the impossible by taking unnecessary risks, inevitably increasing the likelihood of catastrophic events.

When senior managers set near-impossible goals and over-ambitious key performance indicators, they effectively encourage their subordinates to distort and falsify information in their reports, and convince their bosses that their departments are achieving their targets. How else can they appear to achieve all the development goals?

FEAR AMONG SUBORDINATES AND CONTRACTORS THAT THEY WILL BE BLAMED AND PUNISHED FOR REPORTING A PROBLEM

(this factor was identified in 70% of cases)

Pressure from senior management on employees to implement a high-risk corporate strategy often includes severe penalties for anyone who fails to meet their targets. Senior management may have little or no real interest in how subordinates will actually achieve the goals demanded of them. In such a punitive culture, why would employees try to warn their managers about the safety problems inherent in implementing an over-ambitious corporate strategy, raise issues relating to their area of competence or request assistance or additional resources? Instead, as a rule, subordinates are left to try and solve any problems that arise without assistance from their seniors. This can easily lead to a situation where the only way an employee can be seen to be meet the targets set for them is to violate safety regulations and falsify reports. Therefore in many cases, subordinates prefer to keep quiet about the problems in their area of responsibility when they report to their superiors, insisting that everything is under control and going according to plan. Lower-level workers are afraid of financial penalties; lower and middle managers are afraid of being fired.

When accidents occur, organizations generally focus on mistakes by specific employees instead of looking for possible root causes of the problem. Much might be learned, for example, from analyzing the impact of corporate goals on the work of those employees, or investigating whether they had sufficient resources at their disposal to adequately manage possible risks. But during incident investigations, senior managers very rarely admit that the risk escalation was a result of over-ambitious corporate goals, lack of resources in the field, or other weaknesses in their organization that left individual employees feeling isolated, and seeing no option but to violate safety regulations in order to deliver what was demanded of them.

All these factors ultimately lead to a culture of fear, where most employees are reluctant to disclose risks and problems to superiors in their area of competence.

INEFFICIENT STATE REGULATION (including, in some cases, corruption of government officials)

(this factor was identified in 65% of cases)

Over the past few decades, politicians financed by private business have tried to convince voters that reducing government involvement in the regulation of private business should be the order of the day. The arguments they use are as follows: cuts in budget expenditure for government officials can free up additional resources (for example, for social programs); free from state control, private business can develop more dynamically, creating new jobs and increasing tax revenues; and less control by officials means less corruption.

In addition, several other factors have contributed to the convergence of political and business elites' interests: active cooperation between authorities and private business in the development of state economic policy; the widespread practice of employees moving back and forth between private business and public service; legitimate corporate financing of election campaigns; and finally, on occasion, outright corruption of specific government officials.

As a result, under the pretense of cutting spending on bureaucracy—and with strong support from private business and voters—politicians reduce the salaries of officials responsible for overseeing regulation of critical infrastructure. Due to a decrease in funding and the reduction of their powers, regulators cannot attract highly educated and experienced employees for the key roles of ensuring quality control and regulation compliance.

In the accidents under consideration, the reduction of effective state regulation allowed the managers of private business, with the approval of their owners and shareholders, to focus exclusively on chasing short-term, ambitious financial goals. This behavior led to frequent security breaches, which jeopardized the long-term sustainability of the business. The accidents reviewed in [7,8,9] demonstrate that companies who actively lobby for weakening regulation measures and public accountability of their activities often do themselves a disservice: they lose the input of objective external controllers, who could prevent the development of critical events by prohibiting or modifying risky and reckless management decisions. Competent regulators can: (I) impose additional legal restrictions on the operation of critical infrastructure facilities under extreme conditions; (II) require a company to provide comprehensive solutions to control existing risks at industrial sites; and (III) significantly strengthen its own emergency response services to reassert control as soon as potential accidents are identified.

---

[7] Dmitry Chernov, Didier Sornette, Man-made Catastrophes and Risk Information Concealment, Springer, 2016, https://www.springer.com/gp/book/9783319242996.

[8] Dmitry Chernov, Didier Sornette, Critical Risks of Different Economic Sectors (Based on the Analysis of More Than 500 Incidents, Accidents and Disasters) Springer, 2020, https://link.springer.com/book/10.1007/978-3-030-25034-8.

[9] Dmitry Chernov, Didier Sornette, Giovanni Sansavini, Ali Ayoub, Don't Tell the Boss! How poor communication on risks within organizations causes major catastrophes, Springer, 2022, https://link.springer.com/book/10.1007/978-3-031-05206-4.

In addition, the legislative reduction of fines for violations in the field of industrial safety has allowed the leaders of some private companies to disregard the regulatory framework, preferring to pay penalty fines: even when repeatedly incurred, fines work out cheaper than making serious investments in the facility to eliminate risks, modernize equipment and prevent further breaches.

Ineffective state regulation, reductions in regulatory resources and powers, and even outright corruption have all encouraged some critical infrastructure managers to disregard warnings from their subordinates about significant risks at the site that could lead to serious accidents. In some cases, employees feel it is their civic duty to prevent a critical development of events, but they cannot take their concerns about safety violations at work to regulatory authorities—because they know that their own bosses have close links with members of those authorities. So why bother? No remedial action will be forthcoming and all that will happen is they will get fired. In other words, they choose to remain silent about the critical risks they know are there, and keep their jobs. Often the cozy relationship between private business and regulators only comes to light after a serious accident—when investigators question hundreds of workers, site managers, and regulatory officials, and begin reconstructing the whole sequence of events.

FEAR AMONG SUBORDINATES AND CONTRACTORS OF APPEARING INCOMPETENT IN THE EYES OF THE MANAGEMENT

(this factor was identified in 60% of cases)

Whenever an organization sets short-term, ambitious development goals, subordinates are afraid to appear incompetent in the eyes of superiors and colleagues when it comes to implementation. If someone in an organization fails to achieve the goals set by owners and shareholders, managers will automatically accuse them of incompetence. Managers expect their subordinates to demonstrate excellence in achieving successful outcomes—even when there are insufficient resources, or when it is physically impossible to achieve them without violating safety regulations. Employees are understandably afraid to appear weak and useless in the eyes of their superiors. On the contrary, they want to demonstrate their competence, efficiency and resourcefulness in order to justify the trust their leaders have placed in them. Sometimes, this can only be achieved by hiding negative information and embellishing reality when reporting progress to senior managers.

PERMANENT "*RUSH WORK*" CULTURE

(this factor was identified in 60% of cases)

Adopting short-term targets based on maximizing corporate profit and production, as well as tight commissioning schedules, can foster a culture of haste for both managers and their subordinates. As a result, everyone involved in a critical infrastructure organization is always in a hurry. Never having the time to work out best practice and quality solutions means employees will inevitably make mistakes and poor decisions. The pressure on managers to meet deadlines means that they do not want to hear about problems that might cause delay, so employees simply do not report them to their superiors. Instead of honestly admitting the reality of the situation, and pointing out to managers that it is impossible to complete the work safely within the specified time frame, many employees prefer to send reassuring reports to their managers and colleagues. It is easier to just say that everything is going according to plan and on time, rather than warning your supervisor—let alone the site manager—about unrealistic targets and the risks of always working in a hurry.

"*SUCCESS AT ANY PRICE*" AND "*NO BAD NEWS*" CULTURE

(this factor was identified in 55% of cases)

Ambitious business development goals create a very high bar for everyone in an organization to achieve certain goals or demonstrate continuous improvement. Leaders often create a climate within an organization in which goals must be achieved at all costs. Therefore, they do not tolerate subordinates who bring them bad news, wishing only to hear reports of success. In response to this, employees feel they have no choice but to solve the problem on their own, and only let management know after a successful solution has been implemented. In some organizations, a real dread of senior management can develop, and it becomes almost impossible to admit any professional mistake or uncertainty for fear of the sanctions or punishments that may follow. Some employees cannot stand this pressure and, in order to maintain the illusion of success, they begin to falsify their achievements. Unwilling to hear anything but good news from their intimidated employees, executives are happy in the illusion that everything is going well on their watch and there are no serious problems or risks. It is only when a critical incident does finally occur that the real situation in an organization finally becomes clear to the management. Until then, the truth has been hidden—because employees felt compelled to conceal it, to meet the inflated demands imposed by their senior executives.

IGNORANCE ABOUT RISKS AND WISHFUL THINKING/OVERCONFIDENCE/SELF-SUGGESTION/SELF-DECEPTION

(this factor was identified in 50% of cases)

Self-deception on the part of those involved in the receiving, processing and reporting of risk information is one of the main obstacles to quickly identifying critical situations and communicating this to other stakeholders. Instead of analyzing the situation, studying the facts, looking for primary sources and objectively evaluating the information received, many managers choose to believe what they want to believe. Even in situations where a cautious and critical attitude would seem eminently sensible and necessary, a significant proportion of managers prefer to rely only on the calming reports they receive from subordinates. Reassured that all is well, they can avoid the anxiety of having to take a critical view of their own earlier management decisions. This kind of wishful thinking from the top can lead to a group mentality developing, where everyone is eager to convince everyone else that all is well and any risks are under control. Self-deception inevitably leads to a faulty perception of reality, which can clearly compromise an organization's ability to respond effectively to existing and mounting critical risks.

WEAK INTERNAL CONTROL WITHIN AN ORGANIZATION

(this factor was identified in 30% of cases)

For executives seeking to achieve impressive results in a short time, the relaxation of internal control within an organization could appear to support these goals. A professional, efficient and independent control department, which collects information about all activities of both staff and managers and produces impartial assessments, constitutes a dangerous witness that can be exploited by regulators and government investigators in the event of disaster. It is therefore not surprising that, in some accidents examined in the study, internal regulatory departments had either been abolished or were staffed by incompetent or under-resourced employees who failed to perform their duties adequately. If employees are aware that their leaders are unable or unwilling to exert appropriate control on the ground, they are much more likely to delay or withhold the truth about risk concerns in their area of responsibility.

2.3 Views of Practitioners Managing Critical Infrastructure About Why Managers Are Reluctant to Receive Risk-Related Information, and Why Employees Are Reluctant to Disclose Risks

Between October 2018 to June 2021, the present authors conducted in-depth interviews with 100 senior managers, technical managers, middle and lower managers, and occupational health and safety managers with leading industrial companies in Western Europe (41% of all respondents), Russia (32%), North America (10%), Middle East (9%), Africa (5%) and Australia (3%). These were drawn from the following sectors of critical infrastructure: power industry (40% of all respondents, including nuclear, thermal, wind and hydro generation, and electricity distribution), oil and gas (35%), chemical and petrochemical (9%), mining (6%), metallurgy (6%) and other industries (3%). Some of the respondents had previous experience as representatives of state regulatory bodies in the field of industrial safety.

It was important for the present authors to hear directly from practitioners about the factors that affect the poor quality of risk information transmission within traditional hierarchical companies. Practitioners manage critical infrastructure on a daily basis, constantly analyze technological risks, are immersed in occupational health and safety issues, and regularly participate in internal investigation of incidents.

All interviewees were first asked about: (I) "*Why subordinates are sometimes reluctant to inform managers about problems within an organization (e.g. fail to report problems with equipment, errors that have been made, or the impossibility of achieving corporate goals, etc.)?*" and (II) "*Why managers are sometimes unwilling to hear bad news from subordinates about observed risks and problems in an organization, and about additional investments such as equipment upgrades that are necessary to create a safer production process?*".

2.3.1 Who Creates an Internal Climate Within an Organization Where It Is not Acceptable to Talk About Problems?

Most of the interviewees, when asked why employees might hide information about the problems of an organization, soon moved on to talk about the responsibility of the leaders themselves. By their reluctance to hear about problems within an organization, leaders can discourage employees from raising these issues in the first place. Therefore, the next question to all respondents was: (III) "*Who bears more responsibility (managers or subordinates) for creating an atmosphere within an organization in which discussion of problems is not welcome?*". 97% of interviewees responded by placing the majority of the blame on managers. 2% of respondents argued that the responsibility is equally shared by managers and subordinates. 1% of respondents believed that the reasons for such an internal corporate atmosphere lay mostly in the personal qualities of individuals and their relationship with colleagues, and not in their organizational roles, whether manager or subordinate. None of the respondents placed the main responsibility on employees. However the head of HSE department of a mining company, at the beginning of his interview, categorically stated that he believed employees have a tendency to conceal information about their activities from their superiors. However, on further discussion around the actions of the owners and senior managers he worked with, he changed his point of view and concluded that managers bear most of the responsibility whenever an organization distorts information about risks.

Delving deeper into the points of view of some of the interviewees provides an interesting perspective on the matter.

The head of HSE department of a gold mining company cited Deming, who said that most quality problems at work are due to system errors in management, which put employees in a position where they are forced to make defective goods. Deming believed that 96% of all organizational problems are due to managerial errors and incorrect processes, while employees and other factors influence only

4% of cases.[10,11] The respondent maintains that it is the same story with incidents and safety violations. Ordinary employees are not suicidal. They feel uneasy about the risks involved in running critical infrastructure, and so are ready to talk about problems and safety issues. However, the culture fostered by top management makes it very difficult for employees to air their concerns. In his opinion, the silence of employees about safety issues starts with managers who do not want to hear them.

The head of HSE department of a fertilizer manufacturer agrees that a significant part of the blame for the widespread practice of concealing problems lies within top management. As the Latin proverb puts it—"*piscis primum a capite foetet*"—"*fish rots from the head*". The way employees behave is predetermined by the unspoken position of top management, who do not want to hear about bad news.

The HSE head of an oil company also believes that the behavior of employees depends on the corporate settings that managers determine. Employees are afraid to send bad news up the hierarchy because management do not want to hear about problems. Too many leaders respond aggressively to any negative information, assuming that the bearers of the bad news must be responsible. As a result, employees shut down and decide not to bother this kind of manager anymore—knowing that, if they carry on, their well-intentioned honesty may well threaten their careers. Sometimes a leader like this will not say a word, but his demeanor will make it perfectly clear that he is extremely dissatisfied. In general, until senior managers show their subordinates that they want to receive information about problems, and will put their time into solving them, the transmission of risk information in an organization will not improve.

A safety consultant and former HSE director in mining and metallurgy shared the following experience. When he advises top management in industrial companies, he starts by asking them what they want to hear after an external safety assessment at a production site. 1/3 choose the option "*only good news, we are not interested in hearing about bad news*". 2/3 choose the option "*if there is bad news, we want to hear it, along with good news*". Clearly one should not expect much appetite from managers in the first group to change things in their companies, even if they have serious safety problems in the workplace. If managers only want to hear good news, then they are not committed to change: they are comfortable in the fictional world that they have created around themselves. But the second group of managers is focused on change. The fact that they want to hear bad news shows they are ready to take action and allocate resources to stop the most critical problems, although in this case too there is still a question of breaking down priorities. It is very important to understand that employees will adapt to the settings determined by top management. Employees are only the executors of decisions made by the authorities. They play by the rules that exist in a company and are established by the leadership. If the CEO and his

[10] BW (Ben) Marguglio, Human Performance Improvement through Human Error Prevention: A Comprehensive Implementation Guide for Protecting Employees and Maintaining Cost Efficiency, CRC Press, 2021, p. 17.

[11] Lawrence P. Leach, Critical Chain Project Management, Third Edition, Artech House, 2014, p. 281.

deputies make it clear that they have no wish to concentrate on the negative, or discuss issues, and constantly turn the conversation to positive news and achievements, their subordinates will get the message: if they want to be respected and build a career in this company, their reports to the management must be a success story. The way ahead is to solve problems yourself, without disturbing the leaders. The responsibility for creating a culture where discussion of problems is not welcomed in the company lies solely with senior management and the owners.

A regional manager in the power industry, responsible for the operation and maintenance of turbomachinery, agrees that executives are the ones who play the greatest part in fostering a corporate culture of "*no bad news*". Leaders are those who set an example. If managers do not actively promote dialogue and open communication with subordinates, they are helping to maintain the existing atmosphere of silence about problems.

The head of an oil production facility believes that what causes employees to hide problems is a corporate culture of silence, i.e. the unspoken rules established by senior management that discourage employees from reporting bad news. After all, the attitude of the average senior manager will be something like this: "*We must find the culprits who allowed this problem to escalate to a critical level, and they must be punished so that others will see and not repeat their mistakes*". In fact, what "*others will see*" is that reporting a problem will just lead to a search to identify putative perpetrators; they had better make sure they only send good news upstairs, so that they look positive in the eyes of the leadership and continue to make progress in their careers.

A critical infrastructure manager also maintains that the key reason why employees are silent and only give good news to the top is that managers are reluctant to hear about problems, reluctant to understand what has caused them and reluctant to allocate resources to subordinates to solve them. According to the respondent, employees are ready to talk about problems—the key question is whether managers are ready to listen.

The HSE manager of a production company managing a large number of hazardous chemical processes agrees that managers are chiefly responsible for this situation. By superimposing the organizational structure of the company onto the company's "*risk pyramid*", he showed that up to 80% of all critical risks are supervised by a board of directors. Top and middle managers deal with 10%, and 10% are left to shop floor managers and ordinary employees. The priorities set at the very top of the company determine how risks will be managed at the very bottom. If a board of directors and senior management are focused on financial results and do not want to hear about safety and technological problems, they will create an atmosphere in which delivering bad news that impacts profitability will not find support. On the other hand, if leaders prioritize the long-term maintenance and development of production assets, they will want to hear about production safety issues and react to risks proactively. Employees will respond to this and inform managers if they see any problems.

The HSE manager of a metallurgy company believes that, when employees violate safety requirements, they often do this because of external pressure. In the investigation of most industrial accidents, it turns out that employees were under unspoken pressure from management to complete the production task faster and without involving additional resources. It is important to understand that the head of a company is responsible for the atmosphere that prevails there. It all starts and ends with the boss. Moreover, leaders can pay lip service to the need for openness in discussing risks, but if nothing really changes, and instead such openness just seems to be punished, then employees will continue to be economical with the truth.

The HSE head of an oil company describes the chain of logic as follows: when subordinates come to senior management with a list of operational problems, they are told that solving these problems is their job because management do not want to be bothered with such minor operational issues. But when these same operational problems cause an emergency at a production site, leaders will reproach the site manager: "*Why didn't you say anything? You need to understand when not to disturb headquarters with small things, and when you need to ring the alarm bells!*". In reality, senior managers are creating and reinforcing a tacit system so that employees do not bring them any bad news. And it is a vicious circle: managers do not want to hear from subordinates about problems, so their employees keep quiet and do not inform them. The respondent identified two primary reasons for employees concealing information about problems: corporate goals and priorities dictating over-ambitious production and financial targets, and the behavior of leaders who say things like "*I don't want to hear any more bad news*", or "*We trusted you and delegated power, but you couldn't do it*". The respondent did not see silence or concealment stemming from the personal motives of ordinary employees. Most employees are simply forced to adopt the established and unspoken corporate rules, and comply with them in order to make a career; those who openly disagree with the rules are forced to leave.

The vice president of a gas pipeline construction and repair company is very skeptical of the idea that employees, on their own initiative, will disclose information about risks if the leaders do not want to hear about them. If managers are not interested, then employees will certainly not volunteer. Even when managers want their employees to tell them about problems, it takes years for employees to believe that they would not be punished for doing so. Therefore, if senior managers want to get an objective picture of the situation on the ground, they must be very persistent. Employees need to see real positive experience for many years before they stop being afraid of disclosing risks.

The head of a power plant shares the belief that the leadership, not the workforce, are responsible for the poor feedback and the concealment of risks in a company. It is the expectations they set that determine how employees will behave when they encounter a problem or a critical risk. If managers create a system of punishment whenever negative aspects of the operation are revealed, then employees will avoid sending honest feedback to the top. If managers guarantee that the prompt reporting of accurate information will meet with approval rather than punishment, then employees will be happy to inform the authorities.

A psychologist and consultant in the field of organizational behavior identifies the following deep-seated fears that employees tend to carry: (I) loss of one's own worth, of respect or love; (II) loss of career prospects; (III) loss of certainty of the future (knowledge/ignorance); (IV) loss of recognition as a professional. The question is how executives manage these fears to reassure and empower employees. Punitive and intimidating behaviour by executives can increase fear among employees, in which case they will keep quiet. Nevertheless, executives can also reduce fear, by increasing openness and explaining the logic of management decision-making in a transparent way, in which case subordinates are more likely to share their concerns. If a company has a negative track record in responding to feedback, punishing employees for reporting to the head and fostering fear in the workforce, then employees will try to avoid giving feedback. The kind of environment created in a company depends on management. If the pervading environment is one of fear, then the fears of employees will be exacerbated. In an environment of fear, an employee's response to a threat may take the form of avoidance, paralysis or aggression. If an environment of respect and openness is created, employees will be reassured and their fears will decrease or completely fade away.

The HSE head of a metallurgy company believes that it is important that senior management understand that bad news brought today can help to solve a problem proactively, and prevent it developing into a catastrophe tomorrow. Discussing bad news is a two-way process: it involves leaders, who are responsible for decisions, and employees, who are responsible for reporting problems. Of course, it is up to senior management to make the first step. Without the sincere desire of managers to solve the problems that employees are reporting, nothing will work. If managers do not want to deal with problems, and punish employees for bringing bad news, then of course employees will be silent for the sake of their career prospects. If the upper levels do not want openness and communication, the lower levels will never feel safe to participate in it. If leaders want to change the situation for the better, the higher up they are in a company the more likely it is that positive change will happen.

The chief HSE officer of an oil service company thinks that the reason it is a cliché to say "*everything starts from the top*" is that it is true. Experience shows that it is the top of the corporate pyramid—the CEO and the board—who are responsible for the corporate culture that develops in an organization. The base of the pyramid—middle and lower-level managers and rank-and-file employees—will follow the tone, orientation and culture set by those at the top. This applies to attitudes around safety just as much as other aspects of the corporate climate.

In this regard, a consultant in nuclear safety with long experience in nuclear power plant operations cites a simple example from everyday life. There is a family of two parents and three or four children. Who is responsible for creating the family culture? Of course, it is the parents! It is the same in an organization: senior managers are responsible for creating a culture of openness when discussing issues, and for ensuring that this culture is correctly interpreted at all levels of the hierarchy.

An HSE/EHS consultant specializing in manufacturing makes the following argument: history shows that all great battles have been won and lost because of the decisions of generals. Subordinates follow the rules set by those in power. Therefore, those who manage by setting goals and giving instructions to subordinates must always be held accountable for their decisions.

A safety expert and consultant working mainly in chemical and steel manufacturing considers that 85% of the internal corporate culture is formed by top management. If senior management has a clear idea of how it wishes to develop the organizational culture, then this can be rapidly achieved.

An HSE manager and consultant with experience in nuclear power and construction believes that a reluctance to discuss problems within an organization will only change when senior management turn to subordinates and say: "*We need to know what is happening! Please help us with this*".

The senior vice president managing HSSE (Health, Safety, Security and Environment) for the asset operations of an international electricity company believes that the corporate culture around risk communication is created by senior management. Employees in a company tend to imitate the leadership style used by their superiors, even if this is often subconscious. If senior management want to make a good impression on shareholders and the media by issuing a glossy annual report, then employees will be encouraged to emulate this behavior at their level by reporting good news up the hierarchy and not revealing the problems in their area of responsibility. Employees will not consider this shameful or dishonest: they can see that senior management are doing the same with shareholders and the media, apparently with impunity. Thus, the behavior of senior leadership can create conditions that encourage the concealment of serious incidents, and this unhealthy intra-corporate atmosphere prevents the prompt reporting of problems. Left unattended, those problems will only get worse, and eventually lead to a serious accident.

The CEO of a consulting company in human performance, with wide experience in power generation, agrees that it is senior managers who create the internal organizational environment. If an unspoken culture that does not want to acknowledge or discuss problems is established as the norm, it becomes so ingrained that even a change to senior managers do not alter the situation. Change will only happen when senior managers encourage a more open atmosphere, and show their employees how to behave differently.

The HSE head of an electricity production company considers that a manager who does not want to hear the views of subordinates on the problems of an organization does not meet the standard of a good and effective manager as generally defined in most countries.

2.3.2 Reasons Why Leaders Do not Want to Hear About Problems from Their Subordinates

In addition to the text, cartoons will illustrate the relationship between owners, senior management, middle managers, junior managers and ordinary employees.

1. **TACKLING REPORTED PROBLEMS WILL BE COSTLY, AND OWNERS AND SHAREHOLDERS ARE IMPOSING STRICT FINANCIAL AND PRODUCTION TARGETS**

58% of the respondents' answers about why managers do not want to hear about problems from their subordinates because the costs of addressing any serious issue in a critical infrastructure company will be very high. In addition, owners and shareholders are often imposing strict financial and production targets on their senior managers already. Reports from employees about any serious safety and technological problems may threaten the implementation of these plans, as well as negatively affecting the career and the earnings of senior managers.

Essentially similar opinions are shared by leaders at various levels in a wide variety of industries. The fact that their views are so similar shows how serious and how prevalent the problem is, occurring worldwide and across different industries.

The vice president of a company building and repairing gas pipelines postulates that in the market economy, the sole criterion by which an organization's performance is measured is profit. The effectiveness of an organization's senior managers is also evaluated according to the profits they deliver to shareholders, and their bonuses are calculated accordingly. When employees come to them with information about the risks of operating equipment, addressing the situation requires expenditure, which will cut profits in the short-term. In truth, many senior managers already understand the main problems faced by the enterprises entrusted to them, but they cannot take the action necessary because of pressure from the shareholders, who are only concerned about the short-term profit that management will deliver. This is probably the main reason why senior managers are reluctant to receive information from employees about known problems.

The HSE head of an oil company points out that responding to the existing problems of a large industrial company requires huge resources to modernize equipment. These additional costs are unlikely to be agreed by owners and shareholders if they see their involvement with the company as a short-term financial venture rather than a long-term strategic investment. These "*short-term profit*" owners, and the managers hired by them, fail to understand that investments in safety are profitable—but only in the long-term, as they maintain the value of the asset and allow losses from accidents to be avoided. In their view, the key factor is the size of the annual profit and not the long-term growth of the value of the asset. Therefore in companies owned and operated by these people, the emphasis is on the growth of production, not the safety of production. Priority is given to profits, not investment in modernization. In these companies, talk of rising safety costs and the necessary investment in infrastructure upgrades is bad news that neither shareholders nor senior management want to hear. They only want to hear good news about how profits and productivity have risen and costs are falling, how employees are embracing the owners' and shareholders' austerity, and so on. If someone dares to reproach the senior management for their focus on short-term profit, or tries to point out the existence of a fault or dangerous practice within a company, their response is to get rid of that person so that they cannot set a "*bad*" example to the remaining staff. In essence, questions about safety lead to costs, and costs are unwelcome news in these companies.

An HSE consultant working mainly for oil and gas as well as air traffic control believes that the most important thing to remember is that senior managers are always accountable primarily to their shareholders. Ultimately, shareholders and senior managers focus mainly on the profits that can be generated by the production process. The costs of mitigating any identified serious risks can be very high and are liable to result in a drop in production, at least in the short-term. All this can negatively affect the implementation of the agreed production plan. Therefore, senior managers are constantly seeking a compromise between production, finance and safety. In many cases, finding a workable compromise solution is difficult. Hence, many no longer want to be informed about problems, with the motto "*what I don't know doesn't bother me*". As a result, small issues are ignored and left unsolved, and risks accumulate, eventually leading to the development of major problems which are likely to be much harder and more costly to put right.

The head of HSE at a mining and metallurgy company believes that the reason some critical infrastructure executives do not want to hear about workplace problems is the high cost of tackling them. In most companies, key performance indicators for managers are all about maximizing profits, and the additional costs of solving problems will inevitably reduce the profitability of their business. Naturally, this will adversely affect what the owners think of the work performed by their managers, and the managers are likely to lose their bonus. For many executives, it just does not make sense to invest in solving problems that will probably not cause anything disastrous to happen over the next few years. When they look at a potential safety issue, they will assess the likelihood of any negative event occurring—and may then decide to take the risk, and maximize profits by avoiding the costs of modernizing equipment. After all, if senior management do not have the resources to solve a problem, they will have no option but to turn to the owners, and that would mean admitting the unfavorable situation at the production site. Therefore, some senior managers, even when informed by their subordinates of serious issues, will avoid raising them with the owners because they do not want to adversely affect their career prospects. In this regard, it is worth noting the irony that the only managers who are really free to make adequate decisions are those who are wealthy enough, or confident enough in their professional security, not to fear losing their jobs. Job insecurity is no respecter of rank. Whether you are operating production equipment or running the company, you know that your job depends on your performance—and if you are afraid of losing your livelihood and your career, you will be cautious about the decisions you take.

The HSE head of a production company, managing a large number of hazardous chemical processes, observes that any manager has somebody else above them. For senior management, this is a board of directors, and even board members are answerable to owners or a pool of key shareholders. Therefore, if employees bring an issue to senior management that will require significant resources to resolve, senior managers will have to take it to the board or the owners, and admit that there are serious problems in the company. Therefore, in order not to jeopardize their own careers, some executives prefer simply not to hear about problems from employees. The employees, of course, get the message that they should only bring good news to the boss. Ultimately, instead of hearing about a potential threat in advance from their subordinates, leaders will only find out when there is an emergency and the situation gets out of control. The staff on the ground knew about the risk all along, but were afraid to inform their superiors.

The executive vice president of sustainability and HSSE at an international electricity company believes that senior managers are in the same situation in relation to shareholders as their subordinates are in relation to them: no one wants to tell their bosses about bad news, everyone wants to shine and the bosses are mostly told only about the good things.

The CEO and chief nuclear officer of a nuclear power operating company thinks that a poor corporate culture typically starts with the behavior of a board of directors or senior executives and then trickles down from the top. Everyone wants to please the person above them and, as a result, an organization will become a reflection of the values and behaviors of its most senior echelons. If the person at the top is all about results, does not want to hear about problems, and is focused on profits, then everybody else in the company must adopt the same values in order to survive. In this scenario, negative information about problems that might lower profits has no chance of reaching the board of directors or senior managers.

The HSE director of an oil company points out that even senior management are subordinate to shareholders and governments, because critical infrastructure is part of national security. If employees inform them that for any reason it will not be possible to meet the targets they have set, the senior managers will be in a difficult position: they have no choice but to go to the shareholders and the state authorities and admit that the goals they have agreed on cannot be achieved. This will not go down well. At the corporate level, shareholder returns will already have been determined. At the state level, the company's projected production output will already have been integrated into national development programs, and budgets will have been set up according to the expected tax revenues. Desperate not to seem incompetent in the eyes of shareholders and the state, the leaders of an industrial company will try at all costs to prevent the disruption of their production and financial plans. If any subordinate tries to warn them that their plans cannot be safely fulfilled, they will make it clear to them that plans must be implemented by any means necessary. Leaders know they can get away with admitting some minor problems, but anything fundamental must be resolved one way or another before they can report it to shareholders or the state. If it seems impossible under any circumstances to do so and still achieve the agreed production and financial targets, then some managers will attempt to prove to shareholders that the plans could not be implemented without unacceptable risks. However, if the plans seem feasible to managers at what they feel is an acceptable level of risk, they will do what it takes to achieve them, despite the objections of employees.

The head of the Nuclear Design Department of a multinational electric utility company believes that in addition to financial and economic demands, leaders of large projects are under political pressure to develop their infrastructure. The owners and senior management will often have made overly optimistic promises to politicians about the progress they expect to make in developing their facilities. Therefore, they do not want to hear from their subordinates about problems which will cause delays and additional costs during the commissioning of new critical infrastructure.

The HSE manager of a petrochemical company reports that some executives believe that, if they are informed of a serious problem—which even they do not have the resources or expertise to solve—they will have to go to shareholders and ask for additional resources to tackle the situation. This puts them in the same situation as their subordinates, who could not solve the problem on the spot without involving their superiors. Most managers feel that staff members should solve any issues they encounter, but that if their subordinates cannot do so for any reason, the managers themselves will have to deal with it. If a senior manager has to ask shareholders to allocate additional resources to deal with a problem, shareholders may well reconsider their decision to appoint that manager in the first place. After all, most shareholders of large assets have the same attitude as their managers take to their subordinates: "*We hire people on a very good salary, and we expect them to solve a company's issues without having to draw our attention to the problems that arise*". Naturally, senior managers do not want to give shareholders a reason to doubt their ability to hit their profitability and production targets and make effective decisions. Many company executives, especially in finance departments, believe that if a figure is set for expenditure per year, it is unacceptable to exceed this estimate. The last thing managers want to hear about is an unforeseen problem, when they know they cannot find funding to solve it within the approved annual budget.

The head of the HSE department of an oilfield service company has a similar opinion: shareholders hire top managers to show a positive result in the form of growth in profits and output volumes, and they expect them to deliver. Therefore, the ambitious financial and production plans set by shareholders dominate the work of senior managers. When they get a message about a serious production problem, they know that it will require enormous resources to solve—stopping production, lowering profits, and increasing costs. And as soon as senior managers respond to the message, they are on record as knowing about the problem. They will now need to contact the shareholders and explain that they can no longer guarantee that financial and production targets will be met, and that they will need to increase costs in order to solve accumulated problems in the workplace. Obviously, the shareholders will not welcome this news, because they expect senior management to solve any problems that come up and still meet their targets. Some senior managers will therefore simply ignore warnings from subordinates, hoping that they will find a way to solve their problems independently. Managers can very easily make it clear to employees that they do not want to hear about problems, but only want to hear about positive results.

The head of HSE at a steel company has also been a consultant for dozens of large industrial companies, and has come across several cases where senior managers knowingly ignored negative information from subordinates about their company's operations. In every single case, those managers were working on a very ambitious development plan for a company, imposed on them by shareholders. Senior management refuse to respond to problems, even those constantly reported by various employees, if tackling them might stop them meeting targets set by the owners. They are afraid that the owners or shareholders will question their professional competence if they seem unable to independently solve complex problems and achieve the planned production and profit levels. With neither the resources nor the motivation to change anything, they would rather not react in any way to messages about problems coming from below. In this situation, the lack of objective feedback in an organization is primarily related to the goals being set by shareholders and the resulting actions of senior management, rather than any reticence from employees. In other words, the concealment of information about risks does not stem from the bottom of the corporate pyramid—between ordinary employees and lower or middle managers—but primarily arises from the communication between top management and shareholders, and between the middle and top tiers of management. Key performance indicators (KPI) and management bonuses also tend to be tied to the production and financial plan of a company, which has been defined by shareholders. Anxious to meet the annual performance targets set by shareholders, managers prefer not to respond to problems. The head of a workshop at this steel company once told the respondent in a private conversation that the only people who survive in the company are those who solve problems on their own, and do not raise them to the level of their superiors. With a corporate set-up like this, workers will do their utmost to deal with problems independently and complete the tasks they have been set. They will only inform bosses at the last possible moment if there are problems that threaten the financial and production plans approved by management. For their part, managers do not want to hear about risks, because of pressure from owners, shareholders, and aboard of directors. It is incentives and priorities set at this top level that make senior management unwilling to hear about problems being raised from below.

The HSE head of a mining company says that, in recent decades, many owners and shareholders of critical infrastructure companies have begun to view their assets as short-term investments, from which they can squeeze the maximum profit in a short time without making investments.[12] That is why employees and managers who object to such short-term opportunism are driven out of an organization, while

[12] The authors of the handbook consider that such short-termism may be due to (I) the trend towards financialization of economic activities since the 1980s, and/or (II) a growing uncertainty about regulations as well as geopolitical developments and/or (III) the sustainability of operations involving

the rest are expected to dutifully take whatever risks are required to achieve the owners' or shareholders' aggressive corporate goals. Employees who are willing to go against the compromising tide—to defy the leadership at all levels of a company, and jeopardize their own career and livelihood in the process—are extremely rare. Most staff prefer to tow the company line and take the production risks they are tacitly expected to. With no honest feedback from their subordinates, owners and top management impose more and more ambitious and unrealistic production plans, either ignoring or quite possibly having no idea, that their employees on the ground are already taking dangerous risks to fulfill existing targets. In the opinion of this respondent, the position of owners and senior management thus has a key influence on the practice of concealment/disclosure of risks by middle and lower management, as well as by employees on the shop floor. If the owner and the executives do not want to know about problems, they will not hear about them. At every level of management, all the way up to the boardroom, the information passed on will be "*filtered*".

The SSE manager of a gold mining company notes a pattern that he observes when preparing the annual work plan of industrial companies for the next year. Managers at different levels always try to plan their activities based on the most optimistic forecast for profitability and cost reduction. This automatically means that the bar of corporate achievement is constantly rising, affecting all employees, and that the senior management's perception of the real situation within an organization is distorted. At the end of the year, a company is inevitably faced with the reality that it was not possible to fulfill its overly optimistic plans. Disappointment and fatigue from the "*failed*" year sets in, but the problem was that unrealistic goals had been set from the beginning. According to the respondent, senior managers should be ambitious, but realistic.

One lower-level manager, more directly responsible for the operation of a critical infrastructure on a daily basis, shared his company's existing practice. Every year, plans are set for 2–5% growth on several business indicators. Accordingly, senior management set ambitious goals, and no one at the top wants to hear from subordinates that these goals cannot be fulfilled: "*Every year you need to show a better result than in the previous year. In reality, this cannot be achieved without additional resources or without the threat to other indicators, including safety. But fewer resources are being allocated, and plans and KPIs are becoming more ambitious. We cannot object to this system—if we do, we will be perceived as disloyal employees. Therefore, no one can write at the end of the year that he was not able to fulfill the plans set by the management. In all reports the situation is embellished from what is really happening. However, senior management are happy to receive such reassuring reports, which have little to do with the reality of what is happening on the ground. Going openly against such plans is suicide for the career of any employee. After all, the senior management will not listen to anyone, but simply mark down such an employee as a rebel, a direct threat to the course intended by top management. All this resembles a fairground attraction with a squirrel in the wheel, when every minute*

activities related to fossil fuels that are in the crosshair of politicians with plans to phase out fossil fuels and other polluting activities in the coming decades.

the squirrel is forced to run faster and faster. But muscles have a limit, whereas the company has no growth limit and demands that everything must keep improving year after year".

The vice president of an electricity company and the head of a large power plant both maintain that sometimes senior managers are aware of serious problems that employees have reported to them, but do not have the resources to solve them. This is mainly because the shareholders who appointed senior management consider their investment in this asset as short-term, and as such, all they want to hear from senior management is positive news: rising profits and measures to reduce costs. As a result: (I) senior management know that shareholders will be extremely negative about any requests for increased production upgrade costs to solve operational problems; (II) senior managers cannot admit to employees that they lack the resources to eliminate problems and thus in reality are unable to solve them, as admission of their helplessness to subordinates will undermine their authority as managers; (III) when resources are limited, senior management must be creative if problems arise, and make difficult decisions to find resources within an enterprise to solve them. Therefore, senior managers will do everything they can to avoid hearing bad news from employees, shutting down the communication of negative information and delegating the solution of any arising problem to the middle tier of production site managers. But middle managers also lack the resources to fix problems. They too try to avoid receiving bad news about problems that they cannot solve, preferring such problems to remain at the level of lower management. These lower level managers do not want to hear about problems from their subordinates either, and insist that they are solved by the workers who have spotted and reported them. In the end, the only resource available to deal with risks is the ground level staff at the production site, who come to managers of different levels and warn them of these risks. It turns out to be a vicious circle: when employees complain to management, who have extremely limited financial resources because of the position of shareholders, they put themselves in conflict with management. Ultimately, by blowing the whistle, the workers operating the hardware take the burden of solving the problem on themselves, because they are in fact the management's only resource to tackle it. This is how feedback collapses within a large critical infrastructure company: information about risks and problems remains at the grassroots level. The entire management hierarchy is in a state of euphoria from all the positive news about the growth of financial and production indicators, the transmission of which is welcomed. However, the risks are not being tackled and the company is moving towards a catastrophe, which will come as a nasty shock for senior management and shareholders as a result of their approach to the allocation of resources. Such short-term financial target-setting turns out to be a false economy: when disaster strikes, dealing with the consequences will require hundreds, if not thousands of times more resources than the funds that would have been required to reduce risks when they were first noticed, often many years before the emergency, as already mentioned.

A safety consultant with experience in various industries notes that when he works with supervisors and rank-and-file employees in the workplace, people complain that higher and middle managers ask them to carry out a production plan, but none of the senior managers asks about the risks associated with its implementation. In other words, managers often directly push those operating their facilities to violate safety standards to implement the plan, and do not want to discuss how to deal with the risks involved. This is a one-way (monologue) communication system, where orders come from above to achieve certain results, but there is no feedback channel that would allow employees to express their views in an honest dialogue with their superiors.

In this regard, one thermal power plant director (middle management) relates an interesting example of the feedback he once gave to senior management. Headquarters had asked the power plant entrusted to him to increase power generation, reduce costs, and increase productivity to improve the annual financial performance of the energy company. In response to this, he told senior management at headquarters that he would be able to ensure a multiple increase in the revenue of the power plant and the productivity of his subordinates in the very short-term, but only by cancelling planned repairs of equipment and laying off most of the employees. He continued by stating that, in the first year, the plant would show the phenomenal growth of efficiency they were asking for, but most likely, it would be the last year for the plant to produce electricity. There would be an exponential increase in equipment accidents and failure, likely leading to a major incident, which would destroy the plant's equipment. As a result, such a short-term strategy of maximizing profits and minimizing costs would result in a sharp increase in financial efficiency over the short-term, but lead to a long-term depreciation of fixed assets. Having listened to this argument from the director of the plant, the senior management at headquarters stopped sending ultimatums to the power plant to increase production and the productivity of workers by any means necessary.

The HSE head of an oil company expresses the view that responding to the problems of a large industrial company requires huge resources to upgrade equipment and train people to behave safely in the workplace. However, owners and managers hired by them do not understand that investments in safety in the long-term are profitable, that they will preserve the value of the asset and avoid losses from accidents. Priority is given to profits, not to investments in modernization or safety of production. In such companies, talking about rising safety costs and necessary investments in infrastructure modernization is bad news that neither shareholders nor senior management want to hear. They only want to hear good news about how profits and productivity have grown, how costs have been cut, and how the workforce approve of the course of austerity set by owner and shareholders. If someone starts to reproach senior management for following such a course, or tries to point out the presence of very dangerous practices in the company, they will be silenced and most likely dismissed. In general, talking about safety involves cost, and costs are bad news in such companies.

According to the HSE manager of an oil company, many rank-and-file employees subconsciously divide their activities into two components: meeting production targets (drilling indicators, volume of oil, etc.) and managing industrial safety. Of course, management will approve of employees who help them achieve industrial indicators and results. Low accident rates and other actions to improve industrial safety are also approved by managers but, in many industrial companies, this is perceived as a secondary task in relation to achieving production indicators. If employees offer management something that will further improve industrial performance and financial results, these tips are seen as innovative ideas and rationales ("*good news*") and get quick approval from senior management. Information on industrial safety risks and problems, however, tends to be seen by senior management as a headache ("*bad news*") as it will require additional financial resources and time to solve the problems identified. Subconsciously, managers want to hear good news from employees, not bad news, and employees instinctively know what senior management want to hear. Being part of the good news story about achieving industrial performance can make a career within a large company, so many employees invest in this area of work. The same goes for priorities: employees for example prefer to focus on drilling and extracting more oil than filling out STOP [Safety Training Observation Program] cards on the observed risks or taking a leadership course in industrial safety. It is natural that they prioritize these because they know their managers are paying attention to production achievements, not to safety issues. Unfortunately, this leaves most employees with a feeling that industrial safety issues are secondary to production results. It all stems from the attitudes determined by senior management. Accidents are rare—many employees have been working for years and never faced serious emergencies—and managers constantly demand more progress towards the industrial indicators they have agreed with the shareholders. Inevitably then, the focus of communication between employees and senior management is on results and achievements, not on problems. Moreover, employee salaries are generally tied to performance, and safety issues are seldom integrated into pay systems in the same way. Every factor motivating employees is pushing them to neglect technological risks in favor of meeting performance targets.

2. **MANAGERS ARE AFRAID OF BEING SEEN AS INCOMPETENT IF THEY TAKE RESPONSIBILITY FOR PREVIOUS BAD DECISIONS THAT HAVE CREATED CURRENT PROBLEMS**

38% of respondents believe that when employees inform managers about any serious problem or risk, they are highlighting the bad decisions and mistakes made by managers in the past that have led to the problem developing in the first place. So rather than admitting that they may have made a mistake, managers try not to hear about or respond to current problems.

The vice president of a gas pipeline construction and repair company points out: "*Problems don't just come from nothing!*". In other words, every problem is the result of a series of poor managerial decisions.

The HSE head of a petrochemical company says that some executives consciously support the myth that they are infallible, so that their credibility remains unshakable in the eyes of colleagues, subordinates, and shareholders.

An HSE manager of a steel company observes that a problem will often have developed in the first place because senior managers in a company made a bad decision. As such, acknowledging the problem is seen as an admission of guilt.

The HSE head of a mining company takes a similar view. When an employee informs management about a problem, it turns out in many cases that the problem stems from previous erroneous management decisions: as the fable puts it, the emperor has no clothes. For many managers, such a message from a subordinate is tantamount to their own public humiliation and an accusation of unprofessionalism. As a result, in order not to lose face in the future, some managers decide to remove the "*offending*" employee from an organization as soon as possible. It is enough to publicly dismiss one such detractor: after that, most employees will think twice before criticizing the executives. Rather than speaking unpalatable truths, they will send only reassuring and inspiring reports. Many executives do not have the wisdom to make sure they have an honest adviser—someone they can trust to give them an objective assessment of the situation in a company and, when necessary, an unflattering evaluation of their own decisions. In the end, employees just give up reporting to their seniors about any problems that management had a hand in.

The HSE director of a steel company cites some of the tactics employed by senior management when the question of previous mistakes is raised. Especially in organizations that use a system of punishment for mistakes, they will try to prevent any information on the issue coming out. For example, if a company decides to build a new workshop without proper consultation on environmental risks, and a problem later emerges there, management will do all they can to cover up. After all, solving the problem will require huge costs: it may be necessary to resettle residents, scale down production, or upgrade expensive and possibly recently bought equipment. Not wanting to take the blame for this extra expenditure, managers will simply not tell shareholders about the problem. Leaders are very afraid of looking incompetent, so they try not to associate their name with any decision that turns out to have been a mistake.

A project manager in oil and gas exploration notes that in some organizations, executives are promoted before the consequences of their poor decision-making in previous managerial positions come to light. This includes occasions where subordinates had raised objections to a manager's plans but had been ignored, even though this subsequently led to a problem for an organization. Many managers who were involved in the genesis of these problems in their previous position use every possible strategy to prevent senior management eventually finding out, in order to avoid being marked down as incompetent and quite possibly dismissed.

An HSE manager at an oil company thinks the problem is with the egos of some executives, who despite their best efforts to achieve good results, end up failing. Leaders do not want to admit that they did not succeed. They do not want to admit they failed.

The HSE manager of an oil company draws attention to the fact that many modern executives may have training and experience in economics or finance and administrative management, but often have no idea about the specifics of a manufacturing process or the management of industrial sites, let alone industrial safety systems. Therefore, they delegate responsibility for decisions in these areas to subordinates who have competence in them. In order to make adequate decisions, executives need the support of a team of high-quality technicians who can assess risks and make informed technical decisions. Many executives with an economic background simply do not have such a team, so they try to avoid having to deal with technological issues by not hearing about them. For executives who lack the experience to understand the technical aspects of an industrial enterprise, leadership on HSE issues is very onerous. Rather than having to tackle them, they relegate issues of production safety to the background. If senior managers are not willing to communicate on these issues with employees, the employees will not communicate with the managers. If executives do not travel to production facilities or communicate with ordinary employees on a regular basis, employees will not risk passing information up the hierarchy, and no one will report new or existing problems to higher ups.

3. **SENIOR MANAGEMENT ASSUME THAT ONCE THEY HAVE BEEN TOLD ABOUT A PROBLEM, THEY WILL NEED TO SOLVE IT**

36% of respondents think managers are afraid that, if an employee informs them about a problem, they will then have to solve it.

Some leaders think: "*Why is he telling me about this—it would be better if I didn't know anything*". They do not want to get involved in solving the problem, so would rather have the employee find a solution, and come to them with a plan. Or better still, the employee would solve the problem without even notifying senior management, and only inform them about the successful solution. There is an interesting analogy for this: when an employee tells a manager about a problem without solving it, it is as if the employee is passing a monkey sitting on his shoulders over to the shoulders of the manager. Some leaders may feel that the monkey belongs with the boss, not the employee, but if all the "*monkeys*" of a whole department end up sitting on one boss's shoulders, the boss may collapse under the weight. This is what some leaders fear.

4. **SENIOR MANAGERS EXPECT EMPLOYEES TO SOLVE PROBLEMS INDEPENDENTLY IN THEIR AREA OF RESPONSIBILITY**

28% of respondents note that some managers prefer not to pay attention to warnings from employees, because they believe that employees are well enough paid and should be able to deal with problems that arise independently, without involving them.

The manager of a production site at an oil company (middle managerial level) observes that some senior managers do their best to make sure difficult issues are resolved at the level of their subordinates, to whom they have delegated the responsibility and authority to make decisions. They tell their subordinates "*we pay you a good salary to solve problems at your level and not bother us with operational issues*".

The HSE director of a mining company describes the problem as follows. Some senior managers believe that middle and junior managers are paid a decent salary precisely so that they should solve problems in their area of competence, and look for ways to mitigate risks with the equipment entrusted to them, without troubling senior management. Employees working under managers like this are forced to conceal the true situation when reporting to them, in order to survive in an organization and protect their career prospects. The constant unwillingness of managers to listen to employees over a long period of time can only result in one thing: a culture of lies at every level of an organization.

A risk management consultant in the oil and gas industry notes that many management and leadership books promote the creation of an internal corporate culture of only good news. They recommend that senior managers give the following directives to their subordinates: "*Don't come to senior management with problems—only come with problem solving*". Many years later, when their subordinates become senior managers, they will repeat these instructions to the workforce: "*Don't bring me problems, just bring me solutions*". These attitudes are deeply rooted in modern management philosophy in the form of the so-called "*can-do*" culture. However, this encourages employees to hide serious problems if they cannot suggest or implement a solution. They are likely to think: "*Well, if I have a problem without a solution, I'd better not say anything*".

5. **SENIOR MANAGEMENT PREFER NOT TO KNOW ABOUT RISKS, IN ORDER TO AVOID BEING HELD RESPONSIBLE (INCLUDING LEGAL RESPONSIBILITY) IF THINGS GO WRONG**

27% of respondents feel that some managers do not want to hear about existing risks from their employees because they do not want to be held legally responsible for an accident or emergency. Some managers—irrationally but perhaps understandably—believe that, if risk information does not reach management, the responsibility for the onset of an emergency remains entirely with their subordinates who are managing the facility involved. This has some basis in experience. During investigations in several countries following major accidents in critical industries, some senior executives were able to avoid criminal liability because they claimed they had not been aware of the problems that ultimately led to the accidents—while their subordinates, unable to plead ignorance, were punished.

One HSE manager of a steel company postulates that as soon as a manager knows about a risk, they share the responsibility for mitigating it, but as long as the manager is in ignorance, the employee carries sole responsibility.

The board director and site manager of a chemical company points out that from an organizational point of view, if managers do not know about a problem, then they are shielded from the obligation they would have if they knew to inform their superiors or their subordinates about it. From a legal point of view, the authorities cannot punish managers if they did not receive information about the risks that led to an accident. All this motivates managers to remain deaf to bad news.

The head of HSE of an oil company believes that, for many managers, receiving information from subordinates about a problem inevitably means that: (I) they should be involved in its solution; (II) they will have to immerse themselves in the problem and it will take up their time; (III) managers do not always have the competence to make a good decision, and therefore (IV) they will need to spend time learning new facts or skills so they can make an informed decision about the problem; (V) they will be held accountable, both for their commitment to tackle the problem and for solving it. Summarizing all this, it can be said that some managers would rather not know about problems or be part of their solution, but would rather leave solving any problems that come up to their subordinates or other managers. That way, if the problem cannot be solved, they can point the finger at others and not be held responsible. In the respondent's opinion, this is a universal corporate "*disease*".

The senior manager of an electricity company puts the situation more specifically: often the "*system of silence*" about the risks within an industrial company is actually created by senior management. This is primarily because senior managers want to reduce the extent of their responsibility in case an accident occurs at production site, especially if it results in major damage or even death. In organizations where this "*system of silence*" prevails, as the situation progresses upwards through each level of management, real-life information is adjusted to improve the assessment of existing risks. This means that the responsibility for risk prevention moves further down the managerial hierarchy. When an emergency occurs it is the grassroots employees, who independently took risky decisions, that are found legally liable. Meanwhile, their senior managers can say that they did not receive any warnings from them—and in support of this claim they can duly provide embellished reports, previously sent up the corporate hierarchy by their subordinates under the unspoken "*system of silence*" about problems.

The HSE manager of a gold mining company confirms the existence of this practice in many companies: it is in the interest of managers to shift the responsibility for risk management down to ordinary workers. In the investigation following any incident, few people pay attention to the underlying system of over-ambitious tasks set by senior management, which puts ordinary employees under pressure to violate safety rules to comply with unrealistic targets. If employees disagree with the guidelines of their leaders, they will be fired. Understandably, most employees dutifully follow the orders from on high, even though they do not agree with them. Fear of losing their jobs forces employees to keep their heads down, ignore their better judgment and violate safety regulations.

The HSE head of a chemical company believes that, in many companies, managers do not even think they are responsible for safety. Instead, this accountability is delegated to the HSE department, and the HSE director must single-handedly steer the safe operation of a company. Senior managers do not believe that they should be responsible for safety, so are unwilling to receive information about risks. In their opinion, these should be addressed to the heads of HSE units, leaving them unscathed by any bad news so that they look better to the owners and shareholders. This arrangement suits the managers, because they can avoid responsibility in case of an emergency at work.

The HSE director of an oil company cites an example of how she tried unsuccessfully to persuade a senior manager from the parent company to visit the company's foreign facilities. This executive preferred to limit his supervision to the work of a limited number of facilities known to him in one country, even though he was formally accountable for facilities in several other countries. The executive did not officially admit to making no foreign site visits, but instead avoided them under various pretexts, mostly that he had been too busy. However, the facilities entrusted to him were in poor technical condition and the probability of serious accidents was high.

The HSE head of a mining and metallurgy company has repeatedly confirmed that during their trips to production sites, some managers do not allow any new risks identified to be documented. Instead, they insist that subordinates make oral reports. That way, in the event of an emergency, there will be no legally binding documents that could indicate that the risks that ultimately led to the emergency had already been identified, and that senior management were perfectly aware of them.

In some countries, the legal liability in the event of an accident falls on site foremen and shop supervisors. In other countries, the industrial directors or senior managers bear the primary legal responsibility for accidents. In the latter case, senior managers have far more motivation to immerse themselves in the problems of the production sites. They work hard to get information about risks from their subordinates as early as possible, and long before these risks become uncontrollable.

6. **LEADERS DO NOT WANT TO STEP OUT OF THEIR COMFORT ZONE TO SOLVE COMPLEX QUESTIONS**

26% of respondents note that some leaders do not want to step out of their comfort zone, change their routine or take on extra work to respond to problems their employees have warned them about.

The HSE head of an electricity company points out that on hearing bad news, senior management should act. However, some managers do not want to leave their comfort zone to immerse themselves in solving a problem. The last thing they want to do is leave their cozy headquarters, and rush to some remote region to deal on the spot with trouble at one of production sites.

The HSE director of an oil company makes the same point: some leaders prefer not to leave their "*ivory towers*". As long as the bonus comes in, they would rather not know about problems on the ground, and thus avoid responsibility for their solution. The main reason why a company hides information about risks is the attitude of the leadership, which for many years has punished the bearers of bad news and made it clear that only positive and comfortable news is welcome. He attributes this to the reluctance of managers to leave their comfort zone: they would rather "*sit it out*" in the hope that the problem will somehow "*blow through*" without personally affecting them, or upsetting the beautiful positive picture they have built around themselves.

7. **LEADERS ARE PEOPLE TOO—LIKE ANYONE, THEY WOULD RATHER HEAR GOOD NEWS THAN BAD ONES**

24% of respondents point out that one should remember that leaders are just humans underneath, and it is just human nature to prefer good news rather than bad.[13]

[13] Dmitry Chernov, Didier Sornette, Giovanni Sansavini, Ali Ayoub, Don't Tell the Boss! How poor communication on risks within organizations causes major catastrophes, Springer, 2022; subchapter 2.2 Results of other research on the challenges of voice and silence in an organization, https://link.springer.com/book/10.1007/978-3-031-05206-4.

A psychologist and consultant in the field of organizational behavior shares his professional opinion: there is more fear at the top of the corporate hierarchy than there is at the bottom. The higher up the ladder you go, the greater the entropy. Uncertainty causes fear, and managers have to deal with greater uncertainty and have greater responsibility in making decisions. Managers, like their subordinates, want a safe environment around them and want to feel safe themselves, as a result, they prefer to hear good news. Bad news strikes fear into managers because it reduces the feeling of security. Subconsciously, managers do not want to work in a dangerous or threatening environment, so they do not want to hear about problems that could upset their inner peace. When they are faced with a threat, they have the same instincts as their subordinates: avoidance, paralysis or aggression. This is why managers do not want to hear about the problems their subordinates bring to them. This can easily create a vicious circle where fearful executives respond with scorn or aggression to messages from their employees, which in turn raises employees' fears. Employees stop passing negative information upstairs, because they are terrified of the consequences for their own careers.

The HSE director of a mining company said that everybody wants to hear mostly good things. Leaders are people too. So when the manager makes a call to a production site with the question: "*How are you doing?*" he expects to hear "*Everything is fine*". Naturally enough, subordinates want to please the boss, so if the unvarnished truth is likely to be upsetting, they tend to embellish the situation to make it more palatable. Many employees do this quite subconsciously. And of course, many managers are tempted to surround themselves with employees like this, who will dutifully follow the implicit instructions of their superiors and say only good things.

The prevalence of this practice is confirmed by the HSE manager of a metallurgy company: many executives like to surround themselves with people who tell them "*sweet*" stories about what great managers they are, how well they are handling everything and how wonderful everything is in a company under their leadership. However, when things go wrong in the areas of labor protection, fire safety or industrial safety, the message is clear that the company is not doing quite so well after all. But even then, with "*don't upset the boss*" as company practice, many employees will avoid bothering their superiors with inconvenient questions about the causes of accidents, as they do not want to spoil the cozy world picture the managers have built around them. In companies like this, the blame for accidents or "*close shaves*" will fall on specific employees in a specific area, thus avoiding the unpleasantness of looking at systemic shortcomings in the work of the organization as a whole.

The head of the well construction and repair department of an oil company believes that you should not ignore the personal strength or weakness of individual managers. Some can calmly assess problems and take fearless action to solve them; others, unfortunately, are afraid to bear responsibility, so they take a formal approach to finding out about problems, avoid initiating any decisive action and try to create bureaucratic barriers in order to protect themselves.

8. MANAGERS SEE ISSUES REPORTED BY EMPLOYEES AS UNIMPORTANT

23% of respondents express the opinion that most of the problems employees bring to senior management are insignificant from the managers' perspective. As a result, some executives are reluctant to hear about the concerns of rank-and-file employees and do not want to have to respond, as for them these are minor issues. But with this approach, there is the chance that vital information about a critical risk may be overlooked.

The HSE head of an oil company sums up the situation: senior managers tend to think that most of the problems subordinates face are local issues that the employees should deal with themselves, rather than wasting management time.

According to the safety consultant with experience in various industries, some managers have a pretty low opinion about their employees and think there is little point in getting feedback from them. They assume that all they will get from their employees is complaints, and do not expect to learn anything useful.

The head of a power plant has a similar opinion: 99% of appeals to him from ordinary employees are about better wages, reducing the workload, or considering one of the employee's friends or relatives for work at the plant. It is extremely rare that an employee comes to him from the shop floor to discuss the risks of equipment operation or safety issues. The direct channel is intended exactly for communication about operational safety matters, but the issues that concern ordinary workers are things that relate to their material well-being, not to the critical risks of the enterprise.

The HSE head of a fertilizer mining company illustrates the problem with some examples. A group of employees running mining harvesters in a mine will think more about the health risks of working on a particular section of the mine or how to reduce the likelihood of staff injury rather than about the risk of equipment failure. The interviewee notes that, in his experience, very few people working on the shop floor are thinking broadly enough to predict the consequences of equipment failure or consider the critical risks of the entire production plant. A direct line has been set up in the company so that the workforce can warn decision-makers about the risks they have become aware of or suggest measures to reduce them. But instead, employees mostly complain to their bosses about low wages, demand more comfortable working conditions and so on—which, as far as management is concerned, is just more bad news which does not increase the added value of the product. Managers are reluctant to communicate with ordinary employees because the value of such communication for the growth of productivity and profits is almost zero. It simply generates a huge heap of problems, which cost money to solve and do not increase profits. For example, let us take the old steps in the mine. If a person slips on them, he/she may get injured. For senior management, the issue of replacing the steps seems inconsequential. Instead, it is easier for them to just tell the staff to go more carefully. The bosses are willing to spend a few pence to have signs put up warning people to be careful on the steps but will not allocate money to renovate or replace them. It seems obvious that this is a problem that could be solved promptly and that tackling it properly would be in everyone's best interests. But in fact, managers are reluctant to deal with "*small issues*" that do not increase the added value of the product, creating only a wave of routine tasks in the field of occupational health. Replacing all the old stairs and ladders in an enterprise does not lead to improved product quality or speed up production time—"*all*" it does is slightly improve the injury rate but with increased costs and reduced profit margins. It is worth noting that all this is more likely to occur when the law only imposes low fines and penalties on companies for industrial injuries and deaths at work. If every serious injury or death of an employee cost a company a significant amount of money and made them financially responsible for creating conditions that caused such accidents, then managers would recognize that rectifying faults in non-critical equipment would be cheaper for the company than the costs arising from financial penalties and employee absence following injuries.

According to the SSE manager of a gold mining company, there are many small problems that bother ordinary employees at industrial sites every day. Workers disclose these problems to their seniors and anticipate their solution. However, the managers generally reply that these problems are insignificant, and thus not worth dealing with. In this way, the managers imply that they only want to hear about big, serious problems from their subordinates.

As serious critical problems appear quite rarely, this dismissive attitude from senior managers towards minor problems reported by their subordinates may well mean that information about a critical risk does not reach the ears of management in a timely manner. If employees see that managers will not take action to tackle small problems which are personally important to them, they may decide that it is pointless to inform them about more important issues—issues that could eventually affect the work of the entire enterprise.

9. **SHORT-TERM CONTRACTS FOR MANAGERS**

19% of respondents note that the reluctance of some managers even to hear about serious problems, let alone get engaged in solving them, is influenced by their short-term contracts as part of the short-term corporate goals they are expected to achieve. These respondents feel that the short-term contracts of senior managers (up to 3 years) are detrimental to creating a favorable environment for the reporting of information about risks. Leaders feel under pressure to show shareholders a quick positive result and, therefore, they are unwilling to receive bad news about production issues that will require time and money to rectify. Solving serious problems in critical infrastructure companies generally requires sustained effort over many years. The technological cycle of renewal of fixed assets can take many years or even a decade. The design and regulatory approval of a project alone will require at least three years.

The HSE head of a mining and metallurgy company emphasizes that the main reason for the priority of profit over safety is the desire for short-term results. In many companies, corporate strategies are planned over 3–5 years, so the maximization of profit takes precedence, and industry leaders agree to a KPI to produce a given rate of profit within just one year. Few bosses even have as much as a three-year action plan. If a company is only working towards short-term development goals, there is no need to prioritize safety above profitability. Therefore, many executives focus on "*quick wins*" rather than on tackling longer-term issues.

The HSE director of a mining and metallurgy company points out that managers' performance is judged mostly by short-term profitability, and their work contracts are also short-term (1–3 years). As a result, most of them will choose to maximize profits and increase personal bonuses to the detriment of solving serious problems or investing in costly equipment upgrades. At the same time, it is worth noting that most companies do actually have the resources to solve serious problems—but at the level of owners and top management, there is a consensus on the priority of bringing in short-term profits and minimizing costs.

The HSE head of a petrochemical company postulates that the short-term nature of executive contracts is dangerous for companies managing critical infrastructure. If a management team is "*in it for the long haul*", it is not in their interest to leave risks unmitigated, as this could lead to an accident that will adversely affect the success of the entire company. Therefore, it makes more sense for managers operating critical infrastructure to be offered contracts with no end date. This will encourage them to remain with a company for a long-time, to employ forward-looking strategic thinking, and view the success of both themselves and the company as one and the same.

The head of a power plant (middle management) also makes the case for longer-term contracts for senior management. According to him, 3–5 year contracts for executives have recently become more popular. This is an extremely flawed idea because it does not motivate the heads of large industrial companies to deal seriously with equipment problems, which may well have a payback period of several decades. Such short contract terms do not motivate managers to think strategically and develop solutions with an eye for 20, 30 or even 50 years ahead. Understandably enough, they will plan to work out the specified short period of their contract, show a short-term result and then move on to another company. Of course, such managers do not want to take responsibility for solving the serious problems that arise from the operation of critical infrastructure. After all, solving major problems requires significant resources. The bill for these will ultimately be passed on to shareholders, and spoil the rosy picture the senior management want to maintain regarding the success of the company under their charge. Leaders must take responsibility for any decisions they make in solving problems, but the trouble is that solving problems inevitably leads to a decrease in the profit margin of the business. If they go to the owners and shareholders with news like that, they will not be welcomed, even though they are only trying to make the business more sustainable in the long-term by actively investing in equipment modernization. For many shareholders, the ideal senior manager is somebody who does all they can to reduce costs, keep the payroll down, and increase profits year on year. All of this works well in the short-term and this is the role model that most senior managers on short contracts try to follow. That is why many senior managers do not want to hear about the risks and problems faced by production sites. In their world of short-term contracts, such information is bad news, which threatens the security of their position. As a result, senior managers make it clear to subordinates that they do not want to hear about problems, and that lower and middle level managers should deal with risks on their own and on the spot, without disturbing their superiors.

The HSE director of a gold mining company expresses a similar opinion. Some managers tend to focus on quick victories instead of serious changes—but experience has shown that quick victories are often followed by a pushback, as they do not fundamentally change the way a company works.[14] Such a short-term approach means that managers do not focus on working with critical risks, but concentrate on minor problems that they can quickly solve. As a result, many managers prefer to ignore information about the more serious problems that employees bring to them.

According to HSE director of an oil company, executives responsible for production are generally very competent in technical terms and have a deep understanding of the potential life of the equipment, based on their own experience. When properly informed about an operational problem with a given piece of equipment, they can usually assess whether it will continue to run without accident for, say, the next two or three years. However, if they are due for promotion soon, they will be tempted to ignore the warning and leave the problem and the costly equipment upgrades it will bring to their successor. As the saying goes, "*Après nous, le déluge!*" ("*We don't care what happens once we're gone!*"). The point is that the moral character of managers, and their willingness to engage in solving problems, are more important than professional competence in many technical matters.

The HSE head of a mining and metallurgy company stresses that some managers do not want to share responsibility for serious problems that have developed in a company over decades. According to the respondent, this is especially true for those who are going to leave a company. They do not want their names to be associated with the solution of some old problem because solving a problem takes time, and may also cause a conflict that will become public and cast a shadow on the reputation of the manager who was in charge. Therefore, they do their utmost to avoid hearing about problems or participating in their solution.

The head of a safety department at a nuclear power plant cites his own electricity company, which forces a mandatory rotation of managers every 3 years. This model of personnel management has its advantages and disadvantages, but in the end, the company cannot identify who exactly is to blame for this or that problem. For example, a problem was recently discovered at one of the company's sites. The former head of the site is already working at another site, and the new head says that he is not to blame for the problem, because it existed before he joined the company. All these short-term contracts lead to a situation where managers prefer an easy life and are reluctant to plunge into solving the serious problems at their current company, knowing that they will soon be transferred to a new job somewhere else.

According to one of the HSE directors of an oil company, managers with very short contracts will only "*bring to light*" a serious problem in their first six months in post, as this allows them to explain it away as something they inherited from their

[14] This is called "*reversion towards the mean*" and is arguably the most powerful predictor of future outcomes. See e.g. Michael J. Mauboussin, The Success Equation: Untangling Skill and Luck in Business, Sports, and Investing, Harvard Business Review Press, 2012.

predecessors. After this period, identifying any problem can be a threat to the career of a senior manager, as he can no longer claim that the problem has not occurred on his watch.

In general, high management turnover can have a damaging impact on safety. The frequently limited tenure of individual managers affects the whole management team's knowledge of the details of the infrastructure entrusted to them. According to one respondent, it will take at least a year and a half simply for a new senior manager to get acquainted with the peculiarities of a company, such as learning what each subordinate manager is capable of, who amongst the workforce can be trusted, and so on. A manager in charge of a critical infrastructure company for many years has time to get to know the company thoroughly, develop an understanding of all the critical risks and weaknesses of the infrastructure, and have a good instinct for the priorities for investment. In this regard, it is significant that some critical infrastructure companies do not seek external applications to fill senior positions, so that all the managers are long-term employees who have passed all the steps of the career ladder within the same company. Until a few decades ago, many critical infrastructure industries operated like a family, with entire dynasties of workers and executives who linked their fate to that one company or industry. This good practice is now largely lost.

10. **A COMMON CORPORATE LEADERSHIP CULTURE PERVADES THE ENTIRE COMPANY AND INDUSTRY**

15% of respondents refer to the existence of generally accepted rules of conduct for leaders in their industry. These often predetermine the behavior of senior managers in any given company, and in turn foster a pathological corporate culture that discourages honest and accurate communication through the corporate hierarchy.

According to the HSE vice president of an oilfield services company, the industry's accepted "*code of conduct*" and a company's accumulated corporate culture together influence how former workers behave when they become executives. As managers, they feel they should behave in much the same way as they saw their bosses behaving when they were still ordinary workers. If there is an unwritten rule within the company that workers should not bring bad news to managers on a Friday night, then a leader who has come up through the ranks will also not expect to hear any bad news on a Friday night. If a subordinate turns up then with a negative report, the leader will say, "*What can I do about this now? You've ruined my weekend! Why couldn't this have waited till Monday morning?*".

The HSE manager of an oil company notes in this regard that in many companies there is simply no expectation at all that senior management will interact with ordinary employees. The system is designed so that managers never communicate with employees on the shop floor, let alone get honest, practical feedback from them. Managers live by the rules of a traditional hierarchy, where senior executives interact only with their immediate subordinates, and they in turn, only with their subordinate middle managers, and so on down to ordinary employees. In top-down organizations like this, information about the real situation on the ground, which ordinary employees observe from their direct experience, only ever reaches the leaders in a distorted state, if at all—because as it is passed up each successive level of management, things are slightly embellished to make them more palatable to the next manager up. If you introduce some feedback tools—for example, a "*master's day*" when the craftsmen tell the chief engineer about their problems, or the senior management visit industrial facilities to meet with ordinary employees—then more managers will be included in the feedback process. But in many companies, there is simply no tradition of such initiatives, and all communication is strictly hierarchical. It is because of this that senior managers never get to hear the truth about the situation at the bottom of the organizational pyramid. These attitudes have formed over decades, and are deeply embedded within the corporate culture of a company and the unspoken rules that senior management impose. Entire generations of leaders have grown up in a culture in which only good news can be brought to the authorities. Therefore, one should not expect quick changes, even if a company's management genuinely want to radically improve the quality of the information that reaches them.

A regional manager, responsible for the operation and maintenance of turbomachinery in the power industry, believes that in some large companies, a corporate culture of "*no bad news*" accumulates over decades, and dictates the behavior of many executives. If individuals—even at a senior level—try and take a stand for more open communication, the system will reject them, and they will be forced to leave the company. The prevailing culture is cumulative, constantly reproducing and reinforcing an atmosphere in which it is better not to discuss problems. Ordinary managers can do little about this situation as corporate culture is almost always formed at the level of a company's board. As a result, a tacit collective agreement not to raise problems develops between the various levels of a company. Ultimately, it is impossible to blame any particular individual for creating this atmosphere.

An HSE consultant believes that most companies simply do not have a benchmark for a more effective model of organizational behavior. People who work in companies with a corporate culture of "*don't bring bad news to the boss*" have probably never seen an alternative way of working where they can calmly discuss problems with their managers or directors without fear of being punished for simply telling the truth. The prevalent corporate culture has its own inertia and those imprisoned in it have no idea of how it might be possible to work differently. This respondent once worked in a regional division of an international safety consulting company. The company preached all the right values about the priority of human life, the need for dialogue between the employee and the leader, etc. But many clients of the regional unit, after talking with consultants there—who, like the interviewee, had come back to the region after several years of work at headquarters located on another continent—believed that these consultants were "*spoiled*" by the high-flown values of a different culture, which encouraged praise rather than punishment, and treating everyone humanely. Local customers would shake their heads and say: "*These are all fairy tales about a good attitude to the workers! We know that in reality, employees do not work without a whip and tight control*".

The HSE manager of a metallurgy company considers that if a company continues to promote a culture of "*no bad news*", it will eventually collapse. The most principled employees will be the first to leave; the rest will keep reassuring the management and the owner, maintaining their self-deception that everything is going well in the company. One major accident at a critical infrastructure facility can lead to financial disaster, change of ownership at the insistence of the authorities and a complete replacement of the senior management team.

OTHER REASONS

No free time (12%)

Senior managers are extremely busy people who often lack the time to (I) comprehensively study safety and technological problems that employees have reported to them, (II) find an opportunity to visit a given industrial site in order to discuss with other managers possible ways the identified problems can be solved, or (III) make an informed decision about how and by whom the problems will be tackled. One respondent explained that some senior managers are so busy from 8 am to 6 pm in constant face-to-face and online meetings, that only from 6 pm onwards do they have any opportunity to focus on examining documents, reports, and e-mails. When managers are under time pressure like this, they often react negatively to any serious problem that unexpectedly crops up, because in order to deal with the situation they will have to completely reschedule their already busy diary, with the result that there will be delays dealing with other, often equally pressing, matters.

Manager reputation (9%)

Bad news undermines the success story that executives wish to project. Many managers do not want to associate themselves with unpleasant events, so they try their best not to respond to—or ideally, not to hear at all—information about serious problems in an organization. They do not want to be responsible for solving them, and would rather concentrate only on successful projects and good news stories. Unwilling to risk his or her career and reputation by having to go to the board with bad news, a leader will try and arrange matters so that bad news will simply not reach them. One of the respondents relates an example of this from his practice. He knew of one mine manager who was legendary in the industry, with an excellent professional reputation, many state awards, and honor and respect among his colleagues. He was also renowned within the industry for never having had an accident on his watch. However, the truth was that there had been incidents, but they never made their way into official reports. To maintain his image as a senior manager in charge of a safe coal mining enterprise, he had told his entire work team that if there was ever an accident, the whole team involved would be fired, and the same would apply if he found out later that an accident had been hidden from him. To keep their jobs, all the miners' brigades kept quiet about any injuries their workers had suffered, claiming they had been injured in their everyday lives. Officially, the company had a perfect safety record for years—but the truth was that incidents had been repeatedly hidden by frightened employees.

Unwillingness to talk about anything bad and reluctance to discuss what a company will do in the event of an accident (4%)

In some large industrial companies, there is no talk of organizational problems at all. The whole orientation of the company is to show off their successes and achievements, because many leaders insist that "*Everything has to go well on my watch*". No one, especially the leaders, wants to talk about problems and risks. If we look at corporate brochures and magazines, everything is about what is good in a company: achieving production targets, the results in those areas of the corporation that are clearly successful. But at the level of official corporate rhetoric, no one publicly talks about organization's weaknesses. There is silence on any area where the corporation is not winning, no acknowledgment that any organization must overcome difficulties in order to become stronger. Successes and strengths are promoted; failures and weaknesses are not officially discussed. Naturally, this leads to reluctance on the part of employees to bother senior management with negative news. In addition, many companies do not make a point of recognizing and praising safe production. Instead, they want to create the impression that this is a given, that there are never problems with safe working conditions. These days most industrial companies are active on social networks, but public discussion about serious safety issues involving a company are quite rare. Employees pay attention to the trends set by corporate media, and unfortunately, production safety does not get a mention. If safety is off the public agenda of an industrial company then its employees, taking their cue, will not want to bother managers with their fears or technical risk assessments. Therefore, much depends on the stance of senior management—if leaders are willing to openly discuss problems within an organization, employees will follow their example.

Few executives regularly discuss what a company will do in the event of a dangerous incident. Special divisions of a company systematize risks and rank them by probability of occurrence. However, at the highest management level of many companies, there is no system for regular discussion of a company's actions in the event of these risks. There is in effect a kind of unspoken taboo that nothing bad can happen in a company. This has a direct impact on the general reluctance of employees and managers to discuss problems.

This lack of a formal dialogue on problems means that they are only discussed behind the scenes within specific units. If there was a company-wide discussion of difficult issues, then everyone could bring their experience to bear on solving them: a problem coming up in a particular division or region will often have been encountered and solved somewhere else in a company before. But because of this organizational silence, each unit has to face problems in isolation, learn from making their own mistakes and try and find their own solutions. Obviously, a more efficient approach would be to use standard, consistent, tried and tested ways to solve a problem. Best practices for solving problems, and lessons learned from similar incidents in the past, should be shared and replicated.

Some executives consider themselves smarter than their subordinates (3%)

Some managers believe that they know better than their subordinates. They have a big ego, and think their job is to give the orders, not to listen to their subordinates. These managers are "*dazzled by their own crown*": they are not interested in going down to the shop floor to talk to their subordinates on an equal basis about the risks they can see. If they did, they would find that those working at the forefront of the production process have a much clearer understanding of many of the technological issues than anyone else in the organization.

Reluctance to learn the lessons of the past, in case management errors are brought to light (3%)

Many managers will not countenance even an internal corporate discussion of lessons learned from past incidents, let alone a public enquiry. The HSE head of an oil company cites an example of this from his practice, when he had been involved in the investigation of a major fire at a gas field. Headquarters had ordered the field managers to launch the field on time, although many safety systems had not yet been installed—indeed, it would not have been possible to start production safely with the project at its existing level of readiness. Many ordinary employees knew there were problems, and some, including field managers, were aware of a small leak of gas condensate. But managers and workers alike were in a hurry to meet the deadline set by headquarters to get the field into operation, so they ignored these obvious minor flaws. Nobody in the field dared to defy headquarters and report that the facility was not ready for production. Their careers and reputations depended on a timely launch. Smart field managers have certain strategies for working round such impossible demands. If they fail to persuade senior management to delay the launch of facilities due to the existing risks, they will do what they are told by the boss and launch. However, once the ceremonial ribbons have been cut and the company top brass and state officials have left the field, production stops immediately until the facility has been brought to a safe state. But in this case, the facility continued to operate after the official launch despite gross safety violations. Field managers wanted to show their superiors that they would do whatever it took to deliver on time and according to the production plan. They were cutting corners to meet their impossible targets, but nobody at headquarters knew about the dangerous risks being taken as a result. Why would they have known when they had made it implicitly clear that they did not want to know? It is possible that the field managers had tried to warn headquarters earlier, but their concerns were overruled, and they were reminded of the deadlines for putting the facility into operation. In other words, the field managers were only allowed to send good news ("*we are on track to meet the deadline*"), and never bad ("*the facility is really not safe for operation yet*"). Fortunately, no one was killed or injured in the fire that resulted from the condensate leak. But it is noteworthy that after this major accident nobody was held to account, and the company was not required by national regulators to change the way they were working. It goes without saying that senior managers were very keen not to raise the issue of what had forced middle managers at the site to neglect safety procedures. They also appreciated that the field managers had been doing their utmost to meet the deadline from headquarters. No one was punished, and in the respondent's opinion, no lessons were learned—essentially because none of the senior managers were interested in discussing what had led to a gross violation of safety regulations, and ultimately to a major fire. Had there been any desire from within the company to understand the root causes of this accident, grave errors at all levels of management would have come to light: the pressure that headquarters had put on the local leaders, and in turn, the willingness of those running the field to allow a gross violation of safety precautions rather than put their careers on the line by demanding postponement of the launch. Conducting such a detailed investigation, which would have called into question the entire decision-making structure of the company, was not in the interest of managers at any level.

A senior HSSE vice president at an international electricity company is accountable for their asset operations around the world. He believes that after an accident, many companies immediately look for culprits in the workforce who violated some internal regulation or instruction. For senior managers, this is the easiest way to apportion blame and much less trouble than investigating the true causes of the incident—which will unearth uncomfortable questions about what motivated junior employees to make errors, and whether they had been provided with everything they needed to carry out their work safely. Honest answers to such questions would inevitably lead to a discussion of the role of the wider organization, and of middle and senior managers, in causing the emergency. Therefore, for most leaders, it is much less dangerous to quickly point the finger of blame at specific culprits within the rank and file, rather than identifying organizational flaws that may be impossible to fix without jeopardizing the careers of many of the senior staff.

Previously reported risks have not led to any incidents (3%)

The HSE director of one mining company observes that sometimes, senior managers hear about critical risks and do not respond to them. A few years then pass, but these risks do not translate into incidents. Some managers begin to believe that the criticality assessments put forward by employees must have been erroneous. Next time a subordinate reports another critical risk to them, they will ignore the warning—they just assume the sender is "*crying wolf*" and overestimating the risk. This approach leads to a dangerous situation where the manager now generally refuses to respond to information about problems, greatly increasing the likelihood of an accident. Based on more than 25 years of experience in the field of safety, the interviewee makes an intuitive observation, not perhaps amenable to logical interpretation: in his opinion, what should go wrong often goes right. There is not always a direct relationship between the information received about a critical risk and its impact any time soon. There are many examples of critical risks existing within a company for many years or even decades without a serious incident occurring. However, after a major incident occurs, things can quickly swing to the opposite end of the scale, where every message about risk is analyzed in detail.

The head of the well construction and repair department of an oil company believes that the problem with the perception of the criticality of risks is that on many previous occasions, ignoring similar risk notifications did not lead to any serious consequences. Therefore, managers chose to ignore disturbing messages from employees, expecting that once again the problem will simply "*blow over*". And indeed, on ten occasions the risk might not lead to a serious incident, perhaps due to a favorable combination of related circumstances. But the eleventh time, a serious problem does arise—quite possibly on account of one changed variable that the manager was probably not even aware of, having refused to receive any up-to-date information about the problems that the field operatives were facing.

State regulation in the field of industrial safety (3%)

In some countries, state authorities monitor the situation with injuries on industrial sites, and if they seem to be on the increase, they will demand that the company concerned take measures to reduce them. But some executives prefer to redefine injuries in order to "*improve*" the statistics for reporting purposes. Legally, they cannot order that injuries be hidden, so they act through a motivation system. For example, senior management can ensure that if there are injuries at a production site, the site manager loses some of their salary. This principle then cascades throughout the company, so that if an ordinary employee is injured, they will declare that the injury occurred outside of work, rather than at the production site. In this way, managers can set specific KPIs to create a climate in which it is beneficial to hide risks. Of course, this affects what it is acceptable to say about risks and incidents within a company. With such a pervasive culture of concealment, it is difficult, if not impossible, to gain any understanding of why dangerous events occurred, to analyze cause-effect relationships, to draw conclusions, or to take corrective measures. Without this key information, it is very difficult to prevent the recurrence of similar emergencies on a company's sites.

It is hard to credit that senior managers could believe it is in anyone's best interest to conceal risks in this way. It suggests a very poor understanding of the importance of safety for the long-term development of an industrial company. If a serious incident is hushed up, it is destined to be repeated, quite probably on a larger scale.

The CEO of a nuclear operating company states that he personally remains open to discussing problems at the nuclear power plants entrusted to him, including when communicating with external audiences and regulatory bodies. However, he observes that some representatives of regional regulators are liable to use such openness from a critical infrastructure company against them. Nevertheless, this senior manager has successfully established honest and trusting relationships with representatives of the national regulator of the nuclear power industry. During their association, a number of problems have come to light, and all have been successfully rectified. Senior management at the company willingly disclosed the problems, actively set about solving them and have thus helped to create a strong culture of safety at their power plants. When the promises of senior management are translated into concrete action and problems at nuclear power plants are effectively resolved, regulators begin to trust the company. In this way, good relations between business and regulator can be built over a period of time, to the benefit of all involved.

Executives hope and believe that a problem will somehow be solved by itself (2%)

The head of risk management at a renewable energy producer points out that some managers naturally prefer to hide bad news and will convince themselves that things will eventually sort themselves out. So rather than rock the boat by telling a board of directors and shareholders of a company, they keep quiet in the hope that the problem will somehow cure itself and go away.

Growth in insurance and fundraising costs in case of risk disclosure (1%)

Divulging information about the deplorable state of an enterprise's assets can have a dramatic effect on insurance fees. The leak of such information can cost a company millions of dollars by increasing the cost of insurance. It may also push up the cost of borrowing from any financial institution. So in preparation for IPOs (initial public offerings on a stock exchange) HSE departments are often tasked with "*polishing up*" industrial sites, even to the extent of sweeping any problems under the carpet.

High-ranking clerks—the gate-keepers at headquarters (1%)

A respondent from the electric power industry states that in his experience, there are many levels of control between those who directly manage equipment (site employees, lower-level and mid-level managers) and those who manage a company at the most senior level. Often the reason that crucial information does not reach senior management is that there is an additional management layer, which is often not officially represented in organizational structures of industrial companies but can be very influential. This layer consists of high-ranking clerks and is located between the heads of industrial sites and senior management. Such clerks include the heads and senior executives of departments at headquarters and represent the third tier down from a company's CEO. These senior figures will often be responsible for filtering risk information, and when reporting to their seniors may distort information to suit their interests. Managers of production facilities (middle management) may not always call their boss (senior management) with their problems, because the CEO of a company is often busy with strategic issues, and it does not always seem advisable to distract them with information about the risks at a particular site. Therefore, the site managers inform senior clerks about their problems, expecting that resources will duly be provided from above to solve them. High-ranking clerks are like a retinue under the king, whose information and advice forms the king's opinion on most matters. These managers greatly value their status as senior advisers and stay very much in their comfort zone. They tend not to be considered responsible for the quality of decisions within a company, as their job basically boils down to passing on information to senior management in a manner that is advantageous for them. Senior managers make the decisions, which will then be implemented by middle management under the supervision of these same senior clerks. As a result, the clerks have great authority within a company, but are not held responsible for their actions. The legal responsibility lies with the CEO of the company (senior management) and the heads of the production facilities (middle management). If the high-ranking clerks are brought in information about a significant risk or problem from a production site, then they will often downplay the scale of the problem to senior managers, tailoring their account to suit their own best interests, or avoid passing it on at all, thus avoiding any responsibility for solving the problem. Sometimes there are letters and reports sent from the production site to headquarters describing safety and technological problems, but the clerks do not reply. If they did, they would have to commit to helping to solve the problems. After all, if the clerk comes to senior management with a problem that has arisen at the site, then they may well decide to make the bearer of the bad news accountable for overseeing the solution. Even worse, the boss can always ask a question: why did the clerks ignore the problem for years, and only now, when the situation has become a threat, come to senior management for an urgent solution? In general, for a high-ranking clerk, telling senior management the truth about a significant production problem can create potential problems for them—both the onerous responsibility for solving it, and the threat to the clerk's career if the boss accuses them of taking no action on it for years.

This interviewee cites another example. Production sites inform headquarters every year about the resources needed for the balanced development of sites over a given period: equipment modernization, investment in the construction of new facilities, etc. Meanwhile the senior clerks are cutting back on resources for these same sites. Often, senior officials and their deputies do not look too deeply into the problems of production sites, trusting their senior clerks to develop optimal solutions for equipment modernization and other production issues. And if a company's shareholders have given senior management targets to reduce costs, then the clerks will know what is implicitly expected of them, and implement a company's policy of cost reduction. For example, with production sites, they will cut the resources allocated for repairs and updating of fixed assets. Moreover, in many companies there is an unwritten rule: do not go over your boss's head. Therefore, when the senior clerks "*slash*" the allocation of resources needed for the routine safe operation of production facilities, the site managers (middle management) cannot complain to senior management about what the senior clerks are doing as this would be seen as insubordination, and would "*expose*" clerks to senior management. If the middle link continues to complain to the top link, then they risk making enemies of senior clerks, with whom they may have to work for many years and who may well threaten their own career prospects. So middle managers prefer not to disagree with the decision of high-ranking clerks and accept as a given their decision to cut resources for the production sites. In their day-to-day operations, middle management just have to try to manage the issue of safe site operation with scarce resources.

In general, senior clerks help senior management to achieve cost savings, and middle management see senior management as innocent, uninformed victims of the ruthless senior clerks (the "*bad cops*"). But to use the analogy coined already, clerks are the *"king's retinue"* helping the *"king"* implement the cost-reduction strategy set by the shareholders. When middle managers gain an audience with senior management, senior managers will play the role of *"good cops"* who are ready to listen and take action, punishing the *"bad cops"* (the senior clerks) and allocating resources needed to solve the problems. Not every mid-level representative is ready to seek an audience with the senior officials of a company. Many are worried that their perceived insubordination will damage their careers if they are viewed at headquarters as upstarts, rebels, or employees who expect special treatment. Instead, many middle managers accept the resource cutbacks imposed by the clerks at headquarters and try to solve existing problems on their own. In truth, senior managers often understand the situation perfectly well, and are working together with their clerks to force the middle management to deal with local issues without asking for help.

If an emergency occurs, no big decisions can be expected from the senior clerks. They will scrupulously avoid taking any initiative and distress messages from the site about the escalating crisis will remain unanswered. Instead, the clerks will try to push all decisions down to middle managers on the ground or up to their bosses. At all costs, they will avoid endangering their own careers by making any decision that would implicate them in the responsibility for dealing with the emergency.

Low-level shift work and lack of collective responsibility (1%)

Some lower-level managers, team leaders and shift supervisors live in a shift-work paradigm and are only willing to take responsibly for what happens during their shift. If you are working with potentially dangerous industrial machinery and you report a risk to the head of the next shift, this lower-level manager may refuse to work the shift and demand that you take responsibility for initiating an emergency shutdown. Disclosure of risks also attracts the attention of senior managers and threatens to worsen their perception of the manager concerned. In organizations with a widespread system of punitive sanctions, a halt in production could have negative consequences for the career of an honest and proactive manager. Therefore, some lower-level managers prefer to keep quiet about the problem, leaving its solution to their successors.

Fear of competition from employees with an active civic position (1%)

In companies, as in the wider world, there are some people who are "*active citizens*", full of ideas for how to change things for the better. A confident and forward-thinking senior manager might see natural leadership potential in such employees and want to encourage them. Nevertheless, some managers do not want to promote subordinates with this kind of reforming energy, let alone bring them to the attention of a board of directors or shareholders. They are afraid of competition from such proactive employees: at the bottom, they fear for their own jobs. If key owners or representatives of a board of directors were to meet such a workplace reformer and be won over by his/her ideas, they may decide that the current bosses are not dynamic and effective enough and invite the reformer to take their place.

2.3.3 *Reasons Why Employees Are Reluctant to Disclose Risks to Their Managers*

1. **FEAR OF BLAME AND PUNISHMENT FROM EXECUTIVES: SUBORDINATES ASSUME THAT THEY WILL BE HELD RESPONSIBLE FOR THE OCCURRENCE OF ANY PROBLEM THEY REPORT TO THEIR MANAGERS**

63% of respondents believe that employees are afraid that if they raise the alarm about a problem, management will accuse them of having caused or exacerbated it through their mistakes. In most cases, problems do not arise from nothing. They usually develop partly because of poor decision-making by managers (for example a refusal to approve adequate resources to keep facilities running safely) and partly through the actions of employees who may have been forced to violate operational safety to meet production targets dictated by managers.

The HSE head of a fertilizer mining company sees the situation like this. Employees reluctant to report problems upstairs are often afraid that they will be accused of allowing what was initially a manageable situation to develop into something so serious that senior management have to step in. In response to such an accusation, they may say that they were not given the resources to prevent the escalation of events, or that they requested resources but were refused or ignored. Senior managers will likely reply that the employee was not insistent enough and did not make clear the critical consequences that could follow if resources were not allocated. But making such persistent complaints and demands can threaten an employee's career, as management begin to associate them with the problems they have highlighted. It becomes a vicious circle: people are afraid to talk about problems, because they may be reprimanded for not tackling them earlier and nipping them in the bud, but if they do have the initiative and integrity to bring issues up earlier, they may well be punished for doing so.

The head of a production facility at an oil company describes the situation as follows. When something goes wrong at a facility, the management often look for mistakes or negligence on the part of employees, rather than flawed production processes, outdated equipment, or their own previous decisions. In the traditional paradigm, poor production indicators mean that some people or teams have not done their job properly; someone has made mistakes. The managers tend to believe that the fault must lie with their subordinates, not with production processes, equipment, or the allocation of resources. Hardly surprising, then, that employees are afraid to tell management about problems and risks when they know that the first people the bosses will look to blame are those delivering the bad news, and anyone else involved in letting the situation escalate to the point that it has been brought the attention of the management. In other words, the instinct of managers is to *"shoot the messenger"*.

The HSE manager of a petrochemical company also notes that employees are often afraid they will be accused by managers of allowing a problem to get beyond a stage where they could manage it themselves. This suggests that a critical infrastructure company has precedents when employees were found guilty of mismanaging problems, publicly reprimanded, and maybe demoted or dismissed, despite the fact that they had taken the initiative and voluntarily disclosed information about problems to managers. In fact, the fault lay with senior management, who did not want to understand or tackle the problem, but simply shifted the blame for its occurrence to the employees who had been proactive and courageous enough to talk about it. If senior management had reacted differently to the risks being disclosed—for example, by providing resources to solve the problem rather than searching for subordinates to blame—then employees would be more willing to inform managers about risks when they first encounter them.

A low-level manager describes how most incidents on critical infrastructure in his company are investigated. The company searches for specific employees to blame, rather than changing the corporate goals, conditions or business processes that created a situation where employees were not able to cope with problems in their area of responsibility. Even when analyzing incidents resulting from external factors, the system still looks for shortcomings in the work of specific employees of the company. According to him, the company's senior management are so skilled at conducting internal investigations of incidents that employees have no choice but to admit that they were to blame for the onset of risks that led to emergencies. During these investigations, the management would never acknowledge that a lack of resources from the company had anything to do with the risks escalating. Instead, a scapegoat would always be found. After one such incident investigation, five leaders at various management levels were fired. With this sort of precedent, it is hardly surprising that employees are afraid to get involved in emergency situations or to discuss risks honestly, as they naturally fear that the investigation will end with their dismissal. Clearly, this interviewee's company has an unspoken rule: nothing bad can be reported upstairs and everyone tries to avoid mentioning problems to their seniors. If they do, they know they will be asked: *"What did you do in your role to prevent these risks/problems/mistakes?"*. Everyone is under pressure to meet targets and deadlines, and there is a repressive system of disproportionate penalties for any shortcomings at work, so people are afraid to admit that something is wrong. The interviewee summarized it by saying, *"At the heart of everything is fear. Employees are afraid of material punishment, managers are afraid of dismissal"*. He gave a further example of his company's practice. Years ago, if there was an emergency at a production site, the regional head was called to headquarters. As a result, when the scapegoating of middle managers began, mass concealment of any incidents at production sites became more common. All employees in the regions, including managers, took care not to *"expose"* themselves or their colleagues by disclosing information about risks and incidents to headquarters. Local workers simply stopped reporting incidents to avoid being held responsible and punished. Who is to blame for this situation? Of course, senior management—because they have repeatedly scapegoated specific employees rather than looking for the systemic problems that forced the workers to violate safety rules in the first place. And of those systemic problems, the most significant is the lack of adequate resources, which senior management should have been allocating to the regional enterprise to keep the equipment operating safely.

The HSE manager at a utility services company believes that the problems of a critical infrastructure company can be divided into two categories: problems with equipment and employee errors. In the first case, employees do not feel responsible for the failure of equipment and therefore they are not afraid to say: *"This machine should be fixed"*. In the latter case however, most employees are very reluctant to say: *"There is a problem because I made a mistake"*, because of the possible negative consequences for them if they admit this to their superiors.

The head of HSSE-Q (Health, Safety, Security, Environment & Quality) at an electricity company has an example. It is very hard to go to your boss and say, *"Look, there's a problem because I messed up"*. It is like you are confessing that you are a failure, useless, and can't be trusted.

2. **EMPLOYEES ARE AFRAID OF LOSING INCOME AND RUINING THEIR CAREER PROSPECTS BY LOOKING INCOMPETENT IN THE EYES OF THEIR BOSSES**

48% of respondents expressed the view that, for many employees, the fear of losing earnings and damaging their own career prospects stops them from reporting serious problems and risks.

According to respondents from several different industries, workers are fearful of losing their income, their jobs, and even their whole career. Many have children and a mortgage to pay, so are dependent on having a well-paid job. All this could be lost overnight if a problem arises in their area of competence that they cannot solve with their own resources and experience and have to take up to the management.

Employees are afraid to seem incompetent in the eyes of their boss. If they cannot solve something, they are afraid to let managers know that there are limits to their capabilities. Senior managers may then question their ability to manage risks and will think twice about promoting them up the career ladder. In the traditional paradigm of managerial perception, employees should do everything possible at their level to prevent negative developments. If a problem has to be brought to the leadership, the assumption is that their subordinates failed to show due diligence and have not justified the trust placed in them. Any demonstration of incompetence can have several bad consequences for employees: (I) it will threaten their career progress and maybe even their job; (II) it will irritate the boss, who now has to engage in solving the problem; (III) the boss will be less inclined to trust the subordinate, who may previously have been delegated significant authority. Therefore, employees are afraid to appear weak or lacking in resourcefulness. On the contrary, they want to demonstrate their competence, effectiveness, and ability to resolve any issues, to be seen as "*a safe pair of hands*" and justify the trust of their managers. Sometimes this is done by hiding negative information.

A safety consultant with managerial experience in chemicals, mining, oil and gas believes that if subordinates send a report to superiors that draws a seriously problematic picture of the state of affairs at a facility, then the whole operation at the facility may be threatened with closure, and people will potentially lose their jobs.

A senior technical professional from the oil and gas industry also believes that people are reluctant to be honest about the serious hazards they face in their jobs because know that the costs of correcting them may be considered excessive by shareholders and a board of directors, so that instead of investing they decide to close the entire facility and make lots of the employees redundant.

Competition among work colleagues to keep their own jobs can motivate both front-line employees and middle managers to be very cautious before revealing problems in their area of responsibility to superiors. After all, if the heads of nine facilities tell senior management that everything is fine and under control, but the director of the tenth is constantly informing them about various problems and shortcomings in his enterprise, senior management may well begin to question the competence of that manager and consider replacing them. Senior management will look for a new candidate to promote among those employees who have previously been able to prove that they can solve problems without requesting the attention of their superiors.

3. INERTIA OF CORPORATE CULTURE

43% of respondents consider that most employees will go along with the corporate culture that exists in their company. If that culture dictates secrecy about problems, demands only good news and punishes staff for the presence of problems on their watch, then most employees will simply not inform the authorities. Over time, a continual unwillingness of managers to listen to employees when they raise concerns inevitably results in an ingrained culture of lies at every level of an organization.

The CEO of a nuclear operating company believes that a "*Don't Tell Me Bad News*" organizational culture arises when senior managers lack the skills to create a safe and supportive work environment, where employees are encouraged to share issues and concerns about their work on a regular basis with their superiors. It is clearly the responsibility of those at the very top of the hierarchy to establish and maintain a framework that supports honest and open communication at all levels.

4. **FEAR OF DESTROYING RELATIONSHIPS WITH COLLEAGUES OR LINE MANAGERS**

32% of respondents believe that many employees do not disclose risks to their bosses because they think this will ruin their relationship with their colleagues, or with their immediate supervisor.

The head of HSE at an oil company puts this starkly. The fact is that if risks are disclosed to senior management, many people in his company will face punishment. Management can punish an individual employee or even an entire unit for flagging up a serious problem. Instead of finding solutions, they are more likely to search for specific perpetrators, thereby jeopardizing the careers of many people. To avoid this, many junior employees and some middle managers will put pressure on any would-be whistleblower and silence them.

The heads of HSE departments at two different metallurgical companies describe how a team can influence employees who want to stir things up by warning management of problems. Shop floor employees tend to operate in small teams, who work together to avoid pressure from their bosses. A golden rule for teams like this is "*we do not surrender our own*", or "*never snitch on your mates*". So consider the situation when a proactive employee wants to tell top management about a serious issue. In effect, they will be disclosing the fact that a dangerous risk has been present at the production site for a long time but has been ignored or tolerated. To do so could cost the career of the whistleblower's immediate boss, and probably some of their colleagues as well, because once the risk is exposed, it will be obvious that many workers have known about it for ages but failed to report it. This then raises questions about the competence and trustworthiness of those employees who: (I) coexisted with risks before they were disclosed; (II) did not report them to management; (III) did not take any measures to reduce them. To prevent these awkward questions being asked, the immediate supervisor or colleagues of the employee who wants to "*snitch*" may threaten them: "*Do you think you're so much better than the rest of us? You want to betray us all to get in with the management? Whose side are you on?*". As a result, even those who know it is their moral duty to try and stop dangerous safety violations are afraid that if they do, it will be make their lives very difficult. There will be conflict with colleagues and punishment by the authorities, not only of themselves but possibly the entire team. Most employees, once they realize the consequences of whistleblowing, decide that it is safer to keep quiet about problems rather than risking communicating with their superiors. This peer pressure creates a mutual understanding not to communicate risk information from the production site to senior managers.

One respondent has some experience of "*cracking*" ordinary employees in incident investigations. Without exception, they initially do not want to talk about any safety violations in the team before the incident occurred. It is only when the HSE manager presents them with facts that have already come to light, meaning their careers are already on the line, that some ordinary employees will eventually share information about what actually happened before the incident and any violations their colleagues were guilty of. People are extremely reluctant to immediately inform the authorities about such violations. To do so would break the widespread but unspoken rule for working in teams, that "*shopping*" your colleagues is taboo and punishable with expulsion from the group. If managers do get full information from an employee about the details of an incident, therefore, they are advised to maintain the anonymity of the source to prevent such retribution.

One of the interviewees has an answer to a rather complex question: "*Why does the 'stop-the-job' system not work in so many companies?*". It is widely understood that if an employee shuts an operation down, this will immediately signal to senior management that there is a serious problem. The bosses will then punish the site manager for failing to take steps to tackle the risk sooner. In turn, the low-level managers will look below them and punish the proactive employees who stopped the suspiciously functioning equipment in the first place. Once again, we have a vicious feedback circle: if you reveal a risk, however morally right you were to do so, you will inevitably be punished for it in the end.

A risk management consultant in the oil and gas industry has analyzed the tasks facing employees and low-level managers at the bottom of the corporate hierarchy. Ordinary employees are focused on doing a specific job within designated time and quality parameters, and in return, they receive the remuneration owing to them and go home. The foreman of a work team has a different task: to get the team of ordinary employees to finish the work on time and to the requisite standard and then report to the higher boss that this has been achieved. The task for lower-level managers and the director of a production facility is to coordinate work across all operational areas to meet the output and timeline targets of the production program that have been set by senior management. When someone highlights a problem that requires a halt in operations for it to be fixed, the interests of all workers at the bottom of the corporate hierarchy are affected. A single worker choosing to be an active citizen and blowing the whistle interferes with the whole team's operations. Therefore, it is not always a lower or middle manager who will discourage employees from reporting issues. Just as often, a team colleague will say: "*Don't upset everything now by blabbing. We all need to make sure we get the job finished*". The respondent suggests using the following process in cases like this. Come down to the shop floor and ask employees, foremen, and site managers a simple question: "*Will your day get better or worse if you notice a problem in your operations and report this to higher managers?*". It is obvious in most cases that the answer is their day will get much worse. They will be unable to achieve their production goals, risk suspension from work, and have to prepare various reports explaining exactly why the problem that has stopped production has occurred. Finally, those individuals who disclosed the risk to their superiors in the first place may come under attack from their own colleagues for disrupting teamwork. Therefore, hiding information about risks is not always due to the actions of lower and middle managers. Sometimes it is a result of the desire of shop floor employees to complete their own tasks, knock off from work on time, or simply to keep their heads down, focus together with the rest of the team on the job in hand and not go around creating problems that nobody wants. Anything for a quiet life!

5. **FEAR THAT EMPLOYEES WILL BE EXPECTED TO SOLVE ANY PROBLEM THEY REPORT**

27% of respondents think employees are afraid that if they report a problem to senior management, they will be left to deal with it themselves with no extra resources to do so.

Tackling a major safety issue with no management support or resources will require huge extra work beyond their normal responsibilities. The typical reaction of many managers to negative news will be something like: "*You need to show more determination with problems like this and find a way to solve them on your own*". Employees are understandably cautious about taking on this kind of extra responsibility. They have seen too many managers who do nothing to solve a problem themselves, but instead leave it to the employee who dared to report it in the first place, often without allocating necessary extra resources. Having seen the negative experience of their colleagues, many employees prefer not to inform managers about problems in their area of competence, so as not to be saddled with even more work.

Lower and middle managers are usually aware of most of the risks that may arise with the infrastructure entrusted to them, but they know it is unlikely that headquarters will give the resources to mitigate them. To get these resources, they would have to send the bad news up to senior management level and prove that the situation at the facility is critical. But bringing a critically dangerous situation to the attention of the management can jeopardize an employee's career. As a result, lower and middle managers prefer to remain silent about problems, dutifully accepting the meager resources allocated to them against a likely background of ever diminishing budgets. Consequently, there is no investment for measures to prevent risks escalating, and issues are only tackled when they develop into an emergency. Then, at last, the whole company is forced to pay attention to a critical weak point in the infrastructure, even though these risks were known about long before disaster struck.

The HSE head of an oil company has an example from their own practice. A conventional oil company, working in production regions where the industry has been operating for decades, usually has outdated equipment that needs to be fully replaced. This would obviously require additional investment and impact the financial profitability of the production sites. Senior managers know that the equipment on some sites is outdated, and are willing to consider programs for its replacement, but only if the profitability targets they have agreed with middle management will still be reached. They will tell middle management at the site: *"Do what is necessary to update the equipment, but keep financial performance within such-and-such parameters"*. In effect, headquarters leave site managers with the impossible task of tackling the equipment problems without adversely affecting the budget. There is little point in middle management informing headquarters about these risks and problems: they know they will not be given the resources needed to mitigate them, but instead expected to find some creative solution on their own. By informing senior managers about the risks at the site, middle managers only create problems for themselves. They still will not have the resources to tackle the risks, but now that their bosses know about the problem. As a result, the site managers will be put under additional pressure to solve the problem one way or another. Of course, impossible demands like this from their superiors understandably make middle management increasingly reluctant to report their problems to headquarters. In addition, in some countries senior managers are not held legally responsible for emergencies. Here middle management are stuck between the *"rock"* of relentless production and financial targets and the *"hard place"* of maintaining safe production, knowing all the time that they are likely to be held legally and criminally liable if things go wrong. Production site managers are constantly juggling two conflicting priorities, and having to hide the logic of decision-making from the management. The respondent describes the conflict of purpose and priorities in many critical infrastructure companies. On the one hand, most oil companies prioritize raising oil production and other industrial indicators. Getting results is the overwhelming focus for senior managers, who constantly hammer the importance of this into the heads of their subordinates. On the other hand, managers know they must also give attention to safe production. In many companies, these priorities are in direct conflict. For example, an order was sent around one large oil company: *"The field must be launched on time, but safely"*.

It is worth noting that the deadlines set for commissioning a new field are often very ambitious and based on the most optimistic scenarios. Such messages have a deep impact on employees. They recognize they are being given conflicting goals, but as both priorities are combined into a single message, they are left unsure as to which should take priority. Such conflicting goals are irreconcilable and bring employees to a standstill. When they set such impossible production goals, managers do not absolve employees of responsibility for the safety of their work. In other words, managers do not formally give permission for unsafe work, but they make the true priority perfectly clear: first and foremost production performance is to be achieved, with safety relegated to second place. All too often production and financial indicators prevail over safety because these will bring in the money, whereas safety improvements will lead to extra costs and lower profits.

The head of a power plant describes a typical scenario when senior managers do not have the technical training or experience to make informed decisions on operational matters. Over the decades of his career, the respondent has seen many senior managers—highly experienced and trained in operating energy facilities—leave the power industry, to be replaced by managers with only financial and economic experience. When employees approach one of this new generation of managers about a technological problem, they will often not get much help in solving it. The new bosses simply do not understand the technical aspects of the problem, and instead their concern is avoiding any extra costs that would reduce a company's profitability. Employees soon learn that there is no one to consult with at a senior level, no one who can be relied on for help, and no one who will give them the resources that are required. It is no surprise, then, that middle management give up asking for assistance and try to solve the problem alone. In future, middle managers will avoid informing their bosses of risks and problems as they know it will not solve anything, but instead create additional difficulties.

6. **EMPLOYEES DO NOT FULLY UNDERSTAND THE RISKS THEY ARE RUNNING, AND LACK THE TRAINING OR EXPERIENCE TO ASSESS THEIR CRITICALITY**

22% of respondents point out that sometimes employees do not realize the risks they are encountering in their day-to-day work, so they do not inform their superiors about them. If people are not aware they are taking risks, then they are unlikely to think that they are doing anything wrong, so see no reason to inform anyone. It will generally be employees who are unqualified or inexperienced who fail to recognize or assess risks. Sometimes employees only care about their own area of work and do not want to look at the risk picture across the entire production process. In this case, they are unlikely to report problems outside their own limited area of competence.

Several respondents believe that some employees find it difficult to assess the criticality of the risks in their area of responsibility, so do not report them to their superiors because they cannot be sure they are significant. They do not want to be *"the boy who cried 'wolf!'"*, repeatedly warning management of risks that never develop into serious incidents.

According to the HSE head of an oil company, it is rarely possible to calculate exactly where and when an accident will happen in the future. An accident occurs through a combination of factors that are difficult to predict in advance. For example, if there is aging pipework throughout an industrial plant, it is impossible to judge exactly which sections are most likely to fail and how soon without a very thorough ultrasonic inspection of the entire infrastructure. Therefore, it is very difficult for field staff to rank risks accurately and pinpoint critical areas of potential failure. After all, it is impossible to eliminate all risks: there simply would not be enough resources for this, nor would it be justified. As a result, production staff do not inform managers of every deviation in the equipment they are running. If they did, it would create an unmanageable flood of information from the production site to headquarters.

The HSE director of a mining company states that some employees are gambling on the basis that the risks they have been aware of for many years ago have never led to serious incidents. The people operating the equipment come to believe that deviations in the way the operation is running are the norm, and that there is no need to report them to the management.

The safety manager of a nuclear power plant believes that subordinates will delay communicating certain risk information to managers, for a short period of time, if they believe that the likely impact of the risk on the operation of the enterprise is not serious. By constantly postponing the communication of observed risks from day-to-day, subordinates become used to tolerating and coexisting with these risks. This situation soon becomes the norm, and it becomes habitual for them not to report any risks to their superiors.

7. **EMPLOYEES FEEL IT IS POINTLESS TO REPORT RISK INFORMATION BECAUSE MANAGERS FAILED TO RESPOND TO SIMILAR MESSAGES IN THE PAST**

21% of respondents point out that some employees see little point in informing their superiors about problems or risks, because there has been no response to previous warnings and the problems have remained unsolved. Frustrated by this lack of action, some employees simply stop telling their superiors about problems, assuming that their efforts will be futile.

The vice president of a gas pipeline construction and repair company says that sometimes an organization lacks the resources to maintain and repair equipment properly. In this case, when employees approach senior management with a list of problems that will require funding to address, they may be told that there is simply not enough money to cover all the requests, so the list will have to be drastically reduced. The next time there is a problem, the employee will have learnt that there is little point contacting the management because their organization does not have the resources to prevent and mitigate risks. Making another request will only serve to put the boss in the uncomfortable position of having to admit that there is nothing management can do.

This is how the HSE manager of a mining company describes it. Employees have tried on many occasions to inform the company hierarchy about critical problems. What happened when they did? At best, nothing. No one reacted and no one allocated resources to rectify critical problems. At worst, employees were punished for revealing this information. When workers see another potentially dangerous situation, this information will no longer be sent upstairs. As a result, problems inevitably accumulate, increasing the likelihood of a serious accident.

The HSE head of a metallurgy company believes that existing corporate cultures, built on decades of silence on risk information, often make it impossible for employees to go to the management with complaints or disclose information about risks. There is a widespread assumption that "*nothing will change*" and even if they report serious safety issues, none of the bosses will do anything about it.

The HSE head of a gold mining company gives another example. Sometimes, there are design flaws or errors in the construction of technological facilities, which cost a lot of money to correct. Employees hesitate to make aware managers of these shortcomings, thinking that they are unlikely to allocate resources to rectify the situation, especially for low probabilities of a catastrophic event: "*If nothing will be done anyway, what's the point of bothering the management?*".

An HSE consultant shares the experience of how senior managers sometimes respond to suggestions from employees on how a company's processes and efficiency can be improved. There are a significant number of employees who frequently offer innovative ideas in their area of competence. Some companies have competitions from time to time where they invite employees to submit proposals for possible improvements. Often, proactive employees participate in and win these competitions. However, after the competition the winner has to face the reality of the company's internal bureaucratic systems, which means trying to get their proposal implemented becomes a massive headache, frustrating the winner's initiative and enthusiasm. After such encounters, innovative employees are less likely to offer their ideas again.

8. **FEAR OF BEING SEEN AS DISLOYAL TO A COMPANY, OR AS A REBEL WHO WANTS TO "*ROCK THE BOAT*" OR LABELED AS A "*BAD NEWS GUY*"**

20% of respondents believe that when employees start to "*ring the alarm bells*" and draw attention to problems, they will be perceived by their superiors as rebels, "*black sheep*", troublemakers who want to "*rock the boat*" or "*go on the rampage*". Most managers are afraid of such potential disruption, which could lead to earlier management mistakes coming to light. As a consequence, they will often berate would-be whistle-blowers: "*All your colleagues are quite happy, but you always seem to have a problem with something. You think you're special and you want to wash our dirty linen in public*".

According to the HSE manager of an oil company, for some middle managers hitting certain production and financial indicators is an absolute priority as their careers depend on it. If these targets cannot be reached, then questions will be raised about their competence. To prevent this, middle managers have to try to find a balance, choosing between hitting production targets and operating safely. However, this balance is often impossible to achieve, and it is the production and financial targets that usually come out on top. It is difficult for managers to make a successful career if they keep bringing bad news. Senior management might perceive them as troublemakers, who are always bothering managers with problems. If middle managers tell their seniors that the production and financial indicators they have set are unrealistic or unsafe within the limits of the condition of current equipment and allocated resources, they are effectively saying that senior management have made a mistake. This looks as if they are being disloyal to senior management, and trying to go against their decisions. If middle managers create that impression with their bosses, then their career in a company will be compromised. Senior managers are looking to promote subordinates who are ambitious but loyal and willing to take responsibility for getting results, even if that means taking risks and discreetly violating production safety regulations. So in order not to jeopardize their careers, most middle managers simply do not mention risks to senior management. They believe it is more important, and certainly more advantageous, to tell senior management the news that they want to hear about steadily improving production and financial indicators, for which they know they will be rewarded. No middle manager can expect to earn praise and promotion for saying: "*We can't hit these targets without jeopardizing production safety*". Therefore, they dutifully accept the production and financial indicators imposed from above, striving to meet them without a corresponding increase in injuries or accidents. However, they are trying to achieve the impossible. Sooner or later, the situation will catch up with them, resulting in accidents.

The HSE manager of a metallurgy company ties this in with the punitive system operating in many companies, where there is an ambitious production plan and employees will be punished if they do not deliver. This creates pressure to violate safety rules, as this is the only way to hit the targets. If these strictly enforced, over-ambitious production targets are followed for many years, a culture of tolerating safety violations will inevitably develop in an organization. In such a climate, an employee who objects to violations will become an outcast in the eyes of both management and colleagues. Most workers prefer not to stand out like this, so they go along with the prevailing practice of negligence towards safety matters.

The HSE head of an oil company says that many organizations have created a corporate culture in which any criticism of the system or discussion of problems will lead to dismissal. Hence, no employee dares even consider going over the head of their immediate supervisor to complain about a problem to a more senior manager. This kind of system demands unconditional loyalty to senior management, and anybody who expresses the slightest doubt about the current course, or tries to initiate a discussion about problems, is immediately ostracized. Even in companies that have set up a direct hotline expressly for workers to communicate problems directly to senior management, any employee who actually uses the hotline can expect their line managers to get rid of them. It does not matter how well-intentioned they are or how serious is the risk they have raised: they are deemed to have demonstrated disloyalty by daring to suggest that a company with such wise leadership could be making a serious safety error.

The HSE head of a petrochemical company perfectly sums up the system of absolute loyalty that is fostered in some industrial companies. Senior managers generally like those who defer to them. Those who stand out, who are independent and have their own opinions, will be ostracized and removed from the company. In this context, loyalty to the company does not mean loyalty to what the company stands for, to its genuine long-term mission and values, but loyalty to its current leaders and senior management—demonstrated by implicitly endorsing a company's management policy. It is "*loyal*" employees in this sense who survive in large corporations; indeed, they may never have made a significant contribution to the development of a company, but their compliant attitude at least makes life more comfortable for the executives.

The HSE head of a production company which uses hazardous chemical processes explains that employees are often too afraid to challenge or contradict their superiors. He gives the example of the head of the construction department of his company, who was asked by the owners and senior management about the possibility of building a chemical plant for US $1 billion. This head of the department replied that he could build the plant with the funds allocated. Senior management then came back to him to ask about the possibility of building a plant for $500 million. The construction director confirmed that this would also be feasible. In doing so, he was meekly falling in line with the management proposal—despite the fact that many safety systems would have to be scaled back or removed, only the cheapest building materials could be used and other corners would have to be cut, which over the longer term would lead to very high costs to keep the plant operating safely. In reality, to build a high-end chemical plant that will operate reliably for decades would cost $2 billion. Many subordinates are not ready to volunteer such information unless specifically asked to do so, because they are afraid to put the superiors in an uncomfortable position by telling them their chosen course of action cannot be safely implemented.

The HSE manager of a metallurgy company offers his own simple test, by which a senior manager can measure the safety culture at an industrial facility in just a few minutes. It is sufficient for senior managers to enter a production workshop without personal protective equipment and walk around, unprotected, next to dangerous equipment. If a company has a strong safety culture, an ordinary employee will immediately approach the senior managers and warn them that they are violating safety regulations. If a company has a weak safety culture, nobody will dare challenge the senior managers and everyone will ignore the violation. If employees challenge bosses when they see them flagrantly violating safety rules, it shows that they will not overlook unsafe practice in an organization. Hence, if a serious risk is discovered, management can feel confident that their employees will not be afraid of drawing the issue to their attention. However, if employees ignore a relatively minor violation of safety regulations by senior management, then they are hardly likely to challenge their managers about serious safety and technological problems.

A safety consultant with managerial experience in oil and gas, chemicals and mining believes that the last thing employees want is to become known as a "*bad news guy*" in an environment where the bosses tell the workforce: "*Your job is to fix problems, not create them*".

9. INDUSTRIAL SAFETY PERFORMANCE INDICATORS AND REWARD SYSTEMS ENCOURAGE CONCEALMENT

Corporations use many key performance indicators (KPIs) to manage their productivity. 15% of respondents point out that some of these metrics can incentivize employees to downgrade an incident, and under-report equipment problems or anything else that could stand in the way of hitting ambitious corporate targets. When executives set almost unreachable goals and KPIs, they put pressure on employees to distort risk information in their reports. How else can they possibly hit all their targets?

The head of an oil production facility gives the following example. If we look at the common corporate KPIs in the field of safety, most relate to the number of industrial incidents. Naturally, employees will therefore do their best to reveal as few incidents as possible, to achieve the best possible ratings. Again, the answer to the question of why employees hide information about risks at work is that it comes down to overall company targeting, where management have set performance criteria that motivate their employees to hide the real state of things. Industrial site managers know that the more incidents happen at work, the smaller their annual bonus will be. This hardly motivates them to give management the real picture of industrial safety at their facility. On the contrary, it is directly—and financially—in their interest to embellish the situation. If injury statistics were not included as part of their bonus structure, then managers would be more likely to report them accurately. It is not at all straightforward to create an effective bonus structure, one that motivates all employees to work towards a real world decrease in injury and accident rates, without tempting them to falsify reports and statistics on the safety situation.

The HSEQ director of an international oil company gives the following example. The president of the oil company has a significant part of his bonus linked to safety performance. This means that safety is an important concern for both the president and the entire management team. The problem with this approach is that subordinates are aware that a safety mistake on their part can affect the size of the senior management bonus. Therefore, subordinates may try to hide mistakes so as not to upset their superiors by reducing their reward. In this company, the bonuses of all employees and managers are partially linked to safety performance. The presence of rewards for progress in the field of safety helps to show everyone in the company how important this issue is for senior management. However, because of this reward system, employees have begun to conceal certain safety problems as they know they could create problems for executives of the company.

The HSE head of a chemical company ascribes many of the problems in their organization to its ambitious goals. When directors set almost unreachable goals and KPIs, they put pressure on employees to distort risk information in their reports. In order not to spoil their KPIs, employees will downplay cases of work-related injuries, accidents, and incidents as less significant events—anything to make sure that they and their team continue to be seen to be delivering good results.

A senior HSE advisor and human factors specialist in the oil and gas industry also confirms that cover-ups can occur in more than just health and safety matters. Distortion can affect several KPIs that are set by senior management to evaluate the work of production units. In order to achieve these, people on the ground are liable to underestimate the criticality of any incident and distort information about errors and equipment failures.

The head of the sustainability and systems department of an international electricity company believes that current executive bonus systems are often closely tied to accident and safety statistics. This may motivate executives to underestimate these figures. This reward system has also a negative impact on honest and open intra-corporate communication about risks, failures and negative incidents. To correct the situation, the respondent recommends replacing these inaccurate and under-reported figures with leading indicators, such as the number of events identified where prompt action prevented an incident occurring (near misses) and the number of safety improvement proposals and new initiatives.

10. SOME EMPLOYEES ARE CONFIDENT THAT THEY CAN SOLVE THE PROBLEM ON THEIR OWN

13% of respondents note that employees can sometimes be overconfident in their own capacity to solve a problem. However, when it comes to assessing risks in the operation of equipment, there is a blurred line between where an employee's competency zone ends and that of managers begins. Sometimes in their efforts to make a good impression, employees can be tempted to tackle higher-level issues beyond their competency and pay-grade that should really be the responsibility of their superiors. If subordinates have a strong sense of ownership, they will be tempted to solve problems by themselves and then report their success, rather than reporting the problem to the manager and waiting for them to come up with a solution. In doing so, they may overestimate their capabilities and convince themselves that there is no immediate need to inform their superiors about it. This can lead to an accident at a production site that takes senior management completely by surprise.

The head of risk management at a renewable energy producer believes that some subordinates are overconfident, convincing themselves that they can solve a serious problem on their own, so there is no immediate need to inform the authorities about it.

The head of HSE at a nuclear power plant believes that when subordinates are competent and responsible, they should be encouraged and supported to try and solve problems independently in their area of expertise. This way they can gain credit by reporting their success to their superiors, rather than immediately reporting the problem to their manager and waiting until they come up with a solution. The respondent recalls a situation like this at the nuclear power plant where he works. The plant had a skilled specialist in the field of ventilation system maintenance. It was only when he left for another position that managers and other employees realized just how many problems there were with the ventilation system, and how reliant they had been on him for his expertise in carrying out repairs, all without bothering his superiors. Some subordinates keep their managers ignorant of many problems because they take professional pride in their abilities and draw satisfaction from dealing with issues that they have the skill to put right, quickly, efficiently and without a fuss.

OTHER REASONS

Subordinates do not want to create extra problems for managers (9%)

Some respondents believe that employees who bring negative news will cause emotional stress for managers, and negatively affect their well-being. Leaders are people like everyone else, and most subordinates do not want to upset or burden them by putting them under any avoidable extra pressure.

In many companies, there is an unspoken culture among subordinates not to create problems for their managers. Employees will do anything to avoid the anxiety and distress that can arise if they communicate bad news to their superiors. They will keep quiet until they can bring them some good news. They would rather try to solve problems themselves and then come to the head with a positive result.

Lack of trust between superiors and subordinates (9%)

Problems will be concealed when there is little trust between superiors and their subordinates. Naturally, if employees on the shop floor do not have confidence in their line manager, they will not feel safe to share difficult information in an open and honest way. Instead, they will hide things to try and make sure they are seen in a good light.

In order to have frank conversations about new or existing problems in the workplace, subordinates must have trust in the integrity of their superiors and be confident that they will not face sanctions or punishment if they bring bad news about a production risk or problem to their manager.

Wrong ideology for investigating incidents (8%)

Several respondents point out that some state and internal corporate investigations into incidents tend to focus on the search for specific perpetrators—usually rank-and-file employees—instead of trying to identify systemic flaws in an organization or holding senior managers accountable for the bad decisions they made which meant ordinary employees were forced to violate the regulations.

One interviewee describes the situation as follows. In some countries there is a state policy concerning labor protection and industrial safety, which aims to punish specific employees instead of revealing the systemic or root causes of incidents. Safety regulators send inspectors to industrial facilities to assess the state of labor protection. Most of these inspections are not actually aimed at improving labor protection practices at all, but rather at finding fault with the performance of individual workers who are then blamed and faced with a range of possible punishments. All workers and managers are subject to the requirements of labor protection legislation. The emphasis of regulators on identifying specific perpetrators, rather than systemic flaws in occupational safety issues, leads employees and managers at all levels to take a common approach to hide information about shortcomings in organizational operations. This state ideology, therefore, has a major negative impact on the corporate culture of many industrial companies, discouraging the open transmission of risk information and compromising the safe operation of their facilities. In this regard, another respondent gives the example of an oil company from such country. If they have an incident, in order to conduct an investigation that will actually provide some useful conclusions about how to improve the safety of their operations, the company prepares two incident reports. One, for the regulators, identifies a specific culprit and documents the disciplinary procedure and punishment, as required by local legislation. The other is an internal report, establishing the systemic or root causes of an incident, and setting out measures to exclude the system errors that led to an incident so as to prevent its recurrence.

A risk management consultant in the oil and gas industry has an example of how a company's good intentions in the field of accident investigation can still encourage employees to hide information about incidents. The respondent cites the experience of one oil company with a large fleet of trucks. When an incident occurred on any of their sites, all drivers were automatically checked for drugs and alcohol. There were several cases at one oil refinery when, after even minor road incidents, the company's security guards physically removed all the drivers from their cabs and escorted them across the site in front of all the other employees, to a medical unit, where over the course of half an hour, a nurse tested them for drugs. None of the drivers tested positive. However, the result of this over-the-top and unwarranted procedure was that no employees were willing to report any further road incidents at the refinery because there was an understandable reluctance among the workforce to be subjected to such humiliating and unpleasant treatment. As a result, the management of the oil company eventually changed the criteria for mandatory drug tests and excluded minor incidents from this list.

Lack of official channels for reporting risks and problems (8%)

One of the respondents believes that few companies have an official channel for shop floor workers to communicate directly with management about risks. Any employee who does have the moral courage to report something dangerous is forced to find more informal channels to do so, which may well create a threat to their job security.

It is natural for people to try and make themselves look better and avoid blame (8%)

Several managers express the view that a factor motivating employees to hide information about risks in their area of competence, or when incidents are being investigated, is simply human nature. One of the respondents said that the problem of withholding important risk information exists at all levels of an organization, because it is in human nature to lie. Like many people, some employees habitually lie in small ways. When talking to management, they instinctively distort and massage the facts to make themselves look better. During an investigation into an incident, managers and employees at a site will often manipulate their account of events to avoid implicating themselves so they cannot be blamed for what happened. Another manager who mentioned this tendency described it as a common human practice of "*self-censorship*".

The head of risk management at a renewable energy producer believes that people naturally prefer to share happy and cheerful things in conversation with others. If someone does say something negative, the other person may feel distressed, discouraged or criticized. People feel they should not burden others with troublesome or difficult issues, and this can mean employees avoid sharing risk information with managers.

Social stereotypes (6%)

An HSE consultant suggests that social stereotypes influence employee behavior. Most people involved in the management of critical infrastructure are men. In general, men are taught from childhood to control and hide their emotions, not to show weakness, not to admit defeat. When they make mistakes or encounter problems, their instinct is to try to deal with the challenge alone, without asking for help from their colleagues or superiors. Many employees are reluctant even to acknowledge that a problem might exist in their area of responsibility—to do so would detract from their aura of male strength, resourcefulness and competence. Managers also praise employees who try to solve problems on their own. This male social stereotyping fosters machismo and a brave attitude at all levels of the hierarchy within critical infrastructure companies. The respondent cites the example of an oil company where there is a long-term problem with workers putting themselves at serious risk in their efforts to try and save expensive equipment—up to the point where an employee died attempting to save a tanker truck which had fallen into an icy river or lake. For employees like this, their sense of themselves as strong and honorable men simply cannot allow them to admit—to their superiors or themselves—that they have failed to look after the equipment entrusted to them. They see themselves as hardcore professional men, and all-action heroes. It took a great deal of time for HSE executives to convince ordinary employees there that the company valued their lives far above any piece of equipment.

Human laziness, or slackness in responding to observed risks (4%)

Some employees are simply too lazy to act: they do not want to take any initiative. They are always very reluctant to tackle any risks in their area of responsibility, regardless of the seriousness or how easily the problem might be rectified.

Belief that the problem will somehow solve itself (3%)

The head of risk management at a renewable energy producer believes that some subordinates are very good at convincing themselves that a problem will eventually solve itself, so there is no need to take any action and they can put off telling their superiors indefinitely.

The sad fate of previous whistleblowers (2%)

Many employees in large industrial companies have seen how a corporation will use a variety of tactics to destroy the careers of workers who are brave and principled enough to disclose risks to senior management or board directors, even when it is against the wishes of their teammates and immediate supervisors. Not wishing to suffer the same sad fate, it is understandable that employees prefer not to risk their futures by going over the head of their own managers, and divulging risk information to the leaders of a company.

There is no monitoring of the actions of employees (1%)

One mid-level manager of an oil company points out that the lack of employee monitoring systems can encourage the concealment of industrial incidents. For example, if there is no video monitoring system installed on an oil rig, the entire team can easily hide minor incidents and even some work-related injuries from their superiors, and so maintain good safety statistics. Wherever production processes are not properly monitored, inappropriate behavior or dangerous working practices are more likely to become the norm.

Shift work (1%)

Some employees believe the most important thing is that nothing untoward happens during their shift, so that they cannot be held personally responsible for any incidents. If something goes wrong at the facility on someone else's shift, then that is someone else's problem.

How the Problem Can Be Solved

Chapter 3
Recommendations for Owners and Senior Management: Ten Practical Ways to Improve the Quality of Risk Information Transmission Within Critical Infrastructure Organizations

The previous chapter has presented the insights of 100 practitioners about the causes of existing problems with internal risk transmission (answers to questions I–III are presented in Sect. 2.3). The same 100 practitioners were then asked the following questions about how to improve the quality and speed of risk communication within organizations:

IV. How, in your opinion, should an effective process be built for transmitting information about risks from bottom-line employees to executives? Please outline your perfect picture of how such a process could be established.

V. How in practice is it possible to ensure implementation of an emergency dialogue between employees and executives, where the hierarchy is temporarily leveled in order to facilitate a discussion about critical risks in an organization? Please suggest mechanisms and tools for such an emergency dialogue. What are the obstacles to introducing temporary hierarchy alignment within an organization? How could these obstacles be overcome? (*This question was asked to the first 30 interviewees*).

VI. How can one include middle range managers (directors of production sites) in the process of discussing an organization's emergency risks, when the hierarchy is temporarily leveled? (*This question was asked to the first 30 interviewees*).

VII. Is it possible in practice to prioritize the safety of operations over financial results and production targets during the operation of critical infrastructure? Please outline your perfect picture of how to implement a priority of safety over other corporate goals. Please share your point of view on obstacles to the implementation of such an approach. How would such obstacles be overcome?

VIII. How can one build the trust of employees in executives?

IX. How can one motivate senior management to visit key production sites on a regular basis and demonstrate openness in discussing organizational problems with bottom-line employees? What are the obstacles and fears facing

D. Chernov et al., *Averting Disaster Before It Strikes*,
https://doi.org/10.1007/978-3-031-30772-0_3

executives in initiating active communication with bottom-line employees? How can executives overcome these obstacles and fears?

X. What kind of tangible and intangible rewards for employees can be offered for disclosure of existing risks?

XI. Is it possible to automatize completely the process of gathering information about the operation of critical equipment without the participation of engineers (humans)? What pros and cons do you see in such a total automatization of monitoring and control over critical infrastructure? What are the prospects for artificial intelligence in this context?

XII. What development potential do you see in the idea that companies could provide insurers with full access to internal information about all critical risks, in exchange for lower insurance premiums?

XIII. Do you see national and cultural differences in the reporting and discussion of risk? (*This question was asked only to interviewees with international work experience*).

The information received from 100 practitioners in industry, combined with the results of an earlier study of the reasons for concealing risks in major industrial accidents, provide clear practical recommendations for owners and managers of large industrial companies who want to fundamentally improve the transmission of risk information within their organization, in order to prevent serious industrial accidents.

3.1 Recommendation No. 1: Owners and Senior Management Should Be Willing to Give Up Short-Term Profits in Exchange for the Long-Term Stability of Critical Infrastructure

From the analysis of previous incidents (Sects. 2.1 and 2.2 of the handbook), the prevalence of "*short-term thinking*" was the reason most frequently cited to explain why managers ignore early warnings reported to them, and why subordinates hesitate to report risk-related information internally. It is clear that company owners, shareholders, the business community and even some state authorities all set short-term goals, thus pushing managers to achieve ambitious production and financial targets. This was apparent within both capitalist and socialist economies. Firstly, the priority was always to ensure better financial results over a short period; secondly, there was relentless pressure to increase production, to fulfill or ideally exceed the agreed or imposed targets. Over-ambitious short-term plans for growth in production and revenue negatively affect the corporate atmosphere because any discussion of the risks being taken to achieve them is discouraged.

Based on interviews with 100 practitioners (Sect. 2.3 of this handbook), 58% of them believe that the main reason managers do not want to hear about problems from subordinates is the high cost of tackling them. At the same time, owners and shareholders have imposed strict production and financial targets, which are too tight to be achieved if unforeseen and expensive technological issues come to light. Most large organizations have an approved annual budget for expenses and projected profits. The budget is usually agreed with the shareholders and allows only for some planned expenses. The company's annual profit and personal bonuses for senior managers are calculated based on these forecasts. It is not hard to anticipate the impact on senior management when an employee comes to them with information about a critical risk. Solving the problem will require significant costs and exceed the planned budget. This increased expenditure will greatly reduce the profits they have confidently promised their shareholders. Therefore, for senior management, receiving information about a critical risk from employees is deeply unwelcome, because they will have to explain to shareholders the reasons for the decline in anticipated profits, not to mention the impact that decline will have on their own and their colleagues' bonuses.

Setting short-term ambitious targets for the growth of profit and production, while demanding constant cost reduction, does not foster an internal corporate atmosphere which welcomes the discussion of any risks that might require significant expenditure to mitigate.

Fundamental improvements in the quality and speed of reporting critical risks within a critical infrastructure company are possible only when the owners are willing to focus on the long-term ownership of the company. This involves accepting that the significant costs required to manage existing serious and critical risks may impact short-term profits, but are essential to protect the long-term reliability, sustainability and value of the company. In other words, the owners will make more money on the long-run and with much less risk in case of investing in reliability of critical infrastructure, which increases the long-term return and decreases the short-term risks and volatility.

Senior management follow the requirements of owners

Everything starts from the top: owners and senior management set the tone and standards of behavior, and these spread down and through the whole organization.

Several of the managers interviewed cite the same interesting example of how senior managers follow the corporate goals and priorities that the state and owners dictate to them. In one country, the state had dominant control over critical infrastructure. After the election of a new liberal government, part of the sector was privatized, and some former state assets went to foreign-based transnational corporations (TNCs). In recent times, some of these have promoted a long-term approach to managing critical infrastructure assets, whereby priority is given to long-term sustainability and asset security instead of short-term profit. This approach has arisen from the negative experience of some TNCs in the past, when compromises on safety led to disasters that hit the company's development on a global scale, damaging the capitalization of the business and ruining its reputation in the eyes of governments in other countries where it was operating. Through this bitter experience, some TNCs have come to understand the need to work to the same high safety standards in all regions where they have a presence. Thus when one TNC opened operations in the country in question, the world's leading senior managers and HSE experts came with it to help set up advanced solutions in the field of production and safety management at the newly acquired facilities. The expatriate supervisors who came were strong advocates of a rigorous approach to safety. They actively recruited the agreement of local staff to make safety a priority, and assured them that the company wanted to know about risks to be able to tackle them at an early stage. After 10–15 years of work in the country, the TNC sold its assets to a state-owned company because the national government's strategy for controlling critical infrastructure had changed again. A significant portion of expat executives had left the country, but some who had worked there from the start decided to stay and continue: they had citizenship, had started families and acquired property. Under the new state control, the previously promoted priorities of long-term sustainability and operational safety were relegated to the background. The government pushed to increase production volumes, keen to raise more short-term revenue from the operation of critical infrastructure in order to increase the tax base. They demanded a greater contribution to the state budget and launched new large-scale production projects. It is interesting to trace the conversion of some of the expat executives who had remained. With the new state-driven priorities, many who had previously actively advocated safe working practices began to promote the growth of production and short-term revenue that the new shareholders demanded from their subordinates. Safety ceased to be a priority for the new owner and senior management, so less attention was paid to these issues within the company. At one production site, the expat manager turned from a passionate advocate of safe working practice into an accountant for whom the sole priority was to drive the production and financial plan forward. He was now open to committing significant safety breaches to make this happen—and this metamorphosis occurred in just one year! This example shows that some senior managers are only too willing to adapt to the corporate priorities and demands imposed by owners and the state.

The CEO of an electricity company agrees that everything depends on the ideology that is adopted by senior managers, which in turn is based on the goals that the owners set. If bosses, in response to the demands of owners and shareholders, claim that they really want to improve the quality of their products and give subordinates the resources and authority to achieve this, then employees will fight for quality. The same applies to safety issues. If managers are pursuing a policy of maximizing profits for shareholders and owners, and cutting costs in the short-term, then they will not welcome any information coming from below about problems that need additional investment, because tackling them will cut profits. There is tremendous reluctance to get involved in solving operational safety problems, receiving information from employees about risks, or investing in an effective risk management system.

The CEO of an electricity company described two approaches to obtaining information about risks—reactive and proactive—which are based respectively on whether an organization is focused on short-term or long-term goal setting.

If senior management are focused on short-term results, then they are not interested in the long-term consequences of their actions or the feedback from subordinates about existing serious and critical risks that may in time lead to disaster. When management ignore information about risks coming from below, then ironically, a company's costs for dealing with the onset of serious and critical risks in the long run will be higher than the costs it would have faced to take suitable preventative steps to mitigate risks in the first place. By ignoring early signals about the state of the business, managers are only able to take a reactive approach to crises: they never know what is on the horizon, and every new day is likely to bring unwelcome surprises. Often, senior managers only receive feedback on their actions and begin to respond to shortcomings in the organization's work after the onset of risks (reactive approach). A reactive model is a *monologue*, a one-way communication down the hierarchy: orders come from managers down to subordinates, but managers refuse to hear anything coming in the opposite direction. Such behavior will inevitably stunt the long-term development of an organization.

If senior managers are focused on long-term results, then they consider the long-term consequences of their actions, and take a positive view towards operational feedback from subordinates about the existing serious and critical risks of an organization. This comes from the fact that they recognize this is likely to minimize the chances of a serious incident occurring in the future. The earlier a risk is identified and steps are taken to mitigate it, the cheaper it will be to manage this risk in the long term (proactive approach). A well-informed senior management, that have created a system for working with information arriving from below, can solve many problems before they reach a critical stage. With a proactive approach, risks are addressed in the early stages when fewer resources are required to tackle them, thus reducing costs. Receiving this risk information earlier, before a critical stage is reached, also means managers are likely to have access to a greater choice of effective solutions.

If we take two identical companies at a point in time where they have equally good business performance, it will be the company that responds proactively to risks that has a better chance of thriving in the long term. Proactive management in the modern world is more forward-thinking, because it sets up systems to get information about risks in advance and tries to resolve issues before they turn into serious problems. These systems develop naturally through feedback to senior management in response to its previous actions. Organizations that increase internal feedback are more adaptive and sustainable. A proactive model is a *dialogue* between managers and employees, where senior management are actively interested in feedback from employees.

Whether senior management adopt a proactive or reactive approach to gathering risk information depends largely on the goals set by owners in relation to the terms of ownership of a critical infrastructure company, together with its financial and production goals.

Senior management cannot dramatically increase investment without the support of owners and shareholders

The vice president of a gas pipeline construction and repair company outlines the options available to the board members of a critical infrastructure company when they hear from employees that there are serious problems at one or more industrial facilities. The CEO will assemble the board to present the situation and raise the question of what can be done to reduce the existing critical risks that may require significant expenditure to rectify. The financial director will likely explain that the company simply does not have the funds for a comprehensive modernization—in effect rendering the question as rhetorical, the only answer being "*little or nothing*". In many countries, most critical infrastructure was built decades ago and after years of failing to modernize, the investment now required to bring it up to date is enormous. Sometimes, instead of upgrading old equipment, it is cheaper to buy new or to build a whole new production facility. This kind of capital investment will make a company unprofitable for many years. Obviously, shareholders will not be happy with a course that will wipe out profit and dividends for the foreseeable future. For a start, there will then be no resources to cover large salaries—let alone bonuses—for senior management. Moreover, shareholders will question the professional competence of a management team that cannot deliver decent returns. Faced with such a sobering picture from the financial director, most executives will give up and send critical risk warnings to the back of the in-tray. Senior management will stop asking employees for updates about problems at the production sites and instead of long-term solutions, the managers will opt for the most conservative possible option that will prevent imminent disaster while minimizing extra expenditure. They will invest only in equipment that threatens to fail in the near future, keeping the budget tight enough to still provide for a small profit. They have resigned themselves to the fact that neither radical nor even moderate modernization will have support from shareholders. Despite the long-term reliability—and eventually profitability—that would come from proper investment in new equipment, a period of zero profits and dividends will just not be accepted by the majority of owners and shareholders.

If the business model is all about maximizing short-term profits, there will be no room for the significant extra cost of dealing with serious problems: when expenses rise, short-term profit falls. If profits go down, so will the reputation of the managers who cannot fulfill their profitability targets. Ongoing investment in safety improvements will only find support in companies that are focused on long-term development, whose owners are willing to sacrifice short-term profit in exchange for the stability of the business and the long-term maintenance of its assets.

Owners determine the priorities of critical infrastructure companies

The head of the HSE department of a gold mining company says candidly that, if he were simultaneously the owner as well as the CEO of a large industrial company, the first thing he would try to do would be to reduce his own appetite for profit. It is the desire for ever-increasing profits that underlies many aggressive decisions by senior managers and tempts them to ignore risks. This profit-driven approach from the top puts pressure on middle managers to set barely achievable production goals and do whatever it takes to cut costs. Ultimately, this puts ordinary employees in a position where they are forced to ignore the risks and run after financial and production targets to keep their jobs and obtain career advancement. In most companies, senior management never ask subordinates about risks—they only want to hear about production volume, cost reduction, revenue and so on. The questions that they are choosing to ask make company priorities perfectly clear. In response, these become the issues that subordinates devote their attention to.

The HSE head of a mining and metallurgy company put this very clearly, saying that in some companies, where profit takes precedence over safety due to the priorities of the owners, employees are treated by executives as little more than disposable assets. Obviously, leaders who view their employees like this will pay little attention to safety problems. The respondent cited cases of managers carrying out a cost–benefit analysis, comparing the cost to a company of employee deaths against the cost of installing additional safety systems that could save those employees' lives. Sometimes this even led to the cynical decision not to authorize the huge cost of modernizing equipment, but to accept the far more modest cost of covering employee funerals in the event of an accident. As the respondent put it: "*Everything was cynical, practical, and calculated on a cost–benefit basis*".

The HSE manager of a production company which uses hazardous chemical processes believes that safety ceases to be a priority when it is more important for a company to make short-term profit and deliver an ambitious industrial plan. If however the prime task is to preserve the asset over time and increase the capitalization of a company in the long run, then safety will always be a key corporate priority. Owners will be willing to sacrifice short-term profits, and finance whatever measures are needed to ensure the stability of the company and the safe operation of its assets in the long term. The respondent gives an example from his experience. At the time of the interview, he had been working in the industry for over 25 years. For

22 years he had worked in a company where the owner had a reactive approach to risk management. A few decades ago, this owner gained sole control over what had been a state asset and established a very authoritarian management style. The owner had only one goal: to make as much money as he could, as quickly as possible. Safety issues were only funded on the basis of crisis management—money was allocated only to resolve the most pressing issues. This attitude to risk management only changed after there had been several accidents. By contrast, for the past three years the respondent has been working for a multinational company that has been managing assets for over a century. Over decades, the owners and management of the company have come to understand that the most economical approach to safety is a proactive one. Assets are considered as long-term investments, and the resources are found to maintain their safe operation, even to the detriment of short-term profit. This ensures stability and keeps facilities running safely; it also attracts investors who are willing to fund reasonable expenditure on the development of industrial facilities to yield stable profits in the long term.

The head of a large power plant states that many electric power companies in his country are owned by private investors. Participating in a rush towards widescale privatization, they have little experience of running such utilities. If critical infrastructure is controlled by such private shareholders, they are often tempted to put financial profitability first, to be sure they receive a decent return in the shorter term. The respondent believes that controlling stakes in critical infrastructure facilities should be state-owned. This will allow development goals to be set in a balanced way, to guarantee the stability and reliability of the infrastructure over decades along with a moderately profitable rate of return for the state. Everything related to critical national infrastructure should have a development plan at the state level over a horizon of at least 50 years, because of the very long payback periods for the equipment. Based on his work experience, this respondent believes that owners need to encourage senior managers to "*take the long view*" in the operation of critical infrastructure, and be accountable for their decisions even beyond their tenure as managers. The respondent advocates the idea of making top and middle managers jointly responsible for the state of critical infrastructure facilities over a 10–20 year horizon. To do this, their contracts would have to be set up so that even after the contract has expired, they would still be accountable for the decisions they made during their term with a company. This is especially important in the event of a major emergency on critical infrastructure, when the former leaders of the affected facility clearly share moral responsibility with its current managers, but may not be brought to justice. Changing the legal framework to ensure longer-term accountability would motivate managers to think strategically; it would discourage them from trying to maximize short-term profits by pushing equipment to its limits without making the necessary investment in maintenance and repairs.

The head of risk management of a renewable energy producer believes that, if a critical infrastructure company belongs to a state, safety is often of greater interest to the owners than it would be for private investors. There are some private owners of critical infrastructure who will not authorize significant extra expenditure to improve safety, and expect large dividends and constant growth of stock market capitalization. Moreover, lack of available capital can have a negative impact on senior management decisions regarding safety investments. According to the respondent, a company must be rich to ensure a high level of safety.

The executive of an electric power company believes that many owners and senior managers are making a crucial logical error. In their minds, the concepts of "*safety*" and "*reliability*" are opposed to the concepts of "*profit*" and even "*production plan*". This is because for them, the term "*profitability*" refers only to short-term profitability. Any expenditure towards long-term goals is considered as a cost that does not produce positive financial results in the current financial reporting period. If the calculation period is long enough to correspond to the duration of the life cycle of a material asset, then those longer-term expenses can and should be evaluated as a profitable investment. To make such a calculation requires an accurate assessment of the level of risks over a much longer period. Trying to balance the books only over the normal financial accounting period will not give a fair assessment of costs and revenues over the (much longer) duration of an investment project. In other words, safety will only become a corporate priority when the owner focuses on the ownership of the critical infrastructure in the long term. Then, expenditure on proper maintenance will not be seen as a cost to minimize, but as a wise investment: it will guarantee a profit for many decades, and leave the owner with equipment that is still running reliably and safely, and can be sold for a good price. In the energy sector, the shortest-lived facilities have a design life of 30 years. For hydropower plants the minimum life is 100 years. The respondent was asked what prevents this longer-term priority being implemented in practice. He replied that the obstacles are that: (I) that owners see short-term "*market*" priorities as the only criteria for evaluating effectiveness; (II) that the regulatory framework tends to evaluate the viability of investments and activities on the basis of short-term results; (III) that there are psychological barriers to tolerating short-term losses. To overcome these obstacles you need to: (A) recognize the existence of a multilevel system for evaluating effectiveness; (B) introduce this multilevel system into the regulatory environment; (C) educate executives not only in finance but also in economics; (D) assess the performance of managers against both short and longer-term indicators.

According to the HSE head of a global oilfield services company, there will only be a fundamental breakthrough with safety and the reporting of risk information when owners choose to ensure the long-term sustainability of critical infrastructure to the detriment of short-term financial and industrial results. In order for information on critical risks to be freely transmitted up to the senior leadership of an industrial company, it is necessary to change the way owners see their production assets: priority must be given not to financial and production indicators, but to the long-term functioning of critical infrastructure. If business owners understand that production and financial plans are secondary to the preservation of critical infrastructure assets in the long term, then senior management can create KPIs for the entire company, thus ensuring the long-term safety of production processes and protecting the well-being—and ultimately the lives—of employees. Only then will senior management want to hear about serious problems in the workplace, and feel free to discuss these problems with owners and work together to find acceptable solutions.

In order to radically change attitudes towards the safety of critical infrastructure facilities, owners should tell senior management that: (I) they view the asset as a long-term investment, (II) they are interested in the sustainable development of the enterprise, and (III) they are ready for the additional modernization costs that this development entails.

Owners will then need to work with senior management using these fundamental values to set new KPIs around industrial safety. If owners are willing to allocate resources to prevent critical problems, then their managers will follow suit and begin to pay attention to these safety issues. This changed view regarding safety in the minds of senior management will lead over time to a change in attitudes and working practices throughout a company.

Risk disclosure in public and private companies

Some of the interviewees maintain that it is easier for senior management to discuss critical risks with shareholders in private companies where there are only a few owners than it is in public companies. Large strategic shareholders usually want to know about serious problems or risks in their business that might lead to large losses. Their logic is very simple. "*No one needs a dead cow*": if you own a cow, you want to know when it gets sores on its udders, and treat them so that the cow keeps giving milk in the long run. For an effective ongoing discussion of critical risks, the most convenient situation is when a company has one majority shareholder. Then senior management can stay in touch with the shareholder online, and raise any developing situation to authorize the resources in good time to deal with it. Any critical risks should be confirmed, preferably by an independent examination. The level of risk, possible consequences, and a risk reduction plan—all should be explained in accessible language, especially if the owner is not a specialist in the industry.

For example, one of the interviewees, the HSE head of a large state oil company, previously had experience working in a medium-sized private oil company, with a single multi-billionaire owner. The owner was very interested in preserving and improving his asset. His company had the largest business margins of any oil business in the country. Any major accident involving the company's infrastructure could well lead to the loss of a lot of business, because the state authorities and some competitors had a rather biased and negative attitude towards the owner. The respondent was aware of this situation, and when he was appointed as the HSE director, he made the case for the investment of approximately US $1 billion over several years in the modernization of critical equipment, industrial safety and labor protection. After a detailed study of the proposed safety improvement program, the owner eventually decided to finance it. This shows clearly that, if owners understand that there is a serious chance of losing their assets through industrial accidents, they will be willing to invest significant money in safety.

Several interviewees note that in terms of setting targets, family-owned companies have a longer-term overview in relation to their business, and are not prepared to risk long-term stability just to generate short-term income. Family companies tend not to have annual contracts, there are no short-term bonuses, the turnover of managers is slower, and so on.

It is much more difficult for senior management to disclose the critical risks of public companies where there are a huge number of minority shareholders. These will often be speculative investors who expect senior management to issue a positive annual report, with constant growth in stock capitalization and regular payment of large dividends. If they need to respond to critical risks in this situation, senior management will have to publicly disclose the situation in the workplace to get significant resources from the board of directors to deal with the risk. Highlighting critical risks in this way will likely damage a company's capitalization, attract the attention of regulators, and raise questions about the competence of senior management. However, according to one respondent, if the CEO of a public company is a professional with a strategic vision, he/she will nevertheless publicly disclose risks, look for resources to alleviate problems and take preventive measures to minimize risks. Such a CEO understands that it is better to spend one dollar on risk management today than a thousand dollars on clearing up an accident tomorrow.

Discussion of the objectives and long-term development plans of a critical infrastructure company

The head of a large power plant (middle management) expressed the opinion that for senior management to work effectively, they need to understand the criteria by which their performance will be evaluated by owners over a given period. The first step in creating a long-term plan for the development of a critical infrastructure company is to ask the state and the owners what their goals are for a company. Then they need to ask, "*Where do you want to be in, say, five or ten years' time?*". The objectives will probably include: (I) infrastructure reliability; (II) a certain level of energy output (production); (III) a certain level of profitability; and (IV) some other parameters at certain points in time. Then, within this framework of goals and targets set by the state and the owners, strategic strengths and weaknesses of the existing critical infrastructure can be evaluated. The CEO and senior managers need to explain the consequences of the implementation of certain decisions, showing all the pros and cons and the resources required for each decision. In this way, they will have carte blanche from state and owners to implement specific strategies at given time intervals. Having from the outset established the criteria by which their performance will be assessed, and the targets they need to meet, they can pass appropriate instructions down the corporate hierarchy to their subordinates to set things in motion.

The HSE vice president of an oilfield services company believes that it is enough to change corporate goals and objectives to encourage people to act differently; nevertheless, these new goals must be agreed with the owners. As soon as the goals—and the results expected from implementing them—are clearly defined, many safety priorities will also become clear. It will be understood that the owners have chosen to prioritize the reliable long-term operation of the sites over short-term profits, and that they do not want to make "*dirty money*", or "*a fortune on the corpses of workers*". There is no single list of universally applicable KPIs to achieve this. They need to be developed individually for each enterprise based on the national and cultural characteristics of workers, the level of risk, and the corporate objectives in relation to each production facility. Senior managers should discuss with their mid-level subordinates the new values and goals they have agreed with the owners and shareholders. By discussing the new priorities with managers at various levels and even with ordinary employees, leaders can ensure that all employees understand how these priorities can inform and be integrated into the work of an entire industrial company.

Finding a balance between financial profit, production targets and safety

According to the HSE director of a petrochemical company, the first step towards changing corporate priorities is to introduce management models based on the concept of sustainable development, where the financial, social, and environmental priorities of an organization are balanced.

In response to the question of how, in practice, to put safety above other corporate goals—profit, production plan, and so on—the HSE director of an electricity company maintains that there should always be a balance between safety, finance and production. If one of these three elements takes precedence over the others, there is trouble on the way. For example, if profit takes priority, this can threaten the safety of the production process: equipment may not be as regularly maintained, but aggressively pushed to its tolerance limits. On the other hand, if preventing the over-exploitation and potential failure of production equipment becomes the most important corporate priority, then profit could be lost because production capacity is underused. If industrial safety is the main priority, then some production operations may be halted due to a relatively minor danger to employees.

The task for senior management is to support managers at different levels in conducting daily risk and opportunity analyses to find an acceptable balance of risks. The senior management team can set the example in finding such a balance, by regularly gathering the function managers who oversee these issues and using their knowledge to help reach a consensus between financial targets, production plans and safety. A situation should never arise where the respective managers cannot come to an agreement. Leaders at this level are paid not only to perform their own function in isolation, but also to work together with others towards a unified group objective. Gradually, this practice of reaching consensus will cascade down the hierarchy so that lower-level managers work together in the same way. Ideally, people lower down the ladder will continue to work with their colleagues on the shop floor to find a consensus, without always requiring more input from senior management. To do this, they must be given both appropriate authority and resources. For instance, managers at various levels should have the authority to stop unsafe operations without jeopardizing their career and be able to suggest adjusting financial and production plans if, in their opinion, these cannot be achieved without serious violations of process and occupational safety.

A senior HSE advisor and human factors specialist in the oil and gas industry offers an interesting example from his own practice. He heard an employee say, "*We are not an oil and gas company, we are a safety company*". Colleagues immediately perceived this statement in a negative light and responded, "*Rubbish! We are an oil and gas company, let's be honest about it. The reason we exist is to make money for the shareholders, for the organization and then ideally that filters back to us in the workforce*". According to the interviewee, the idea of operational excellence, whereby all aspects of a company's operations are brought together under one roof, is far more accurate than claims that safety is a top priority for an industrial company. The respondent considers it disingenuous to pretend that operational safety can completely take priority over financial results. The concept of operational excellence recognizes that a company must be profitable or it will fail, and all employees and managers will be out of work. At the same time, a company can manage its operations in such a way as to reduce safety risks to an acceptable level. The respondent believes that companies should be honest in what they promote within the workforce. Therefore, it is better for senior managers to openly admit: "*We are focused on creating an effective organization. This means that we are constantly looking to improve production operations and balancing them against costs. At the same time, we are focused on maintaining excellent standards of labor protection, industrial safety, and environmental care. We are also focused on minimizing lost production time*". Most employees of a critical infrastructure company will accept an honest message like this more favorably than statements claiming they only care about safety and employee wellbeing, when this is clearly not the case in practice.

The Board Director and manufacturing manager of a chemical company gives an example of a balanced model in terms of his company's operating philosophy, which can be summarized as follows: "*our vision is to be one of the safety leaders in the industry and be the preferred supplier for our customers at the lowest possible cost*". The operating philosophy of this company covers three main areas: safety, customers, and cost. Everyone in the organization understands the order of priorities in choosing solutions: first operational safety, then customer service, and finally achieving the lowest possible cost—but without compromising safety and customer care.

The former head of a mining regulator believes that an unprofitable company will not last long. Therefore, there must be a balance between the profitability of a business and the safety of operations. However, there is no justification for focusing solely on a company's profitability. The respondent cites the following example. The annual report of a coal mining company was analyzed to see how much content was devoted respectively to production, environment, and safety. It turned out that 99% of the report was devoted to production and finance! Such an imbalance immediately identifies the business priorities established by that company's board of directors and senior management. This is not to say that an industrial company's financial profitability and production efficiency should be ignored. However, it is necessary to balance these factors against safety issues. A practical example of implementing such an integrated approach is the way some coal mines operate: the day and evening shifts work on mining the coal itself (i.e., production) whereas the night shift deals exclusively with mine safety issues such as ventilation, dust and explosion protection, repair of equipment, and so on.

A risk management consultant specializing in oil and gas also believes that, if a critical infrastructure company reduces its profitability, it will eventually cease to exist, and employees will lose their jobs. Thus in order to sustain an organization, senior management must balance the operational safety against the overall profitability of production. The respondent noted that in a critical infrastructure company, it is essential to make risk management a calculated process, where risk is balanced against profits. If a company does this more successfully than its competitors, it will outperform them, raise company profits and have more capacity to expand.

The HSE manager at a utility services company notes that there are examples of companies with class-leading performance in industrial safety and labor protection that also have better financial performance than their competitors. There are also examples where wealthy companies with smart (and costly) safety policies have the lowest frequency of risk incidents. In the company where the respondent works, risk managers try to explain to all managers that safety improvements lead to improved processes and a reduction in production faults, which positively affects revenues. According to the respondent, this approach is beneficial for everyone, and several key company targets in the field of safety, finance and production are simultaneously achieved. The message that investing in safety improvements can increase the profitability of an organization and reduce losses from accidents should be constantly conveyed to all members of an organization.

The HSE manager of an oil company believes that investment in improved safety standards has a positive impact on a company's share price and reputation, as accidents and incidents are damaging to both finances and reputation. Senior managers are very sensitive to these issues. Therefore, risk managers can use this argument to convince senior managers to make adequate and timely investments in production renewal, industrial safety, and labor protection.

The HSE manager of a company that uses hazardous chemical processes in its production believes that a problem can arise when separate departments in a large industrial company have different perceptions of the safety of the enterprise. For financiers, financial efficiency and stability are the most important for the security of an organization. For production units, safety equates to the adequate operation of equipment. For manufacturers, investing in preventive maintenance and carrying out routine repairs is the basis for safe operation. In the perception of financiers, such costs are not always justified because they impact the financial results and therefore the stability of a company. But in their focus on financial stability, financiers often lack full appreciation of the critical production risks of an enterprise. Senior management tend to listen more to the opinions of financiers. Thus, manufacturers must work hard to explain to financiers the value of timely investment in the repair and modernization of equipment. Financiers should be encouraged to become familiar with the production process and employ risk models to assess the financial damage a company could sustain if equipment fails and fixed assets are lost. This should give them a broader and longer-term perspective and temper their ambition to achieve financial results at any cost. Senior managers should take responsibility for coordinating this interaction between financiers and manufacturers. By organizing regular meetings between the two departments, and demonstrating the decision-making principles they wish to encourage, leaders can bring these potentially antagonistic groups together to listen to each other, negotiate, and find mutually acceptable solutions that will benefit the organization.

The chief risk officer of a national power grid company believes that, if the chief financial officer is too influential in a critical infrastructure company, this can have a negative impact on safety and risk management. If senior managers always focus on costs, this can stop them taking appropriate and timely decisions around safety.

The HSE manager of a metallurgy company believes that it is important for the leaders of a company not only to be in communication with production units, but also to keep track of decisions being made by the departments handling planning, economic and financial strategy, and accounting. Sometimes, their mistakes in setting production and economic indicators put production units under pressure to achieve impossible business targets. Senior executives should harmonize the production plans and financial goals of the enterprise through regular peer-to-peer discussions involving all departments of a company. They should explain to employees how mistakes that seem to be unrelated to the production process could lead to serious problems at a company level. And just as they are promoting the value of greater openness about production risks, problems, and errors, they should urge the directors of the planning, economics, and finance departments to share any uncertainty or error in their area of responsibility. In all sections of the business, workers need to understand that they will be appreciated, not punished, for sharing problems or mistakes. Only by working collaboratively in this way can the issues be fully resolved to the benefit of the whole company.

The HSE head of an oil company explained how they have developed an integrated "*safe production*" program in their company to remove the conflict between financial and safety priorities. Senior managers are encouraged to assess the existing risks of a company and to forecast in detail the likelihood of their occurrence and the possible consequences. If these are not calculated to be significant, then priority is given to production indicators. If, however, the likelihood of a given risk causing significant damage is high, then safety issues are prioritized. It is crucial for the leadership to have such risk assessments operate at all levels of the corporate hierarchy—right down to the shop floor—so that managers at every level have the authority and resources to set priorities and act appropriately to the risk levels involved.

According to the HSE manager of a metallurgy company, metallurgy is a highly competitive industry. Any accidents and incidents occurring within these facilities are carefully analyzed by major customers, who are often worried about the possible disruption of metal or ore supplies if any incidents were to occur at a production site. It can take ten days even to assess the scale of the damage after an accident, and a full resumption of production can take from six months to several years. Therefore, when calculating the costs of potential incidents, it is important to keep in mind lost revenue from downtime and under-production. In other words, organizations that do not invest in safety are not investing sensibly in the business.

Long-term safety and profitability are interdependent. When there is no investment in safety, this is a big threat to continued profits in the long term. However, the only way to guarantee 100% safety is to stop production completely and close the site. On a day-to-day basis, there has to be a reasonable balance between safety and income generation. If you are dealing with an insignificant risk, income can take priority. With a potentially critical problem, it is important to focus on safety rather than immediate income.

The managing director of a gas distribution company believes that, if a temporary safety deficit does not lead to big financial and production problems, then financial priorities may prevail over safety. To inform the decision-making process, it is essential to constantly evaluate the likelihood of serious incidents. Critical risk accidents do generally have a low probability of occurrence, and when an operation has been running for decades without any problems, it can "*blur the vision*" of managers and create the illusion that everything is perfect and nothing serious could ever happen. The challenge is to keep risk assessments as objective as possible, basing them on professional analysis by specialists, managers, and external experts of the likelihood of occurrence and the severity of the possible consequences. If the likelihood of an accident is high, then of course it is essential to prioritize safety over financial and production results.

The head of the HSE department at an oil company believes that safety should not be considered as one among a number of company priorities, as priorities need to often change. It must instead be treated as a core company value, as these are generally fixed. If a company recognizes that safety has inherent value and senior management work to have that value in mind in every aspect of a company's activities, then safety will always remain a focus of executive attention. The respondent also noted that the safest way for a critical infrastructure company to operate is to do nothing at all and stop all production. Obviously, this is absurd, as a company also exists to provide a service or product and ideally to make money too. It is important to keep in sight "*how*" a company operates and not just "*why*".

The vice president of an international oil company expressed a similar view. If safety is a priority rather than a value for a company, then senior managers are likely at some point to shift priorities, at least temporarily, and put financial or operational results first, with safety relegated to an afterthought. However, if a company holds safety as a value, then this will remain as a fixed, fundamental principle even if corporate priorities shift.

A lead safety manager working mainly in oil and gas also believes that using the word "*priority*" in relation to the role of safety in managerial decision-making is inappropriate, and that managers should not talk in terms of "*safety first*". Instead, safety should be part of a company's integral business values. Thus, safety functions are a key operating factor that must be considered every single time any management decision is made. Managers should not have the option of relegating safety issues down a list of priorities. Rather, safety considerations should be built into the process of all managerial decision-making. According to the respondent, there are many examples of the successful integration of safety issues into the decision-making process of critical infrastructure companies.

It is worth noting that most respondents believe that the optimal approach to establishing priorities within a critical infrastructure company involves finding a balance between safety, operations (production) and finances. However, respondents from the nuclear power industry firmly believed that safety should be an absolute top priority for a critical infrastructure company. This is perhaps not surprising, especially given the stringency of regulatory requirements for nuclear safety. Moreover, the potential costs of maintaining a nuclear power plant (NPP) in a safe operating condition are likely to be thousands of times lower than the damage resulting from a major accident, and the subsequent clean-up. For instance, the total damage caused by the Chernobyl accident amounted to US $1 trillion (in 2022 prices).[1] It is estimated that the clean-up after the Fukushima Daiichi nuclear accident will run to more than $200 billion over several decades, and this does not take into account the costs of the immediate forced shut-down of all of Japan's other nuclear reactors (many of which remain closed to date) and the import of fossil fuels into the country for many years after the accident.[2] No wonder then that nuclear industry leaders view operational safety as the top corporate priority of their organizations.

The president and CEO of a nuclear power company believes that in his business, safety should be more important than financial results. However, at the same time, one must be realistic and understand that the corporation has to earn enough revenue to finance its activities.

A consultant in nuclear safety with extensive experience in NPP operations offers another perspective. If safety is considered as an element of sustainability rather than a cost attribute, it radically shifts the viewpoint. The concerns are both the safe operation of the nuclear power plants and the issues of personnel, corporate and financial safety. In the organization where the respondent worked, he went out of his way to associate optimizing safety with maximizing resilience and long-term security. He maintains that associating safety directly with sustainability, rather than with financial costs, is the key to overcoming the previously described obstacles.

A safety consultant with managerial experience in oil and gas, chemicals and mining notes that continuous improvements in the safety of the nuclear power industry could potentially make some NPPs unprofitable to operate. Nevertheless, the respondent considers that nuclear power has very important benefits for the public at large. Thus the issue of nuclear safety becomes not just an economic, but also a wider national political and security issue. This would seem to justify the state subsidizing these companies to ensure the safe operation of NPPs for the long-term public good.

---

[1] Dmitry Chernov, Didier Sornette, Giovanni Sansavini, Ali Ayoub, Don't Tell the Boss! How poor communication on risks within organizations causes major catastrophes, Springer, 2022, subchapter 1.6. Chernobyl nuclear disaster (USSR, 1986), https://link.springer.com/book/10.1007/978-3-031-05206-4.

[2] Dmitry Chernov, Didier Sornette, Giovanni Sansavini, Ali Ayoub, Don't Tell the Boss! How poor communication on risks within organizations causes major catastrophes, Springer, 2022, subchapter 1.16. Fukushima-Daiichi nuclear disaster (Japan, 2011), https://link.springer.com/book/10.1007/978-3-031-05206-4.

Focus on making long-term contacts with senior managers managing critical infrastructure

Senior managers have employment contracts that can affect their willingness to receive bad news from subordinates.

According to some respondents, the ideal tenure for senior management managing critical infrastructure should be at least five years. This allows enough time for a manager to deliver effective solutions for the problems that generally exist within an average company, as well as to develop strategic plans for improvements. The head of a power plant recommends five to seven years as the optimal term for a labor contract for senior management, as this allows them enough time to solve serious problems in a critical infrastructure company. The HSE manager of a metallurgy company believes a period of five to ten years is optimal for senior management to solve the serious infrastructure problems of a large industrial company. The head of the HSE department at an oil company recommends signing an unlimited contract with a manager who operates critical infrastructure, so that the manager does not limit his work plans to a short period, but thinks strategically, linking his fate with this company for years, or even decades. Some respondents agree that in principle an open-ended contract is beneficial for a manager. However, a manager's term of office is rarely indefinite because there is legislation in place in many countries to allow a company to dismiss a senior manager very swiftly if any serious mistakes are made.

Some respondents feel that the annual bonus system easily leads senior managers to set dangerous goals, where profitability and income are prioritized over safety. A better method for encouraging managers to boost the long-term sustainability of critical infrastructure enterprises is to provide their bonuses in the form of company shares that can only be sold after five to ten years. This ensures that the current decisions of senior management are more likely to take account of the long-term success of a company. Some respondents went further, suggesting that the system of paying regular annual bonuses to senior management should be abandoned altogether. Instead, the bonus payments could be spread over time. For example, 30–40% of the bonus can be paid for the current year and 60–70% for the next four years, encouraging senior management to work towards the sustainable development of a company over the long term.

The HSE director of an oil company believes that equal priority should be given to safety as to achieving production and financial targets. He observes that people focus on both what is measured and what is rewarded. If 90% of a senior manager's annual bonus is based on achieving production and financial KPIs, and only 10% is based on meeting safety targets, then this is bound to influence their work priorities. Similarly, if managers believe that a career promotion is likely to follow success with production and financial targets rather than safety achievements, their priorities are likely to follow suit and safety will be pushed down the list.

While this is not directly linked with the question of senior management contracts, the HSE director of a metallurgy company also highlights the problem of short-term contracts for contractors which carry out critical infrastructure maintenance and repair work. If they are retained only on an annual contract, they will be less likely to invest in equipment or personal professional development, because they may not even be working with the company next year. Therefore, this respondent advocates that the optimal contract period for contractors who operate critical infrastructure is between three to five years.

3.2 Recommendation No. 2: Senior Management Should Be Approachable About Problems, and Have the Desire and Resources to Control and Mitigate Identified Risks

Management need to have a genuine desire to hear about problems in production and then solve them

Several leaders observe in their interviews that subordinates work according to how senior management manages them. In other words, the management system in a company determines the actions of employees. Everything comes from leadership. Employees will report problems if managers want to hear them. Managers should want as much information as possible about potential risks. If the management support is not there, all other interventions are doomed to fail. The only way to improve the situation regarding feedback in an organization is if leaders have a genuine desire to hear about risks from their subordinates—and communicate this to them—and then take decisions and allocate resources to stop risk escalation. It has already been established that senior managers should have the necessary support—moral and practical—from owners and shareholders to implement risk reduction measures. Having secured this, they should then take the initiative to implement cultural change, dismantling any system of penalties for reporting risks or incidents, and making it clear that they actively want to hear about problems. Only then will employees, inspired by the evident commitment of their leaders to a safer workplace, be willing to report the risks they have encountered.

Several respondents stress the importance of leaders being honest and authentic. When they claim that safety is one of a company's most important values, they must carry conviction if they are to persuade employees to openly disclose risks and problems. Senior staff should sincerely believe in what they are doing, follow through with their decisions and act with enthusiasm and sincerity, to become a role model for their subordinates. As a rule, managers should behave in the way they want their employees to behave.

The HSE head of a mining and metallurgy company believes that the creation of risk information transmission systems is possible only if senior management actually want to receive the information. If they do not want to hear any bad news from their subordinates, there is little point in investing time and money in setting up a new system.

The HSE manager of a metallurgy company believes that, to start an effective process of reporting risk information from the bottom up, the owner and senior management must want to genuinely hear about problems and have the capabilities and resources to solve any important issues that employees bring up. If a manager simply puts out an order that all employees must reveal risks, few people will be convinced that there is any sincere desire from the top to hear about them. In this regard, the leadership of senior management is obviously very important. The interviewee explains that, if he were the CEO of a company, he would start with himself. He would pursue a policy of open discussion with his subordinates, and assure them that he and his team were not afraid to receive information about problems, and were ready to solve any that were identified as quickly as possible. In addition, he would convey to all employees the possible consequences of concealing risks: a major accident could ensue, leading to the deaths of many people, the bankruptcy of the enterprise and the consequent loss of thousands of people's jobs. He would begin by making regular visits to production sites and creating frequent opportunities to talk with ordinary workers about operational matters. People must see the genuine desire of top officials to understand the problems and solve the issues. He would constantly be asking: "*Why are we operating the equipment like this? What can we do differently to work more efficiently?*". Such questions are necessary to encourage employees to discuss the finer points of the industrial process and indicate the advantages and disadvantages of the current approach to production. No single senior manager can expect to understand all production issues—to fully grasp a problem they need to discuss the details with the expert employees who are running the equipment on a daily basis, and can give a detailed explanation of existing problems and how they could be solved.

A consultant in nuclear safety with long experience in nuclear power plants operations has a similar opinion. He believes that the first step is for the leaders of an organization to show a clear willingness to receive feedback from subordinates about existing risks and problems, which should be seen as a vital source of information to an organization. For this to happen, leaders must be trained to lead: where necessary, they can be offered training on how to change their language and behavior in order to be more effective in their role. Once the behavior of senior managers begins to change, then it is time to move on to do the same for middle and lower managers, and then finally educate the workers. The respondent has experience of implementing new risk management systems at more than 250 industrial sites, in 40 countries and 10 different languages. They worked with senior managers, who then demonstrated to their subordinates the value of this new way of acting and communicating. According to the respondent, when implementing a project to improve the quality and speed of risk information transmission, changes should always begin at the highest level of an organization's management and then gradually work down the corporate hierarchy.

The HSE director of a mining company believes that when managing change, it is crucial that the change is clearly shared and supported by senior managers. Only when managers are actively involved in implementing change will employees be convinced that this is an inevitable and welcomed process. It is essential that managers truly believe in what they are doing and that they are genuinely interested in progress with problem solving. Employees feel it instinctively when leaders are being insincere. For the ideas of change that the leader is preaching to really "*catch fire*", they must carry conviction. However, this is only the first requirement. Many leaders declare that they want to know about the key problems of an organization and believe it is important for them to know. In reality, as soon as solutions to these problems turn out to require a critical reassessment of their own past decisions, or the allocation of huge resources, even leaders who previously seemed committed to change can react negatively: they make it only too clear to employees that they are unhappy about the news, and it is the fault of employees who should have handled the situation independently. Being an adherent is not enough: senior managers need to demonstrate in practice that, even when they hear about the most difficult problems, they will still allocate the necessary resources and be part of the solution. Another instance of the need to "*walk the walk, not just talk the talk*" is that employees cannot be expected to strictly comply with safety rules if managers themselves do not. For example, there was a case when the vice president of one of the world's leading oil services companies refused to buckle up when driving in a corporate car. In the end, he was fired. The board made sure his dismissal was reported to the entire company, to demonstrate that everyone is subject to the same safety rules, and that managers should set the right example for other employees. If the top brass of a company are acting according to newly established principles, their example will communicate the commitment to change better than any statement. Imitation is common among employees as in all human groups—if the leader starts to do something, subordinates will follow suit.

The head of an oil production facility gives another example. Some companies are introducing incentive systems for the detection and reporting of near misses, hazardous actions or conditions, usually reported on so-called "*STOP [Safety Training Observation Program] cards*". These systems are aimed at gradually changing the mindset of workers by having them constantly analyze their actions and those of their colleagues to identify and respond to potentially dangerous actions and working conditions. Unfortunately, in many cases, these systems do not work or are not very efficient. This is mainly because there seems to be little real management support for implementing them. Sure, the leadership verbally support the system—but in reality reported cases are not followed up, and there are no real changes to a company's working practice. Often the implementation of the system is not a high priority for senior management. The system is spread down the hierarchy, slogans for employees are published, but none of the senior managers is ever truly interested in how the system works in practice. For the new system to become ingrained, it is imperative that they respond to the problems flagged up by employees, but senior managers often show little interest. As a result, even though the STOP card system is formally running in a company, it does not actually have much effect in reducing accident rates. If a CEO is not making time to examine the most significant STOP cards from the production sites on a weekly basis, and making decisions based on them, the system will simply never work. If there is no interest from the top, then for the middle and lower management employees there is little motivation to actively participate. They will copy what senior management do and do no more than pay lip service to the scheme.

The HSE manager of a production company which uses hazardous chemical processes believes that a company's management should be willing to immerse themselves in production problems. They must have the resources to make the necessary changes, and make sure all subordinates understand that they want to receive accurate, unvarnished information from the field. Senior management can offer subordinates real motivation—if they receive disclosures of objective information, they will make better and more appropriate decisions. For example, subordinates will be provided with the resources they need to deal with the risks they have reported in their area of responsibility. Senior management can then declare: "*If you let us know the real situation at the site, we will give you the resources you need to solve the problems you have identified*". Subordinates will then have a real interest in disclosing the most critical risks, because management have promised that these risks will be addressed. The respondent has had the opportunity to work under the leadership of three CEOs over the past ten years. The first appeared to welcome all feedback, but in fact made decisions only based on his own experience and opinion. The second delegated all decisions to subordinates and tried not to make any himself. The third tried, whenever he had been informed of a problem, to carefully listen, investigate and then decide about solving it within an agreed timeframe. Unsurprisingly, time shows that the third leader is the most successful and productive in the longer term.

A single team of senior managers

It is important that all senior managers feel part of a single team that shares responsibility for managing a company.

It often happens that the senior management team is very fragmented. Each of them wants to be responsible only for their own functional block. Some do not want to share responsibility for the overall result or for solving problems that affect the whole company. It is important, however, that the whole team understands and is committed to the value of feedback about existing risks, so that they can inspire their immediate subordinate leaders. They, in turn, can then motivate lower management and ordinary employees to give feedback in the same way. There should be complete consensus across the senior management team about the need for change in the way they handle the problems revealed by subordinates.

The HSE manager of a metallurgy company believes that it is important for the CEO to be surrounded by like-minded people: people who will pursue the new policy of openness every day at their management levels, and be ready to shoulder their share of the responsibility for solving the problems that come to light. Subordinates will copy the actions of the management so that after a while, lower and middle managers will also be willing to pursue such a policy of openness, at their own level and in communication with their subordinates.

According to the HSE vice president of an oilfield services company, any alteration of corporate behavior must begin with senior management, and then move to successively lower levels. If senior management do not really want to move on, and instead leave it to external consultants to change the minds of ordinary employees, it is nonsense to talk about meaningful change in a company because people will not believe it. First, senior management need to change—only then will subordinates follow.

In the transition from a reactive to a proactive relationship as regards risks, the communication of risk information improves because all members of every team at every level of an organization are consciously identifying, transmitting, and controlling risks.

Positive examples of owners wanting to hear information about risks, and being ready to allocate the necessary resources to prevent critical situations from developing

The present authors would like to cite two practical examples where owners have been willing to hear about risks in critical infrastructure enterprises, and to make systemic decisions to tackle them.

The first example is from a company whose production involves many hazardous chemical processes. The example was shared by the head of HSE. During an emergency, the senior management were faced with the fact that news of the incident had taken much too long to reach them, and that by the time it did it had been distorted. A fire had broken out at one of a company's key industrial sites in a region remote from headquarters. News of the fire spread very quickly through regional and national press, and was also quickly published in the professional media. This prompt media response was caused not by the company letting the press know but by the actions of the state emergency services. State firefighters dealt with the incident, and reported on their quick response to people living near the production site, the regional authorities and journalists from various media. However, no news of the incident made its way from the site to headquarters. The owners of the company and some of the board of directors found out about it through the professional media, and naturally they had some questions for senior management about exactly what had happened and why. Senior managers were mortified and furious: how could owners and board members have found out about the incident before they themselves—supposedly in charge and certainly accountable for the company—had heard a word?

Unsurprisingly, the whole episode damaged the reputation of the senior management in the eyes of the owners. However, there was a benefit in that it triggered a sea change in the company's approach to communication about incidents and critical risks. Managers came under pressure from the owners to demonstrate that they had full awareness of what was happening in the organization on their watch and, in order to show this, they needed rapid and clear information from their production sites. The owners and the board made it clear to senior management that, from now on, they expected to be informed about any deaths in the workplace—whether the casualty worked for the company or one of its contractors—within an hour of the incident, in any region anywhere in the world. The company called in a specialist who had managed the informational response of state authorities during and after emergency situations, who shared his experience of how information about incidents travels vertically in government bodies. This specialist proposed a keystone for the organization's informational response to emergencies: "*The main thing for dispatchers at the site in the first minutes after an incident is to share information about what has happened with all levels of company management*". In his experience, managers and supervisors who were involved in the lead-up to an emergency tend to distort information about the causes and details of the incident. Therefore, senior management must establish internal communications so that they can rapidly get information about emergency situations from various sources, including recordings from surveillance cameras installed near the scene. All this information should be promptly communicated to managers immediately after an emergency. The consultant also defined new criteria for site managers as to what constitutes a crisis. In the event of such a crisis, dispatchers would simply press a specific button on their control panels that would immediately transmit information about the incident to all recipients included in the contact list, using SMS, email or auto redial—a voicemail that cannot be stopped until the full message is heard. Immediately after an emergency, the information available at the bottom of the corporate hierarchy—workers and dispatchers at the emergency site—began to be transmitted, not only to local specialists involved in the emergency response, but also to company leaders thousands of kilometers from the scene. This helped to reduce or eliminate the distortion of information about emergencies as it passed through the various levels of management, and meant that the company could start dealing with the emergency promptly using the knowledge and resources available at all levels of the hierarchy. The only problem the company now faced was that, for the dispatchers on site, it was difficult to immediately judge the scale of what was already happening or predict how serious the emergency might become. The fact is that at the very beginning of a crisis, dispatchers do not have enough detailed information, and their assessment of the scale of the emergency can very quickly change for the worse because incidents in the chemical industry are unpredictable and difficult to respond to. Therefore, any immediate information that company executives receive about a crisis in the workplace will be incomplete and unstructured. Only when the details are clarified can the true scale of the crisis be determined. Nevertheless, this was a breakthrough for the company. From then on, information about a crisis occurring thousands of kilometers from headquarters has been able to reach senior management within a few minutes of its onset. As soon as managers at all levels know about an emergency, each is tasked to immediately collect information from their subordinates about the developing situation and any measures taken and be ready to report to the senior management of the company within an hour. With executives at various levels reporting upwards in this way, senior managers can gain a more detailed and accurate assessment about the situation, as they are also utilizing a range of alternative channels. These new communication protocols have led the whole company to recognize that distorting information when reporting to superiors is now pointless, because superiors have many other ways of cross-checking the facts. In addition, the company is now committed to releasing a first report about the incident to the media within an hour and it is obviously critical that this public statement is based on the best objective information that can be collated across all the various departments of the company. Before making any statement to the media, senior management should also be advised of how government representatives and media are already interpreting the incident, in order to understand the context in which they are delivering the company's up-to-date account of the emergency.

After the deaths of several contractors' employees at company sites, it was the owners who asked senior management for details of the contractors that the company was hiring. Their logic was simple: if a chemical plant is destroyed, it is equally relevant to the business whether it was a company's own personnel or those of the contractors who were killed, or were responsible. The owners began to insist that senior management should not necessarily select the cheapest contractors, but only those who also have a good safety record. Following this requirement, the selection of contractors in the company was completely transformed. The owners had demonstrated to senior management that, for them, pursuit of short-term profit is not as important as ensuring workforce safety and the long-term security of their assets. The open discussion of the critical risks of the enterprise and ways to reduce them was only possible because the owners and shareholders of the company made this choice. They showed that they were not just paying lip service to the idea of safe and sustainable development—they are ready to allocate significant resources to tackle the critical risks of the enterprise. The whole company very much appreciated this new approach from the owners. Moreover, as soon as they recognized that the owners were genuinely interested in discussing critical risks, many top and mid-level managers began to submit long-standing and extremely important problems to the risk management committee that they had previously held back; the owners now attend this committee on a regular basis. Some long-standing problems submitted to the committee had not previously been resolved because senior management had been in too tight a financial position: they were committed to delivering a certain profit and did not have the authority or resources to go beyond the approved cost estimate. Obviously, most of the projects submitted to the risk committee have a negative impact on short-term profits—but unless risk problems are solved, a company may face huge losses in the event of an emergency at a production site. Now that the owners have agreed on the allocation of additional resources outside the approved annual plan if there is a clear case for them, the company can openly discuss many issues that previously could not even be raised. It is worth noting that only critical risks are brought to the owners for consideration, so as not to create an overload of information noise in the form of proposals to address issues that are not essential for improving safety. When projects are proposed to the risk committee, complex technological issues are explained to the owners in the context of increasing the long-term sustainability of the entire business. Summarizing, this company's experience clearly demonstrates that feedback from employees to senior management on the critical risks of the organization will only be effective if the owners and senior management want to hear it, and are willing to allocate resources to solve the problems reported.

The second positive example was given by the HSE manager of a metallurgy company. This company is an industry leader and the most profitable metallurgy company in its country. The respondent shared his experiences in the company gained over more than two decades. Throughout this time, he has never seen the senior management ignore information from their employees about a serious safety or production issue. The fact is that the owner of the company is focused on the long-term sustainable development of the business. It is more important to him to increase the value of the asset in the long run than to make an immediate profit. Any serious problem at a production site is seen as a threat to the sustainability of the company. Thus, problems need to be quickly identified and comprehensively addressed. The owner has stated many times that, if a serious incident occurs, then it must be the result of systemic shortcomings in the company. He also believes that, if a worker is under pressure to violate safety procedures, then there is something wrong with the organization of the production process itself. Therefore, the causes of any incident should be investigated and established. The company must systematically resolve whatever shortcomings come to light, so that workers are not motivated to take similar risks in the future and incidents do not happen again. The owner insists that, if a safety problem occurs at any of the sites, senior management should raise it at headquarters level in order to systematically resolve the issue throughout the company. He maintains that, if senior managers are not actively looking for safety problems at the production site—and then solving them—they are not meeting their responsibilities. As a result, the management team he has appointed is not afraid to hear about problems and get involved in solving them, because they know the owner is ready to finance whatever measures are necessary to control critical risks. Senior managers at headquarters regularly review the reports of every incident at the production sites. They discuss the causes of incidents not only with the HSE department, but also with middle and lower managers at the site where the incident occurred. In recent years, billions of dollars have been invested in the modernization of equipment and the construction of new production facilities.

Over his two decades of work at this metallurgical enterprise, this respondent has never encountered a site manager (middle management) who did not want to hear about problems, or hid a critical issue from senior management. It is not advantageous for mid-level managers to hide risks: if there is an emergency at the facilities entrusted to them, it could stop production for many months, bring enormous damage to the enterprise and put an end to their career. It is far better for them to identify critical risks and bring them up with senior management and the owner, because the resources needed to solve complex problems are usually many times greater than those at the disposal of middle managers in the field. In this company, at the level of middle and senior management, problems are never hushed up—on the contrary, the main owner and senior managers take pride that problems are identified and resolved quickly. According to the respondent, all this is a result of the values of the principal owner, who has emphasized many times over that he wants to achieve constant improvement of his asset. Accordingly, the management team are set up to proactively resolve issues. It is noteworthy that, when holding meetings at the production site, senior managers start by discussing safety issues, only then moving on to issues of productivity and finances. By addressing the agenda in this order, they are showing middle and lower managers the true priorities of the company.

According to the respondent, at the site where he works, the only organizational level that can possibly conceal problems is the lower tier of managers and some ordinary employees. However, not all employees are the same. Some trust and welcome the constructive attitude of middle and senior managers, and willingly reveal risks; others, perhaps for their own personal reasons, are not always willing to talk about the problems they have encountered.

The respondent summarized the owner's motives regarding his company.

(I) The owner is a professional metallurgist with deep knowledge of the field, who understands production processes very well and is deeply immersed in the details. You could even say that he loves his company and wants to see it develop to the highest world standards. This is akin to the attitude of a fulfilled professional towards their work—when money beyond a certain level is no longer as important as achieving something outstanding in their chosen field, and their reputation as an individual.

(II) The owner is focused on the long-term development of the company and is not concerned with achieving short-term financial gains. Therefore, most of the processes in the company are planned and delivered over years, rather than being a race to hit short-term targets. Billions of dollars are constantly being invested in three areas: (a) equipment modernization, leading to breakthroughs in the mechanization of processes and a sharp reduction in dangerous manual labor; (b) the construction of new production facilities, recently including one of the most modern blast furnaces in the world;

(c) and the environmental sustainability of production, for which the company is a leader in its country.

(III) The owner considers employees as an asset. He actively invests in them through training, career development, high salaries, comprehensive social guarantees, and job preservation during times of economic downturn, to which metallurgy is very prone. In return, employees are very loyal towards the owner: people value their workplace and are happy to remain working at the company for decades. As a result, when an employee sees a serious safety violation, technological failure or critical risk, they do not ignore it, because they understand that the safety of the equipment they operate directly affects their own personal and professional well-being, as well as the company.

(IV) The owner visits each key production site at least three or four times a year and is very familiar with the operational challenges of the enterprise. Representatives of senior management visit key production sites at least once a week. This attention from the owner and senior management allows middle managers to quickly convey critical issues to them, get decisions made and get approval for additional resources to mitigate identified risks.

The respondent also has some reflections about how his company works with contractors. He believes that irresponsible contractor working practices can adversely affect the safe operation of critical infrastructure enterprises. In his company, contractors are treated as partners. They examine all safety indicators alongside their contractors and make no distinction between contractors and permanent employees. Instead, as with staff, the company is willing to invest in contractors and establish long-term contracts with them. Realizing that they are working long-term with the company, contractors are much more willing to invest in the development of their own staff and in the best equipment. They can see that the company values them. In turn, contractors value their relationship with the company and are focused on protecting it in the long term by providing safe, high-quality services.

3.3 Recommendation No. 3: Risks Must Be Prioritized, as It Is Impossible to Manage Every Risk Within an Organization Simultaneously

During discussions with the respondents, the question came up of whether senior management should respond to every risk that employees inform about. When asked why managers are sometimes reluctant to pay attention to the problems that employees report to them, 23% of respondents say that managers tend to assume the issues their employees will be bringing to them will only be minor. As a rule, few managers make employees aware of the risk situation across a whole company.

For the most part, ordinary employees will see risks from their own personal perspective. They are more likely to inform their superiors about problems that impact their own job security and their personal safety, rather than risks involving critical equipment failure. Employees provide feedback to superiors not about what they think might be important for senior management, but about what concerns them and their colleagues. Senior management recognize that most of the information they obtain from employees relates to minor issues, which overload them with trivial information. Senior managers have neither the time nor the detailed on-the-ground knowledge to pay close attention to these minor issues. From their perspective, such issues are the responsibility of subordinates—this is the responsibility and role of lower and middle managers.

The head of HSSE-Q (Health, Safety, Security, Environment & Quality) at an electricity company has an example of this kind of information overload. Several years ago, a number of incidents took place in the respondent's enterprise. The head of the unit where these occurred made it clear to his subordinates that he wanted to know everything that was worrying them. He launched a major initiative to collect detailed information about his employees' concerns—and, as a result, received an avalanche of data. However, he could not do anything with it all, because he was overwhelmed and had no idea of where to start tackling such a huge catalogue of issues and problems. The learning from this is that it is essential to determine in advance exactly what information is being gathered and why. Collect only as much information as senior management can adequately process and provide meaningful solutions to, otherwise the management system will become overloaded, and no useful progress can be made with any of the problems raised by employees.

Having discussed this issue with several respondents, there is a clear consensus that it is necessary to highlight the most critical problems in an organization and direct the efforts of the senior management team to solve those. For effective decision-making, you need to have a system in place to deliver an integrated risk assessment of production processes, where all the key risks of an industrial facility are assessed and then ranked by severity. This will enable an organization to prioritize the allocation of risk management resources.

It is rarely possible to satisfactorily manage all the risks faced by an organization. Resources are always limited and are never sufficient to mitigate every possible risk. Without establishing clear priorities, managers have so much information to handle that they cannot distinguish what is important from what is not. A gradation of risks immediately makes the situation clearer—what further information is required, which risks need monitoring, and which demand urgent action so that "*major negative events*" can be prevented—those that could lead to death or serious injury, major production losses, or environmental disaster, all potentially terminal for an organization. It is vital that critical risks and problems that may threaten the work of an entire enterprise come swiftly to the attention of senior managers so that they can communicate this information immediately to the highest level of the hierarchy, while less serious risks can be delegated to appropriate lower levels of management for further action.

The critical risk structure of a company depends on their corporate strategy

One of the interviewees, the head of a power plant, considers that a production site manager needs to be able to assess the existing technological risks through the lens of a company's strategic plans. For example, if a company's leadership are focused on short-term profit, and the longer-term life of the whole asset is not of interest to them, they will want a realistic assessment of the risks of equipment outage over the short term. If the entire plant will be closed in five years and a new facility built nearby, then the chief engineer of the station will naturally assess the risks of equipment failure over that five-year lifespan. On the other hand, if the facility managers are tasked with delivering stable trouble-free operation for the next 20 years, then the risk assessment carried out on the ground will be completely different. Knowing a company's wider development strategy allows employees to understand the corporate environment they are working in, enabling them to adopt appropriate criteria when assessing the state of equipment. If employees understand how senior leadership view a company's development in the near and longer term, they will be able to better assess which risks may threaten those objectives.

A mining company HSE director, who considers that it is important to know the parameters within which the risks of a company are being assessed, suggests the same approach. If the leadership are interested in identifying which critical equipment may fail during the next year, this generates a specific timeframe for assessing critical risks; if they are focused on a timeframe of 10 years, then the risk assessment picture will be completely different.

It is impossible to effectively manage all risks—prioritization is essential

The head of HSE department of a fertilizer mining company, employing more than 25,000 people and operating millions of units of diverse equipment, gives the following example. Every minute, somewhere at one of the company's sites, there is equipment failure. The most common is a bulb burning out in the production room of a mine or an industrial workshop. Even a blown light bulb is a risk, but does it make sense to inform a senior manager every time this happens? Obviously not: senior management should only be informed about significant risks and incidents that could have a major impact on a company's operations.

The respondent has analyzed the significant risks specific to this mining company. The business is engaged in the extraction of mineral salts from underground, which are then enriched to produce mineral fertilizer. For a salt mine, the worst risk is the threat of flooding due to the breakthrough of groundwater. Mineral salts quickly absorb moisture, and thus most of the mines in the world do not cease operations because everything has been extracted. Instead, they get flooded by groundwater and are abandoned with plenty of mineral reserves still in the ground. Loss of an entire mine is obviously the most critical risk for a fertilizer company. Despite the fact that mineral salt production is highly profitable, the investment required to combat mine flooding can exceed the profit from the mineral sales.

A critical risk like mine flooding can be broken down further. First, it will either be a case of (I) improper exploration or (II) improper extraction. Both areas can be further divided. For example, if the fault lies with extraction, then: (II.a) mining operations may have disrupted the integrity of the mineral reservoir by breaching the saline reservoir. In this case, it will be impossible to prevent water ingress and the mine will gradually begin to flood. By identifying each possible scenario like this, sub-elements of both I and II can be analyzed, and expert estimates made of the impact of each factor on the probability of a critical risk developing. The resulting analysis can be drawn up as a branching schematic of critical risk factors. Comparative assessment allows managers to rank each factor according to its likely influence on creating risk, and thus identify priorities for working to reduce the overall risk. When analyzing the impact of various risks on the work of a large industrial company, the respondent's recommendation is to assess and then rank all the existing risks in terms of the threats they each represent to its sustainable development over a given period.

Identification of owners for each critical enterprise risk

A schematic of identified risk factors should be overlaid on an organization's structure and a set of suitable measures to control specific risk factors should be conveyed to each management level. Then information about each critical risk must be communicated to specific employees, with a request that they inform management about any observed deviations in the operation of equipment or production processes under their control that could lead to these risks developing. For example, if we take the lowest level of an organizational structure—the mining engineer and the driver of a mining harvester—it is critical for them to prevent the harvester from entering the salt medium within the mine. On average, up to 40 mining harvesters work in a large mine. With this critical risk at the forefront of their minds, the harvester operators can monitor the condition of the equipment and the mine, using their practical expertise and experience to continuously assess the situation and be ready to inform senior management promptly about any problems that may lead to flooding—vital information about a critical risk which is obviously of great interest to management. This is the kind of bad news that executives want to hear, since it can prevent huge losses.

Senior management must take the initiative to inform employees exactly what type of risk information it wants them to immediately pass on.

Risk prioritization makes it easier to understand that only a small percentage of the company's personnel—possibly only 5% of the workforce—actually impact the management of critical risks within the enterprise. It follows that senior management do not always need to communicate with every employee working for a big industrial company. In the first instance, management only need to build operational communication channels with those employees who can have a direct impact on avoiding critical risk. For example, the "*owners*" of most critical risks for a fertilizer mining company are the harvester teams that work in salt mines. This is only a small percentage of the company's 25,000 employees. By communicating directly with the critical risk owners, senior management can simulate the most threatening scenario of a mine flooding in order to develop an understanding of what needs to be done to prevent this from happening. No one can give a more accurate assessment of the situation than the mining harvester teams themselves, who go down into the mine daily and see any weaknesses in the flood prevention systems. It is these workers who should be the main source of this critical risk information, if senior management are to receive the most up-to-date reports of potential problems. Realizing their responsibility in preventing this critical risk from threatening the whole enterprise, the mining harvester teams can constantly monitor, and report to the authorities about, any serious deviations in the control measures. If employees understand the coordinated system within which a company manages risks, they can prioritize their actions to reduce risks in their area of responsibility and learn to value their role in helping an organization meet this challenge.

Moreover, if employees know that management will thank them for this information and that they will be rewarded rather than penalized for transmitting it, then everyone in the hierarchy will be encouraged to begin to communicate more proactively. All levels of management will be keen to show senior managers how they are trying to solve critical problems. If it is impossible to solve a given problem at their level, the managers involved will inform their seniors about the need for additional resources. If senior managers demonstrate that they will give critical risks their close attention, the entire management hierarchy will very quickly begin passing this information on to them. Employees raising serious problems and critical risk information will be less likely to say, "*I'm bringing my boss a headache*", but more likely "*I'm saving the company*". Thus the management will receive relevant information on the critical risks of the enterprise, without loads of irrelevant noise about minor problems on non-critical equipment. At the same time, some employees will recognize that they act as a key link in preventing certain critical risks damaging their enterprise. This will positively affect the work of ordinary employees and lower and middle managers, by raising their levels of loyalty, motivation, and professionalism.

Below are the results of responses to anonymous surveys of more than 300 respondents as part of a pilot project to improve the quality and speed of internal transmission of risk information. This pilot project is being implemented in an industrial company that is a world leader in its field (Chap. 4 of this handbook).

Results of responses to anonymous surveys within the framework of the pilot project: *Have you been made aware of the main critical risks of your enterprise?*

	High awareness	Medium awareness	Low awareness	No awareness	Cannot answer	Number of respondents
All survey participants	48.5%	32.2%	10.1%	3.1%	6.1%	326
Senior management, heads of departments and directors of sites (middle managers)	58.5%	39.0%	2.4%	0.0%	0.0%	41
Lower managers: deputy directors of sites, chief engineers of sites, heads of workshops, heads and representatives of HSE services at sites	56.5%	27.8%	8.3%	4.6%	2.8%	108
Engineers, foremen, and ordinary employees who operate critical infrastructure at sites	41.2%	33.3%	13.0%	2.8%	9.6%	177

Interpretation of responses: the lower respondents sit in the hierarchy of the company, the less they know about the critical risks of the enterprise. Respondents' answers indicate that senior management, middle and lower managers are generally more aware of the critical risks of production facilities than engineering and rank-and-file employees. The latter operate production facilities daily and are in an ideal position to personally observe many deviations. However, they may not always understand the consequences of these deviations due to their fragmented understanding of the critical risks of the enterprise. Therefore, the management of industrial facilities need to disclose the critical risks they are facing to their subordinates, so that they operate more effectively within the critical infrastructure and can report promptly to the authorities about problems with equipment function, and so improve the company's overall preventive control of critical risks.

Empowering critical risk owners to manage risks

The head of HSE department of a fertilizer mining company recommends introducing a "*stop-the-job*" system first for the "*owners*" of critical risks, and then gradually expanding it to other employees. It is only where critical risks are being actively managed that employees should be given the power to stop production without prior coordination with line and senior management. There should be a clear rule that employees will never be punished for halting production because of perceived risk, even if the subsequent investigation demonstrates that it was not necessary. In this case, the appropriate action might well be to reassess the way the criticality of the situation is measured. Employees must feel morally justified in stopping critical production. For example, it is better to give all mining harvester teams in a salt mine the absolute right to stop operations because of a significant risk, without the fear of a reprimand for "*crying wolf*". Otherwise, the danger is that employees are too fearful of repercussions from management to halt production, and choose instead to passively stay on the sidelines as an emergency rapidly escalates towards disaster.

In industrial companies with many hierarchical levels, information going up the chain of command has to surmount numerous barriers, and a lot of time can be spent on securing scanty resources. Here it is better to delegate some of the authority and resources to managers and employees lower down the hierarchy. They can then respond more quickly to risks that arise in their area of responsibility, rather than waiting until the bureaucratic apparatus above them processes their request and allocates—or chooses not to allocate—resources to them. People at the production sites are often better placed than those at headquarters to see what priorities should be set in their day-to-day work in order to reduce accidents, and empowering them will increase the reliability of critical infrastructure in the long term.

What to do with non-critical risk information

At the initial stage, senior management need to focus on the reasons for stopping critical production and take measures based on the analysis of each case. Clearly, the first priority is always to find solutions that reduce the likelihood of critical risks. Nevertheless, management should not forget about other less critical risks. Information about these should be directed to delegated managers at different levels of the corporate hierarchy, who should be given the authority and resources to manage them. The critical risks of an industrial company should be managed by top and mid-level management, and the less significant risks delegated to ordinary employees and lower-level managers. Having established a "*stop-the-job*" system with the "*owners*" of critical risks on the ground, the scope of the system can be gradually extended to include less critical, but still significant, risks. It is neither necessary nor appropriate to immediately give the entire workforce the right to "*stop-the-job*". Some people who manage non-critical infrastructure will always be able to find inconsistencies in how they are working when measured against official procedures, and may misuse the system. Quite apart from significantly disrupting production, this would lead to a surge in information traffic because senior management would have to be informed of the reasons for every single pause in production, even on non-critical equipment.

Prioritizing and systematizing risks as the basis for targeted messages to specific risk owners

When employees receive too much information but no guidance on how to use it in the context of their own work, they will simply ignore it. It is more effective for management to inform employees in the field only of matters that have relevance to their own work.

A negative example: in a huge oil company, information about every injury or incident is collected. All of this information is sent throughout the company with no classification into specific context. Understandably, no one reads this unsorted data array or makes any attempt to use the experience of other probably unrelated departments in their work. Thus for example, drillers at remote sites receive information about an accountant's injury during a corporate football match at an oil refinery; or staff at a gas station receive information about injuries on an oil rig.

In other words, the company sends out a mass of unnecessary and unfiltered data that is irrelevant to the recipients, and which encourages employees to ignore it all entirely. This approach likely obscures what may well in fact be useful information for a given employee. Drillers could benefit from information about incidents at other rigs, and a detailed analysis of exactly what happened and what was done. Accidents at refineries will be of interest mainly to employees at other refineries, accidents at gas stations only to employees of other gas stations, and so on. Prioritization, risk selection, grouping of risks by type—all this would allow an organization to convey the relevant specific information to risk owners without overloading them with unnecessary material. Employees are interested in what is relevant to their area of responsibility. This is the case higher in the business hierarchy too, where senior managers are only interested in hearing about the critical risks of an organization and cannot afford to be distracted by minor information.

Risk ranking by function rather than geography

The HSE head of a mining and metallurgy company recommends categorizing risks into critical and less critical, and then grouping them functionally rather than geographically. Often, critical risks in any given functional area of operations also occur across different sites located in different regions, and even across countries.

Engage independent risk assessment experts

The HSE director of a steel company recommends engaging independent external appraisers to assess risks. Their impartial judgment will avoid the manipulation and bias that can distort assessments carried out exclusively within a company. Risk assessment implies a collegiality of experts. In organizations where the reactive approach to risk management is dominant, there is a greater likelihood of distortion and concealment of the real situation.

The HSE manager of an industrial company working with hazardous chemical processes said that regular external independent audits carried out by their insurers help them to establish a list of any equipment that is critically at risk. These experts make their own assessments, and rank the existing production risks across all company operations. The company can then decide on the best preventive action to reduce the most significant risks, and the audits allow them to delegate what detailed response they require at each level of management.

Create a list of critical equipment

One interviewee manages the HSE division at a metallurgy company, in this case called the "*Production Safety Directorate*". At the core of the directorate's work are the classic functions of an HSE department, essentially aiming to change behavior in the workplace to reduce risk and injury. However, an additional element is the assessment of the safety and condition of equipment, in order to reduce the industrial accident rate. The directorate thus deals with both the reduction of injuries (people) and the reduction of technological accidents (equipment). The first task after the "*Production Safety Directorate*" was introduced was the identification of critical equipment at various production sites, the failure of which could result in large losses and harm employees, the environment, or local residents. For example, at the company's leading manufacturing facility, 1,500 units of critical equipment were identified.

Another example comes from the aerospace industry: the safety team working on the Space Shuttle program formulated a "*Critical 1*" list, which included 750 critical elements of the shuttle and the launch vehicle, the failure of which could destroy the spacecraft.

The HSE manager of an oil company gives the following example of identifying critical risks in the industry, through the experience of assessing damage after accidents. An accident at a traditional oil rig will not cause much damage: it is possible to duplicate many industrial operations so that the process of oil production and transportation from an oil field is not disrupted by any single accident. However, if an accident occurs at a major refinery, then a cascade effect is possible, and the loss can amount to billions of dollars. This is because it is impossible to quickly replace such huge industrial facilities and it will likely take many years and billions of dollars before the pre-emergency production level is restored. An accident at an offshore oil field could cause tens of billions of dollars of damage due to the enormous clean-up cost after an oil spill and the regulatory penalty fines it will incur. First and foremost, the senior management of a diversified oil company need to be in close control of the most critical risks: offshore oil production and the operation of large oil refineries.

What should be done about critical risks? Ask the staff!

The vice president of a gas pipeline construction and repair company suggests asking the employees themselves how to control critical risks more effectively. If managers are interested in hearing about problems, they can tell employees: "*Imagine that this company has an unlimited amount of money and resources. Write a program of modernization for the most important and critically worn-out equipment, with a clear justification of priorities, and the senior management will then select the most relevant issues and finance the modernization*". Alternatively, they could stress the need to prioritize: "*There is no extra money or resources, but one of the key goals of the production system is to work without serious accidents. Please provide us with a list of critical equipment and recommendations for its repair, so that this organization can avoid major accidents with minimal additional expenditure*". Despite the difference in emphasis, both approaches are aimed at using the knowledge of employees who are working with monitoring these risks daily, to create a ranked priority list of the most important problems for an organization to tackle.

Explain investment priorities to employees when managing critical risks

The head of a power plant believes that, if shareholders are ready to allocate resources to solve problems, then senior management should identify investment priorities, indicating the most important problems that a company needs to solve. These priorities should be communicated to all technical staff. With a better understanding of the logic of senior management concerning the investment strategy being implemented, each unit should be able to relate their immediate problems to the critical problem list for the entire organization. For example, senior management may receive information about the current risks at all 20 of a company's production sites, but only five facilities would be selected to receive the main investment resources over the next three years, due to the criticality and urgency of the problems identified at those specific sites.

It is also recommended to categorize risks by groups, and then designate those who are to take responsibility for monitoring each group. Risks need to be prioritized so that the most critical problems can be tackled first.

Use simplified terminology when communicating with subordinates

The head of risk management at a nuclear power plant recommends that managers ask employees not about perceived risks, but about observed problems. Employees are not always aware of which issues constitute a risk to an organization, and how critical each of these risks is. Their knowledge is limited to their professional practice. That is why the respondent recommends that managers, when communicating with employees who operate critical infrastructure, replace the word "*risk*" with "*problem*" and ask subordinates the following questions: "*What is the problem? What is it connected with? What factors caused this problem (e.g. problems with administration, with procedures, with equipment)?*". With this information a manager, together with technical specialists, can assess the current criticality of a given risk and decide what action is necessary to manage it.

In addition, managers can advise their subordinates to examine existing problems through the lens of criticality, and then inform senior management immediately about critical and very serious problems they have identified in their area of responsibility.

3.4 Recommendation No. 4: Senior Managers Must Be Leaders in Safety

If senior managers require employees to achieve a production plan, then the production targets will be stipulated in the plan; if they require a reduction in costs, they will get a reduction in costs; if they require employees to meet safe working practices, they will get safe working practices. It is imperative that any initiative to prioritize safe operation of critical infrastructure comes from senior management. In highly hierarchical companies, the example set by the leader is paramount. It is worth noting that most critical infrastructure companies have several management levels and are quite bureaucratic. If subordinates see that safety is extremely important to the CEO, and the entire corporate system makes it a top priority, then most employees will imitate senior management and follow the principles they are espousing. Production site workers will have no grounds for relegating safety down the list of their own priorities, and will be willing to place it first, above production and profitability indicators.

Senior management are responsible for safety too, not just the HSE department

Many companies still have a problem with understanding who is really responsible for process and occupational safety. Some senior managers still believe that only those who have "*industrial safety*" or "*Health, Safety and the Environment*" in their job titles hold this responsibility. In fact, everyone who operates critical infrastructure, or carries responsibility for it, should be prioritizing safety. This applies to managers at all levels, and above all to the senior leadership.

The CEO of a consulting company in human performance with extensive experience in power generation notes that, if an incident occurs in an average critical infrastructure company, the first thing senior management do is to call the HSE director and ask them: "*What happened? Why did this happen? What is the director of occupational health and industrial safety going to do to ensure that such an incident does not happen again?*". When the costs of a project exceed the allotted budget, then senior managers do not call the financial director and ask him: "*What happened? Why did this happen? What are you going to do to ensure that overspending does not happen again?*". They take an active role in resolving the issue. Clearly, senior managers consider financial risk management as part of their managerial role, but often they do not see themselves as being equally accountable for health and safety. Until this corporate situation is rectified, fundamental improvements in health and safety within any critical infrastructure company are unlikely to take place. Instead, this essential function will remain outside the focus of senior management and likely be perceived by them as a bother and a drain on their budget.

The head of the sustainability and systems department of an international electricity company, looking back over the past 20 years of his career, notes that the safety situation in many critical infrastructure sectors has definitely improved. How was this achieved? It is all about encouraging leaders to take an active role in the safety of their companies! In the past, leaders often paid little notice to safety issues and were not adequately trained to make decisions in this field. Managers did not fully recognize the consequences of disregarding safety rules, or put safety on an equal footing with production and financial matters. Therefore, the first step to improve the situation is to properly train all senior managers in safety. The second step is to make safety an essential component of every manager's skill set so that safety is no longer seen as the sole responsibility of HSE, but as an essential concern for every leader throughout the organizational hierarchy. The third step is to successfully balance safety issues with other production and financial goals that managers may have.

The CEO and chief nuclear officer of a nuclear operating company believes that the place to begin the process of involving senior managers in safety matters is to educate them on the best practice and behavior in the field of safety. After 25 years working on this, the respondent came to the conclusion that the key is to ensure managers are fully trained in correct safety culture, so that they have sufficient knowledge and expertise to ensure operational safety across all areas of their leadership responsibilities and hold it as a central value when making managerial decisions.

The HSE director of an oil company cites an example of how educating senior managers can increase their awareness and involvement in safety issues. The respondent developed a training program that involved the CEO and all other senior managers of the company spending two days on location at a special training ground where explosions, fires and other serious accidents that can occur at production sites could be simulated. In addition, actors were employed to take the roles of workers and thus create a range of realistic emergencies, all of which graphically illustrated how a facility can fail and employees can suffer if safety regulations are violated. Participation in such training events allowed senior managers to safely experience what real explosions and fires feel like and the injuries that workers can suffer. This created opportunities to challenge senior management with questions like: "*What would you do to ensure emergencies do not occur on your watch? What would be your role in preventing emergencies? How will you respond if an emergency does occur at your site?*". According to the interviewee, several thousand managers at various corporate levels have now completed this training and have become very different managers as a result, recognizing the huge importance of safety issues in every aspect of their work. This also gave the HSE heads of the company new opportunities to introduce leaders of diverse departments to the challenge of managing site and worker safety.

An HSE consultant working mainly for oil and gas as well as in air traffic control gives another example. In his practice, the respondent presents to senior managers the consequences of accidents at other industrial facilities. For example, he acquaints them in detail with the accidents that occurred at the nuclear power plants at Chernobyl, Fukushima-1 and Three Mile Island. He also examines major railway accidents and serious plane crashes. In discussing these disasters in detail, he encourages the participants to examine the current practices of their own companies and to recognize that similar problems could potentially happen at one of their production sites. It only needs a few safety regulations to be overlooked, a bit of bad luck with several things unexpectedly going wrong at the same time, followed by a couple of delayed or bad decisions by either workers or management (or both) ... and before you know it, you have a full-scale emergency on your hands. To help prevent such sequences of events from happening, senior managers must be constantly immersed in the safety of a company's operations, as well as attentive and responsive to any developing problem.

An HSE manager and consultant with experience in the nuclear and construction industries regularly updates senior managers with reports on significant incidents at other critical infrastructure facilities. For example, in an informal conversation with senior executives in the cafeteria, he might mention that one oil-refining giant has just been fined millions of dollars after a blast wave and a fire led to the death of workers. This respondent then asks them: "*Are we in a situation where something similar might happen in our organization?*". He believes that to convince senior management to treat safety issues seriously, it is necessary to use data, well thought-out arguments and external examples.

An HSE manager who has been working for several oil companies also believes that the best way to increase the engagement of senior managers in safety is to frame it in monetary terms: to obtain funding to implement safety solutions for example, the argument needs to be presented to senior managers in the form of a business case—this is the language that senior managers understand the best.

The HSE manager of an oil company considers that every CEO should begin their working day with a discussion with their senior management team about progress on industrial safety issues. This should include an analysis of the causes of any recent incidents, the actions that are being taken to tackle any current problems, and the steps being taken to minimize any chance of a recurrence. Essentially, safety is like a bicycle. If you stop pedaling and moving forward even briefly, then the bike falls over. This need for "*constant pedaling*" is first and foremost the responsibility of senior managers. They should regularly be asking their team for an analysis of where risks exist in their operations, what are the vulnerable spots, where are the points of possible failure. These discussions should also cover actions being taken to eliminate any identified risks. They should include progress reports on the implementation of decisions already made. It is only when these questions are constantly asked at headquarters that middle and lower management will start following suit and begin to pay proper attention to industrial safety issues. Senior management must constantly be sending the right signals down the hierarchy.

Unfortunately, in some industrial companies, safety is not a priority for senior managers. As a result, the entire corporate management system treats safety issues as a low priority, and the monitoring and control of risks then becomes reactive, with safety policies only receiving attention after major accidents.

In the respondent's experience, most industrial companies define their corporate strategies by first focusing on production targets and finance—and only then, if at all, move on to HSE considerations. A more proactive approach to safety would be to first establish the risk limits that a company cannot exceed, and then develop the industrial solutions that will keep the operation within those limits. The goals set for progress on industrial safety have to be feasible. Trying to achieve a sudden overnight transformation—demanding that employees immediately adhere to ambitious new safety standards—will probably alienate the workforce and leave them feeling under pressure and criticized. This is likely to result in employees falsifying reports, not this time to show they have hit production and financial targets, but to exaggerate their achievements in HSE. Safety must not become another box-ticking exercise. A meaningful transformation in employee behavior around safety, the kind that will save lives and prevent serious accidents, requires persistent application and monitoring, all initiated and maintained by a committed leadership.

One interviewee made a radical proposal. Any lower or middle manager whom senior management consider as a potential future director of production should be required to work for several years in the HSE department, either in a production unit or at headquarters. His responsibility should include direct work with ordinary employees and regular visits to multiple sites to look at risk. According to this proposal, lower or middle managers should only move up the corporate ladder after demonstrating that they are effective leaders in HSE.

Legal responsibility of senior management for safety

A former head of mine regulation believes that, in order to ensure that senior managers treat safety issues seriously, they should hold legal responsibility for preventing accidents and injury to employees. This should be achieved by passing laws that impose this specific responsibility on them. The respondent explains that this proposal was motivated by the situation in his own country. There, the responsibility for safety has traditionally sat not with a company president, but with foremen and superintendents working at production sites, who receive reports of any safety issues or accidents. A coal company president was not even required to have a degree in mining engineering, and could be an accountant or a politician with no understanding of either mining or industrial safety. The respondent believes that the more a country's legal framework absolves senior figures within critical infrastructure companies of the responsibility for safety, the less likely they are to give the issue the attention and resources it merits. Instead, they will try to shift all the responsibility on to foremen, and convince them that, if something goes wrong and an accident occurs, it is foremen who will carry the main responsibility. To change this, it is essential that legislators create a robust legal framework that will hold senior figures accountable for non-compliance with health and safety rules and regulations across a company. Only in this way will senior management be motivated to take an active and ongoing role in safety issues.

In some countries, senior managers can be charged with criminal offences in the event of a serious accident. According to several respondents, this provides a very effective leverage. It encourages executives to pay close attention to safety issues and risks within a company, and to take prompt action when problems arise.

Project manager in oil and gas exploration sees legal accountability as an opportunity to frighten senior management, especially company presidents and heads of finance: it can motivate them to keep a close eye on any risk information being raised by their employees, and to identify and deal with significant risks before they become critical and lead to incidents.

According to the HSE manager of an oil company, industrial safety should not be a matter of voluntary compliance: the responsibility for ensuring safety should be compulsory and be included in the labor contracts of senior managers, with a stipulation that in the event of a major accident, they will be held legally and criminally liable, and that ignorance of the technological risks that led to an accident will not be a valid defense. In other words, it is necessary to ensure that senior managers understand that they cannot afford to be unaware of risks. On the contrary, it is their job to have full knowledge of potentially critical risks and act in time to prevent major accidents. Rather than waiting for the "*lower classes*" to request an audience with the boss, senior managers must take the initiative in establishing effective communication with their subordinates.

An HSE manager of a manufacturing company believes it is important to have candid conversations with senior managers. To persuade them to take a proactive approach to safety, he might say: "*Look, under the law, you, as a manager, can personally be fined up to $5 million if a worker dies in the production process you're in charge of, and you can be shown to have been negligent. I'm telling you this because I'm trying to protect you and protect the business at the same time. To take care of your own interests, you need to take safety seriously and take action*". The respondent is also very clear that he believes industry regulators should be much stricter with the senior management of critical infrastructure companies.

The CEO of a consulting company in human performance with extensive experience in power generation gives an example of how it is possible to explain to senior managers about their legal responsibility should an accident occur, and through this vulnerability encourage them to take active steps to reduce the chances of such accidents happening. Every year in the respondent's country, regulators and the legislature put more and more pressure on senior leaders of organizations that operate critical infrastructure. By law, they can be held criminally liable for accidents at production sites they are responsible for and lawsuits can be filed against them personally if an incident results in the death of an employee. To train senior managers, the respondent takes the opportunity to bring in a law firm that runs mock trials where they stand as the accused. The whole trial process is re-created, with a realistic courtroom, aggressive interrogation by the prosecution and so on. In most cases, the senior managers fare very badly and quickly fail in their defense. They quickly realize that being taken to court is unpleasant and tough, that they have very little control of the process once litigation begins, and that a defense of ignorance around the safety violations that led to an accident is very unlikely to succeed. The respondent also uses real-life case studies of accidents that occurred in various industrial sectors to show the very serious personal consequences for senior managers caught up in the process. Some of these cases resulted in bankruptcy of the companies and senior managers sentenced to prison. Through emphasizing their vulnerability and the serious personal consequences of criminal proceedings, senior managers are highly motivated to pay close attention to what their subordinate managers and line workers are doing—or not doing—in relation to safety protocols and reporting problems and concerns.

Managers must have a unified position on safety issues and be consistent

The HSEQ director of an international oil company believes that leaders should be consistent in upholding safety principles and objectives and be clear about exactly what they expect from their subordinates. When visiting company production sites, this respondent always emphasizes the mantra "*safety above all*" to the workforce, and that operative safety is the number one priority for the senior management of the company. This repetitive messaging from a senior manager makes it very difficult for lower and middle managers to tell their subordinates to ignore safety regulations when performing their duties, and focus solely on increasing production. If one senior manager tells junior employees one thing about safety, and then another says something else entirely, misunderstanding within the workforce is inevitable and subordinates are much less likely to take future messages coming from senior management seriously—with the risk that important messages are ignored. To avoid sending such conflicting messages about corporate priorities to lower-level managers, the leadership must reinforce a company's safety priorities clearly, frequently and consistently throughout the hierarchy.

Consistency is also important when making important corporate decisions. Senior managers must ensure that they allocate sufficient authority and resources to lower and middle managers, so that they can implement these decisions at their level of responsibility. This includes giving employees the right and responsibility to stop unsafe work in their operating area without any subsequent penalty. It is fundamentally important that all company leaders conscientiously apply the safety principles they are promoting to their own day-to-day work, and whenever making management decisions that have safety implications. It takes time to change the culture of a company, and trust can only be built up within an organization through consistent messaging and application of principles. Managers must persevere and be consistent for their messages about the safety priorities of a company to penetrate every level of the hierarchy and to ensure that every employee, from top to bottom, knows what is expected of them.

The senior HSE director of an oilfield service company and the director of operations at an oil company both believe that everyone within an organization must speak the same language to achieve consistency in the implementation of corporate priorities. Over time, a common line of thinking is formed across an organization and making decisions in favor of safety at all levels of management can become routine. According to the respondents, consistency is achieved when decision making and action plans, at every level of management, are always accompanied by the questions: "*Are we always identifying what the hazard is? Are we always controlling that hazard and verifying what safeguards we have? Are we following our own processes and procedures*?".

An HSE manager and consultant with experience in both the nuclear and construction industries gives the following example. The respondent knows one company, which had a very good operational system (the so-called "*heart*" model). The model consists of (I) leadership and (II) systems and processes. These two corporate functions must be in harmony. If a company has good senior management leadership on safety, but the systems and processes are terrible, then a company has no chance of succeeding. If company's safety systems and processes are great, but there is no leadership from senior managers, then the project is doomed. It is essential that leadership and processes operate in tandem if significant improvements in safety are to be achieved. The role of the HSE team is to work both on the leadership skills of senior managers—their behavior, communication, and attitudes—and equally importantly, to help build effective and safe systems, processes, and policies.

Managers have personal responsibility for safety—an example from the oil and gas industry

As part of the studies reported in this handbook, several interviews were conducted with managers at various corporate levels of an oil company. This company has faced many challenges concerning occupational safety and labor protection—their injury and accident rates were higher than other similar energy companies across the world. As a result, they recognized the need for fundamental changes to improve their safety record. Over time, the company developed and implemented a rather unique project to redistribute the responsibility for safer production and working conditions across all levels of management. The interviewees emphasized that it could not be right—or efficient—for the safety of production and the welfare of ordinary workers to be the responsibility of only one managerial level. Better outcomes would come if this was shared between managers at all levels of the company.

First, the concept of zero injuries ("*goal 0*") was adopted at senior management level. As many senior managers had previously worked in other leading energy companies, most of them did not require any explanation of why safety should take priority over production and financial targets. Next, the CEO gathered together key

Employees need to have security guarantees, both for their careers and for their colleagues. Managers must guarantee the security of their sources and take responsibility for solving any significant problem they are informed about.

Having studied many publications on the issues of trust and the transmission of information about risks,[4] the authors of the handbook have found it difficult to settle on a simple and understandable definition of the word "*trust*" in the context of the relationship between a manager and a subordinate. Among 100 interviews with practitioners, one head of an oil production site gave a short, yet comprehensive, definition of what he thought trust was—*the predictability of the behavior of another person based on positive experience of previous interactions with that same person*. According to this manager, trust in a leader develops when employees can predict the leader's behavior based on previous positive experience of interaction with them. If there is a clear logic and predictability to the actions of a leader, there is trust; if this is absent, there will be no trust.

Without trust in the leadership, there can be no high-quality feedback from employees on the problems of an organization. Often, employees evaluate the possible consequences of disclosing risk information based on rumors about how senior management behaved in a previous situation with colleagues. Employees project both the positive and negative experiences of their colleagues onto themselves.

[4] Dmitry Chernov, Didier Sornette, Giovanni Sansavini, Ali Ayoub, Don't Tell the Boss! How poor communication on risks within organizations causes major catastrophes, Springer, 2022; subchapter 2.2 Results of other researches on the challenges of voice and silence in an organization, https://link.springer.com/book/10.1007/978-3-031-05206-4.

executives from both headquarters and production sites and announced that he was now taking full personal responsibility for safety issues across the whole company. His subordinate leaders were then instructed to follow his example and take responsibility for safety in their areas of control. The subordinates of these leaders should follow the example of their superiors, and so on down the hierarchy. The core objective of the project was to make safety issues a priority for managers at all levels of the corporate hierarchy. For the first time in the history of this company, each leader was set his or her own personal safety targets, a key step in the overall task of reducing injuries and industrial accidents. By stating that he was personally responsible for safety across the company, the CEO was sending a message to the entire workforce that this was the number one corporate priority.

To make sure that the goals of this program were shared through the entire workforce, special strategic sessions were held for each managerial stratum of the company: senior managers, key managers at headquarters (heads of departments and directorates), and middle managers from oil production sites and refineries. The sessions began by acknowledging that in real life work environments, it is impossible to maintain zero accidents and injuries forever. Instead, the aim should be to strive to deliver accident-free indicators over shorter periods of time—for example, during a shift or the implementation of a discrete project. From these short pushes over a particular time-period or project, a significantly lower long-term accident rate would gradually develop across the whole facility.

It is insightful to look in more detail at what this oil company did. First, the CEO brought together ten key senior managers. Based on the discussions held in this meeting—which lasted for eight hours—the project's goals, priorities and main areas of work were identified. The functions of each senior manager within the framework of the project were determined. It was important that the project did not rely solely on the HSE department: each of the ten key senior managers were to lead project delivery within their respective areas of responsibility.

Then a further hundred less senior but still key managers were recruited: the immediate subordinates to the ten executives from headquarters, plus middle managers from production sites. Each of the ten senior managers stood up before this larger group and made a commitment to deliver such and such a safety target within a specified timeframe. This was made while acknowledging the personal commitments of the CEO to transform the management of occupational safety and labor protection, with the following measures: (I) to create a new HSE committee under the leadership of senior management; (II) to create a new CEO's Award in the field of safety; (III) in the case that a fatal accident occurs in the workplace, to personally call the head of the industrial site where the accident happened. The ten key senior managers all stated that they too had committed themselves to creating a safer company. They requested that from now on, the top hundred managers should analyze safety problems in their areas of responsibility and make a commitment to their own top one thousand subordinates to solve them.

An essential foundation of the project was that all managers in the hierarchy made firm public commitments to their subordinates about what they would deliver on safety. This had the effect of cascading the responsibility for safety all the way down the management hierarchy and enshrining safety as a critical focus for improvement. This approach broke many stereotypes that middle managers held about their superiors. For the first time in the company's history, the leadership had articulated that safety was an absolute priority.

Here are some organizational details of the project. The HSE department developed a presentation on cascading risk responsibility down each level of the management. At each production site that the project team visited, the presentation was adapted to the specific risks there. Each production team determined its own list of critical risks, in order to focus on monitoring and controlling the most pressing safety challenges of a given enterprise. The HSE director of the whole company, and one of the 10 senior managers, came to each site to conduct sessions on cascading communication.[3] Representatives from company headquarters helped the heads of the production sites (middle management) articulate the commitments they wished to make to their subordinates. The presentation included a special video message from the company CEO, while the HSE director presented the results of the original safety audit that had first revealed the company's alarming safety figures. The head of the given production site (middle manager) informed their subordinates about the priorities set by headquarters. Then the site manager moved on to the risks at their site, and made safety commitments to his subordinates across five main areas. Subsequently, a team of eight specially trained moderators helped the site management team understand the main risks within their responsibility in the five areas identified, and helped them articulate safety commitments to their subordinates. To allow proper delivery of these cascading communications sessions across the whole company, the HSE director set aside more than eight months.

The central moderation team trained moderators at each site to help lower-level managers conduct similar sessions with their subordinates. As a result, the project reached every level of the management from the top brass down, and even included contractors working with the company. The presentations always included the CEO's appeal, a description of the company's plan to work towards zero injuries and an explanation of the key safety rules to be introduced, along with a warning that non-compliance with these rules would be punished.

All this information—areas of responsibility, commitments, plans and regulations—were combined into a single corporate order control system, which the HSE department then regularly monitored to ensure fulfillment of obligations at all levels of management. The project initiators understood that the key thing was to motivate managers to change their habits and develop their skills in the field, rather than focusing on formal indicators of injury reduction. When implementing the system, desired injury reduction targets were not presented as precise percentages. There is good reason for this: if a safety policy states that "*this year we want to reduce the accident rate by 5% or 10%*", there is a good chance that managers will be tempted

[3] Cascading communication is a process that elucidates how all stakeholders will receive aligned and accurate information. It starts with the senior team and is cascaded down throughout the entire organization.

to falsify incident reporting simply in order to meet the target. It is more important that managers at all levels want to genuinely prioritize safety in their decision making, and perform all work safely, so that injury and accident statistics are reduced over the longer term. All the leaders of the company took part in this project, and it involved well over one thousand employees.

Within the framework of the project, the company developed new strategic goals that include, inter alia, changes in organizational behavior. For example:

- managers need to communicate openly with their subordinates;
- managers should spend time at the bottom of the hierarchy in order to understand the real situation on the ground;
- meetings need to be focused and effective;
- two-way feedback channels need to be developed;
- the leadership need to be actively involved in the process.

These goals are a de facto manifesto for good dialogue and communication. They are aimed at teams that have always worked in a rigid corporate hierarchy where the approach to risks and safety issues has been reactive rather than proactive. It is very difficult to alter employees' behavior quickly to embrace this new approach, especially when for decades leaders have been used to shouting at their employees rather than seeing them as potential partners, or in an organization that has previously imposed a rigid system of blind obedience and subservience.

The main achievement of this project was to involve the entire workforce of the company in the discussion of safety issues, beginning with the CEO and spreading out to include ordinary workers at drilling sites thousands of kilometers away from headquarters. It has also helped managers understand that monitoring safety matters is their direct responsibility, not the responsibility of HSE alone. As a result, company managers began asking the heads of production units for progress reports on production safety improvements. Previously, these issues were only handled by the HSE department; production units had only concentrated on the implementation of the oil production plan, regardless of potential risks. A final confirmation of the project's immediate impact: the year after it was launched, the level of injuries in the company fell by 15%.

3.5 Recommendation No. 5: Senior Management Should Build an Atmosphere of Trust and Security, so that Employees Feel Safe to Disclose Risk-Related Information

Many respondents expressed the opinion that, unless senior management build an atmosphere of trust with employees, there is no chance of them passing on high quality risk information and objective feedback. If there is no confidence that senior management will receive it in good faith, workers will not feel safe to disclose it.

Results of responses to anonymous surveys within the framework of the pilot project: *A trusting relationship between a manager and their subordinates is necessary to create an environment where if feels safe to share information about existing problems.*

	Strongly agree	Rather agree	Rather disagree	Strongly disagree	Difficult to answer	Number of respondents
All survey participants	75.4%	22.9%	1.4%	0.0%	0.0%	280
Senior management, heads of departments and directors of sites (middle managers)	86.1%	13.9%	0.0%	0.0%	0.0%	36
Lower managers: deputy directors of sites, chief engineers of sites, heads of workshops, heads and representatives of HSE services at sites	76.0%	22.1%	1.0%	0.0%	1.0%	104
Engineers, foremen, and ordinary employees who operate critical infrastructure at sites	72.1%	25.7%	2.1%	0.0%	0.0%	140

If a leader took an authoritarian stance and punished their colleagues in the past, they will naturally try to avoid a similar fate by not bringing bad news. A few negative examples—where employees were punished for providing objective feedback, and this became common knowledge across an organization—are enough to stop other employees sending information to their seniors.

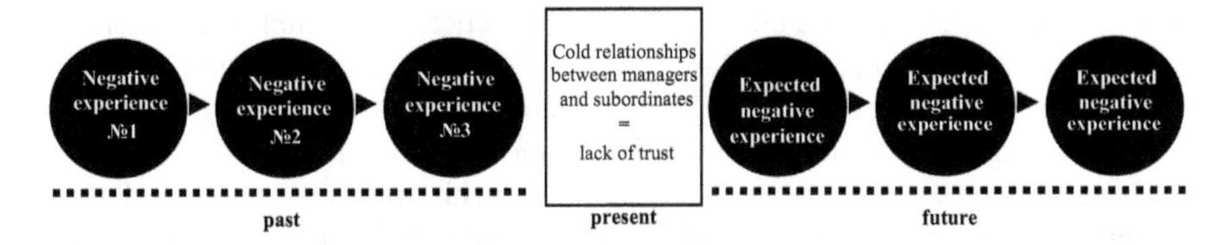

If a leader is perceived as a person who listens, responds positively and then solves the problem, then employees will be willing to tell a leader about any important issues. Just as negative experiences can influence the whole workforce, a few positive examples of feedback that the whole company hears about can dramatically improve the process of transmitting risk information. Employees will understand that senior management want to work constructively with their employees and will expect the same approach if they have an issue to report.

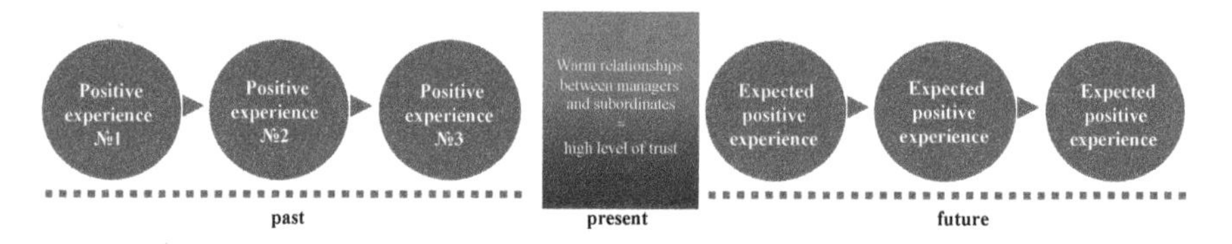

Therefore, if senior managers state that they want to receive objective feedback, they must show appreciation to employees who act on it. Initially, employees who manage critical risks will be afraid to go to executives with bad news. But gradually, seeing the positive experience of their colleagues, confidence in their superiors will grow. Even the most cautious employees will be encouraged to share their assessments of the shortcomings in the work of a company with their managers, and perhaps propose their own solutions to these problems.

A psychologist and consultant in the field of organizational behavior believes that, if there is an atmosphere of trust between employees and management in an organization, employees will feel safe to be honest with management. If an environment can be established where employees do not feel under threat, they will begin to give candid feedback. To increase employee confidence, it is essential to reduce their uncertainty about the actions of managers. Managers need to demonstrate exactly how employees are treated when they give honest feedback. Only through repeated positive responses from managers will it be possible to dispel the common perception that an organization can be dangerous to employees who speak out. The first step is for senior management to make a declaration that feedback is encouraged at all levels of a company. But this is not enough in itself: employees must see the truth of the statement applied in practice, with employees receiving praise and not punishment for offering honest feedback. The message that senior management actively want to hear about problems, and that it is safe for employees to tell them, should come right from the top of the hierarchy. It is important that the CEO and senior executives provide employees with specific examples of their colleagues' positive experiences of communicating problems to their seniors. Rank-and-file employees tend to see senior management through the negative filter of both their natural distrust for people who have hierarchical power over them, and the previous poor experiences of their colleagues' interactions with management. These perceptions will only be improved if employees receive repeated reassurance that managers are not seeking to criticize them or their performance, but genuinely wish to cooperate in order to achieve the shared goal of improving work processes and safety across all areas. It is also vital that managers demonstrate respect for their subordinates, including a sincere interest in their well-being, safety, and progress. If these principles are applied reliably across the board, then even the most cautious employees will gradually come round to the idea that a company is a safe environment, where they can confidently reveal their concerns about the situation on the ground without any negative consequences.

Results of responses to anonymous surveys within the framework of the pilot project: *A high level of employee trust in managers leads to improved transmission of information about risks throughout a company; and conversely, low trust leads to a lack of willingness to transmit risk information to superiors.*

	Strongly agree	Rather agree	Rather disagree	Strongly disagree	Difficult to answer	Number of respondents
All survey participants	62.1%	32.9%	2.5%	0.7%	1.8%	280
Senior management, heads of departments and directors of sites (middle managers)	88.9%	11.1%	0.0%	0.0%	0.0%	36
Lower managers: deputy directors of sites, chief engineers of sites, heads of workshops, heads and representatives of HSE services at sites	57.7%	37.5%	1.9%	1.0%	1.9%	104
Engineers, foremen, and ordinary employees who operate critical infrastructure at sites	58.6%	35.0%	3.6%	0.7%	2.1%	140

It is also necessary to raise awareness among managers and employees of the value of objective feedback in critical infrastructure companies. This issue should be talked about throughout a company, with regular dialogue encouraged between managers and employees, so that each party understands the fears and interests of the other when disclosing information about risks, including those arising from their own mistakes. With a determination to build better communication and feedback channels with managers, employees will gain confidence and security in their position and the process, and gradually begin to speak more openly to their superiors what they think.

Managers should show respect for their subordinates

The executive of an electric power company maintains that for employees to trust senior management, the leader must respect employees at every level and be willing to listen to subordinates. This will reduce fear among subordinates and encourage them to discuss any problems on an equal footing, one professional to another. Managers building a long-term career in a company must be humble and recognize that they cannot know everything. If communication within an entire company at every level is built on a shared belief in honesty, openness, mutual respect, and high professional standards, then a leader can feel confident that they can approach all subordinates working at any particular site or performing any specific role and receive a competent assessment report and useful advice.

Results of responses to anonymous surveys within the framework of the pilot project: *Managers should not offend or humiliate employees, as this can have a negative impact on productivity and damage communications.*

	Strongly agree	Rather agree	Rather disagree	Strongly disagree	Difficult to answer	Number of respondents
All survey participants	51.1%	37.5%	6.1%	1.1%	4.3%	280
Senior management, heads of departments and directors of sites (middle managers)	80.6%	19.4%	0.0%	0.0%	0.0%	36
Lower managers: deputy directors of sites, chief engineers of sites, heads of workshops, heads and representatives of HSE services at sites	51.9%	39.4%	5.8%	1.9%	1.0%	104
Engineers, foremen, and ordinary employees who operate critical infrastructure at sites	42.9%	40.7%	7.9%	0.7%	7.9%	140

Conversely, if a leader looks down on their subordinates, showing contempt and threatening reprisals for perceived misdemeanors, they will become isolated: (I) skilled confident professionals will cease working under such a leader; (II) a company will retain only less skilled employees who cannot find work elsewhere; (III) subordinates who remain will simply execute the orders of the leader without question or comment; (IV) subordinates will falsify results rather than admit they have failed or been unable to execute their orders; (V) nobody will risk reporting the true situation on the ground for fear of provoking anger and punishment; (VI) nobody will raise wider issues about whether a company is heading in the right direction, or the wisdom of their market activities; (VII) a leader like this will be feared and hated—and after their resignation or the failure of a company, despised.

To gain the trust of their employees, a leader must be competent, respectful, willing to listen and constantly looking to learn and improve.

Openness and accessibility of the leader

One of the interviewees is a manager responsible for the operation and maintenance of turbomachinery in an electric power company operating across several countries. He believes that most of the problems between employees and managers stem from the fact that employees do not have easy access to their managers and feel that their opinions do not matter. The level of trust in a leader is dependent on how well they are perceived by their employees. No one can force employees to trust a leader: trust can only be earned if the leader makes himself or herself accessible, welcomes feedback and is happy to hear both good and bad news. Successful managers give their employees ample opportunity to talk freely about the problems they face. Rather than imposing their own solutions, they encourage others to offer up their own ideas and then work cooperatively as a team to find effective solutions. They are always willing to listen to their subordinates, which encourages them to work harder because everyone feels involved and invested in the decision-making processes within a company.

The HSE director of an oilfield services company believes that a company's safety culture is partly a function of whether employees feel part of a team. Effective teams are built on trust. Trust is not easy to build: it must be earned. Leaders should initiate the team building process. Confident leaders should be open when they are in a situation where they feel vulnerable or uncertain. Their subordinates can then see that they are honest enough to show they do not have all the answers and can make mistakes, and when they do, they will openly admit it.

The HSE manager of a fertilizer company believes that, in order for subordinates to feel safe to talk to their superiors about problems and risks, an organization should look to create a sense of family. Leaders should therefore look to act in such a way that their subordinates see them foremost as a friend and fellow worker, rather than as a strict and unapproachable boss.

Security guarantees for employees who reveal risks

The HSE manager of an oil company believes that there is only a small proportion of employees who truly care about their work. It is these people who are willing to step up, fearlessly take the initiative, and disclose risks to superiors or regulators. However, the fate of these "*active citizens*" is often unenviable. Line managers are often afraid of the independent nature of such employees, seeing them as a threat. They are more concerned about making things safe and efficient than about receiving approval, and are willing to report issues and even point out mistakes to senior management. Colleagues can feel threatened by such morally upright workers who are not afraid to speak out even when the rest of the team would rather remain silent. When a rank-and-file employee independently tries to inform senior management about critical issues, they will usually suffer from a negative reaction from their co-workers and immediate supervisors: being seen as an "*informer*", a traitor, a troublemaker, "*a black sheep*", someone who thinks they are special. Often reasons are found (or invented) to dismiss them, but even if they remain in their job, they are often ostracized and find their promotion chances disappear. According to the respondent, some senior managers have sacrificed their moral principles at some point in order to keep moving up the corporate hierarchy. People with a strong sense of civic duty simply cannot ignore their conscience. Therefore, their career prospects will be limited within most reactive organizations. As a consequence, for employees who are willing to disclose information about risks, it can be easier for them if they have some financial security and can afford to be without a job, at least for a limited period. The converse however is also true: when employees are in debt or otherwise struggling with money, they will do whatever it takes to hold on to their jobs and will thus be reluctant to take any initiative with risk disclosure. The respondent believes it is difficult for any company to create a good enough system of protections and guarantees to make most employees feel safe enough to willingly disclose risks, problems, and mistakes to their seniors.

The chief risk officer of a national power grid company notes that subordinates usually pay close attention to what happens to people who disclose problems in their area of responsibility to superiors. The actions of managers towards whistle-blowers tell employees the truth of the situation, far more than any fine words about how staff should not be afraid to disclose information about risks. To encourage honest discussions of complex issues like this, leaders need to create a climate of psychological safety across an organization. This will need to include some cast-iron guarantees: no penalties for subordinates who disclose information about their own mistakes, and no castigation of the supervisors of the workers exposing serious risks. In addition, leaders should commit to expressing public gratitude to these employees who have acted as responsible workers and active citizens, and to highlighting how their actions have prevented a serious accident, improved productivity, saved a company's reputation and employees' jobs and so on.

The HSE vice president of an oilfield services company believes that, if someone willingly risks their career by disclosing risk information, then senior management should do everything possible to protect them from any potential retribution from their managers or colleagues. If an employee is punished by their immediate supervisor for negative feedback, senior management should immediately intervene in the situation in order to protect the whistleblower. The mid-level manager involved should be warned that punishing or harassing employees who have given critical feedback in any way is totally unacceptable. Management must continue to keep a close eye on the situation. In many ways, it is like the kind of witness protection program sometimes required in criminal cases where there is significant threat to the witnesses before or after a trial. Senior managers must also warn middle managers that they are closely monitoring the career of the employee who spoke out and emphasize that it is in their own interest to maintain good working relations with them, rather than threatening sanctions. To get an objective assessment of critical risks from all levels of an organization, it is vital to encourage whistleblowers by doing everything possible to protect them from any possible negative repercussions.

The HSE director of an oil company insists that senior management should give employees legal and financial security guarantees. Employees with critical risk information must be sure they will be safe. The respondent worked in one oil company where there were notices up at all the production sites with a photo of the CEO, stating his own personal guarantees that employees disclosing important safety information would be fully protected. Included were extracts from safety regulations and policies, contact numbers for emergency confidential communication, all covered by the signature of the CEO. There should also be provisions for conscientious employees having problems with their line managers in order that they can receive prompt support and assistance. For instance, senior managers should be ready to give their mobile number to any employee who feels threatened enough to request an immediate transfer to another unit, or any other urgent protective measure.

According to the HSE manager of an oilfield services company, it is never enough just to give security guarantees. The very fact that such guarantees are being issued suggests that the whole company is not willing to discuss risks openly. There is a degree of mistrust present, and most likely a punishment system is in place, executives from both headquarters and production sites and announced that he was now taking full personal responsibility for safety issues across the whole company. His subordinate leaders were then instructed to follow his example and take responsibility for safety in their areas of control. The subordinates of these leaders should follow the example of their superiors, and so on down the hierarchy. The core objective of the project was to make safety issues a priority for managers at all levels of the corporate hierarchy. For the first time in the history of this company, each leader was set his or her own personal safety targets, a key step in the overall task of reducing injuries and industrial accidents. By stating that he was personally responsible for safety across the company, the CEO was sending a message to the entire workforce that this was the number one corporate priority.

To make sure that the goals of this program were shared through the entire workforce, special strategic sessions were held for each managerial stratum of the company: senior managers, key managers at headquarters (heads of departments and directorates), and middle managers from oil production sites and refineries. The sessions began by acknowledging that in real life work environments, it is impossible to maintain zero accidents and injuries forever. Instead, the aim should be to strive to deliver accident-free indicators over shorter periods of time—for example, during a shift or the implementation of a discrete project. From these short pushes over a particular time-period or project, a significantly lower long-term accident rate would gradually develop across the whole facility.

It is insightful to look in more detail at what this oil company did. First, the CEO brought together ten key senior managers. Based on the discussions held in this meeting—which lasted for eight hours—the project's goals, priorities and main areas of work were identified. The functions of each senior manager within the framework of the project were determined. It was important that the project did not rely solely on the HSE department: each of the ten key senior managers were to lead project delivery within their respective areas of responsibility.

Then a further hundred less senior but still key managers were recruited: the immediate subordinates to the ten executives from headquarters, plus middle managers from production sites. Each of the ten senior managers stood up before this larger group and made a commitment to deliver such and such a safety target within a specified timeframe. This was made while acknowledging the personal commitments of the CEO to transform the management of occupational safety and labor protection, with the following measures: (I) to create a new HSE committee under the leadership of senior management; (II) to create a new CEO's Award in the field of safety; (III) in the case that a fatal accident occurs in the workplace, to personally call the head of the industrial site where the accident happened. The ten key senior managers all stated that they too had committed themselves to creating a safer company. They requested that from now on, the top hundred managers should analyze safety problems in their areas of responsibility and make a commitment to their own top one thousand subordinates to solve them.

An essential foundation of the project was that all managers in the hierarchy made firm public commitments to their subordinates about what they would deliver on safety. This had the effect of cascading the responsibility for safety all the way down the management hierarchy and enshrining safety as a critical focus for improvement. This approach broke many stereotypes that middle managers held about their superiors. For the first time in the company's history, the leadership had articulated that safety was an absolute priority.

Here are some organizational details of the project. The HSE department developed a presentation on cascading risk responsibility down each level of the management. At each production site that the project team visited, the presentation was adapted to the specific risks there. Each production team determined its own list of critical risks, in order to focus on monitoring and controlling the most pressing safety challenges of a given enterprise. The HSE director of the whole company, and one of the 10 senior managers, came to each site to conduct sessions on cascading communication.[3] Representatives from company headquarters helped the heads of the production sites (middle management) articulate the commitments they wished to make to their subordinates. The presentation included a special video message from the company CEO, while the HSE director presented the results of the original safety audit that had first revealed the company's alarming safety figures. The head of the given production site (middle manager) informed their subordinates about the priorities set by headquarters. Then the site manager moved on to the risks at their site, and made safety commitments to his subordinates across five main areas. Subsequently, a team of eight specially trained moderators helped the site management team understand the main risks within their responsibility in the five areas identified, and helped them articulate safety commitments to their subordinates. To allow proper delivery of these cascading communications sessions across the whole company, the HSE director set aside more than eight months.

The central moderation team trained moderators at each site to help lower-level managers conduct similar sessions with their subordinates. As a result, the project reached every level of the management from the top brass down, and even included contractors working with the company. The presentations always included the CEO's appeal, a description of the company's plan to work towards zero injuries and an explanation of the key safety rules to be introduced, along with a warning that non-compliance with these rules would be punished.

All this information—areas of responsibility, commitments, plans and regulations—were combined into a single corporate order control system, which the HSE department then regularly monitored to ensure fulfillment of obligations at all levels of management. The project initiators understood that the key thing was to motivate managers to change their habits and develop their skills in the field, rather than focusing on formal indicators of injury reduction. When implementing the system, desired injury reduction targets were not presented as precise percentages. There is good reason for this: if a safety policy states that "*this year we want to reduce the accident rate by 5% or 10%*", there is a good chance that managers will be tempted

[3] Cascading communication is a process that elucidates how all stakeholders will receive aligned and accurate information. It starts with the senior team and is cascaded down throughout the entire organization.

to falsify incident reporting simply in order to meet the target. It is more important that managers at all levels want to genuinely prioritize safety in their decision making, and perform all work safely, so that injury and accident statistics are reduced over the longer term. All the leaders of the company took part in this project, and it involved well over one thousand employees.

Within the framework of the project, the company developed new strategic goals that include, inter alia, changes in organizational behavior. For example:

- managers need to communicate openly with their subordinates;
- managers should spend time at the bottom of the hierarchy in order to understand the real situation on the ground;
- meetings need to be focused and effective;
- two-way feedback channels need to be developed;
- the leadership need to be actively involved in the process.

These goals are a de facto manifesto for good dialogue and communication. They are aimed at teams that have always worked in a rigid corporate hierarchy where the approach to risks and safety issues has been reactive rather than proactive. It is very difficult to alter employees' behavior quickly to embrace this new approach, especially when for decades leaders have been used to shouting at their employees rather than seeing them as potential partners, or in an organization that has previously imposed a rigid system of blind obedience and subservience.

The main achievement of this project was to involve the entire workforce of the company in the discussion of safety issues, beginning with the CEO and spreading out to include ordinary workers at drilling sites thousands of kilometers away from headquarters. It has also helped managers understand that monitoring safety matters is their direct responsibility, not the responsibility of HSE alone. As a result, company managers began asking the heads of production units for progress reports on production safety improvements. Previously, these issues were only handled by the HSE department; production units had only concentrated on the implementation of the oil production plan, regardless of potential risks. A final confirmation of the project's immediate impact: the year after it was launched, the level of injuries in the company fell by 15%.

3.5 Recommendation No. 5: Senior Management Should Build an Atmosphere of Trust and Security, so that Employees Feel Safe to Disclose Risk-Related Information

Many respondents expressed the opinion that, unless senior management build an atmosphere of trust with employees, there is no chance of them passing on high quality risk information and objective feedback. If there is no confidence that senior management will receive it in good faith, workers will not feel safe to disclose it.

Employees need to have security guarantees, both for their careers and for their colleagues. Managers must guarantee the security of their sources and take responsibility for solving any significant problem they are informed about.

Having studied many publications on the issues of trust and the transmission of information about risks,[4] the authors of the handbook have found it difficult to settle on a simple and understandable definition of the word "*trust*" in the context of the relationship between a manager and a subordinate. Among 100 interviews with practitioners, one head of an oil production site gave a short, yet comprehensive, definition of what he thought trust was—*the predictability of the behavior of another person based on positive experience of previous interactions with that same person*. According to this manager, trust in a leader develops when employees can predict the leader's behavior based on previous positive experience of interaction with them. If there is a clear logic and predictability to the actions of a leader, there is trust; if this is absent, there will be no trust.

Without trust in the leadership, there can be no high-quality feedback from employees on the problems of an organization. Often, employees evaluate the possible consequences of disclosing risk information based on rumors about how senior management behaved in a previous situation with colleagues. Employees project both the positive and negative experiences of their colleagues onto themselves.

---

[4] Dmitry Chernov, Didier Sornette, Giovanni Sansavini, Ali Ayoub, Don't Tell the Boss! How poor communication on risks within organizations causes major catastrophes, Springer, 2022; subchapter 2.2 Results of other researches on the challenges of voice and silence in an organization, https://link.springer.com/book/10.1007/978-3-031-05206-4.

whether official or unspoken. Unfortunately, this is common practice in more than 90% of industrial companies in the world. In the respondent's opinion, there is no point in trying to improve the system of guarantees for employees who disclose risks. What is needed instead is a wholesale change to the culture of a company so that people can freely discuss risks without fear of the consequences. The same logic would apply in civic life: it is the difference between arming every household in a city as protection against attack, as opposed to setting up more advanced security systems and tackling the causes of crime in order to reduce the threat in the first place. Clearly, it is better to make the whole city safer rather than telling citizens to defend themselves with guns!

A psychologist and consultant in the field of organizational behavior agrees that, rather than giving guarantees that employees will not be punished or victimized for disclosing information, it is better to tackle the underlying attitudes that make them feel unsafe about speaking out.

The head of HSE of a fertilizer mining company also warns that managers should understand that guarantees usually do not work. Senior managers may assure the workforce that they will not be fired for disclosing information. But if doing so makes the employee an enemy of their direct supervisor, it will be very difficult for them to feel safe and welcome at work, regardless of senior management assurances that they will keep their job. In the respondent's opinion, the only remedy is a gradual change in an organization's values, so that the disclosure of risks becomes a collectively approved action for all employees and managers. In an ideal world, middle managers should feel grateful to employees on the shop floor who catch a critical risk in time to send a warning up the hierarchy, so that the situation comes swiftly to the attention of their superiors with the resources to deal with it. To achieve this, senior management must create a relationship of safety and trust with the middle managers by not punishing them for the existence of a critical risk, but instead giving them the resources to swiftly address it. The focus must be on shifting the values of an organization and building trust between top and middle managers. Without such a change of culture, no guarantees will convince employees that it is safe for them to reveal dangerous practices in a team they have to work with every day.

The HSE director of a metallurgy company agrees that, in reality, no real guarantees can be given on paper. No document can guarantee that the career of an employee who reveals a risk will not be affected. Everything rests on how much trust the employees hold in their boss and managers. Employees will not just open up to senior management overnight. Positive experiences over extended periods of time are the only thing that will have employees believe that risk disclosure will result in the problem being solved rather than their careers being torpedoed. Trust is hard and time-consuming to build, and quick and easy to lose! The assurances of a CEO may lack any credibility if their subordinates are less open-minded and punish or humiliate employees who speak out about risks. All it takes is for news of one such incident to reach the ears of the workforce, and all the guarantees from the top brass will count for nothing. Therefore, senior managers must take great care when choosing their team of middle managers and make sure they will be their allies in risk identification and control. To do this, middle managers—just like their subordinates—need to know that their careers will not be on the line if a risk comes to light on their production site. Then they will have no reason to be vindictive towards the site workers who first raised the alarm.

The HSE manager of a metallurgy company has similar reservations about giving guarantees alone. Senior management should not simply promise the safety of employees, but rather deal with the factors that pose a threat to them. Leaders need to convince middle managers that risks are never solved by removing "*difficult*" employees, but by making systemic changes in a company, so that these employees no longer have a reason to make complaints above their heads to headquarters. Senior management should reassure middle managers that they will continue to progress in their careers even if problems crop up at the production sites where they hold responsibility. It is obvious really: how can the risks and problems of an industrial enterprise possibly be solved by sacking the employees who pointed them out? These conscientious workers were just sounding the alarm bell. If, instead, middle managers work cooperatively with their subordinates to identify and solve problems, then there will be no more unwanted alarm bells. If mid-level managers do not fear for the security of their own jobs every time their subordinates give difficult feedback to the senior management, then they will not see such employees as a threat.

The vice president of an electricity company believes that disclosing and solving risks should be an obligation for the entire workforce of an organization. Leaders need to make sure that this is a continuous and regular practice, for managers and employees alike. It is essential that every time a problem is highlighted, it is resolved as a result. If this is achieved, then it soon becomes apparent to everyone that it is counterproductive to hide risks, because the simplest solution is to deal with the risk with the support of senior management and with access to the resources of the entire company. It no longer makes sense to ignore the risk or try to solve the problem on the quiet with the limited resources available to a local unit. As well as being likely to fail, this approach can look like a cover-up, bringing accusations of fraud and dishonesty with the likely sacking (and/or criminal prosecution) of all involved. Companies should implement a practice of rewarding workers for disclosing information about risks and imposing consistent penalties for concealing them. Incidences of anyone in an organization trying to prevent a significant risk from coming to light will become a thing of the past. In essence, security guarantees are only given to ordinary employees if the corporate value system is at fault. Guarantees are not required when the right priorities have been set at all levels of management, motivating them to identify risks and eliminate them, both for their own and for the organization's good.

Building trust is not a quick process

According to several HSE managers at one oil company, building trust in senior management is a gradual process: you cannot expect to turn the tide in an instant. People hide information about risks for decades. Managers and employees alike have had a mutual agreement (albeit unwritten) about feedback from employees: managers did not want to receive any bad news from subordinates and were best pleased with a positive report and no questions asked. To change this situation will take constant communication between management and employees—a sustained and genuine personal partnership, so that employees understand the motivations of managers and recognize that they are genuinely committed to solving production safety problems. Everyone needs to see successful examples of risks being identified by employees that are then tackled by the whole company working together. In general, a transformative culture shift like this will take years of dedicated effort to achieve and constant reinforcement to maintain.

The head of HSE of a metallurgy company also believes that one should not expect employees to change overnight, and turn up for work one morning suddenly willing to conduct open honest conversations with senior managers about the problems in their area of responsibility. Most employees will need to witness examples of positive outcomes involving other colleagues before they believe that they can risk disclosing critical information and keep their job. Employees need to see that reported problems are actually addressed. This may well take several years.

Thermodynamic model of risk information transmission

As part of the pilot project in the critical infrastructure company, the level of trust of subordinates in their managers, and the level of trust of managers in their subordinates were both measured.

Results of responses to anonymous surveys within the framework of the pilot project: *How do you rate the level of trust of employees towards their managers in your enterprise?*

	Very high	Medium high	Medium	Low	None	Cannot answer	Number of respondents
All survey participants	3.4%	28.2%	48.8%	12.0%	2.8%	4.9%	326
Senior management, heads of departments and directors of sites (middle managers)	0.0%	34.1%	51.2%	9.8%	0.0%	4.9%	41
Lower managers: deputy directors of sites, chief engineers of sites, heads of workshops, heads and representatives of HSE services at sites	3.7%	36.1%	47.2%	7.4%	1.9%	3.7%	108
Engineers, foremen, and ordinary employees who operate critical infrastructure at sites	4.0%	22.0%	49.2%	15.3%	4.0%	5.6%	177

Results of responses to anonymous surveys within the framework of the pilot project: *How do you rate the level of trust of managers towards their employees in your enterprise?*

	Very high	Medium high	Medium	Low	None	Cannot answer	Number of respondents
All survey participants	2.5%	27.6%	50.6%	12.9%	1.2%	5.2%	326
Senior management, heads of departments and directors of sites (middle managers)	0.0%	39.0%	43.9%	17.1%	0.0%	0.0%	41
Lower managers: deputy directors of sites, chief engineers of sites, heads of workshops, heads and representatives of HSE services at sites	3.7%	33.3%	53.7%	8.3%	0.0%	0.9%	108
Engineers, foremen, and ordinary employees who operate critical infrastructure at sites	2.3%	21.5%	50.3%	14.7%	2.3%	9.0%	177

Interpretation of responses: the responses show that in this organization there is a middling level of trust between managers and their employees and between employees and their managers. Note also that the percentage of answers with "*low trust*" is not negligible. This diagnoses a deficit of sufficiently strong trust that would ensure the active transmission of risk information within the company. There is significant potential for improvement and this pilot study confirmed the value of the recommendations presented here.

It is logical that, if senior managers want to improve the quality and speed of risk reporting within an organization, they should focus on increasing the level of confidence that employees have in them as managers. Efforts of senior managers to achieve this can be assessed dynamically: measuring the transition from a perceived low level of employee trust in managers at the beginning of a project, to a high level after concerted long-term efforts to build open, trusting relationships with their subordinates when discussing risks and problems. With low levels of trust, this vital feedback is not effectively conveyed up the corporate hierarchy. Raise levels of employee trust in their leaders, and the speed and quality of information will reliably improve.

To illustrate this, a thermodynamic model of risk information transmission was created. Information about the risks and problems of an organization is like the water enclosed inside a vessel, representing the hierarchy of the company. The ambient temperature affects the state in which water is held—at temperatures below 0 °C, water becomes a solid—ice; at temperatures between 0–100 °C, water is in a liquid state; at temperatures of 100 °C and above water changes to a gaseous state—steam. The proposed analogy is that temperature corresponds to trust. The diagrams below illustrate this analogy between the dynamic transition of different states of water in a vessel under changing temperature, and the speed and quality of the transmission of information about risks within the hierarchy of an organization with changing levels of employee trust in managers: low level of trust = cold relations; medium level of trust = transitional state of relationships; and high level of trust = warm relationship.

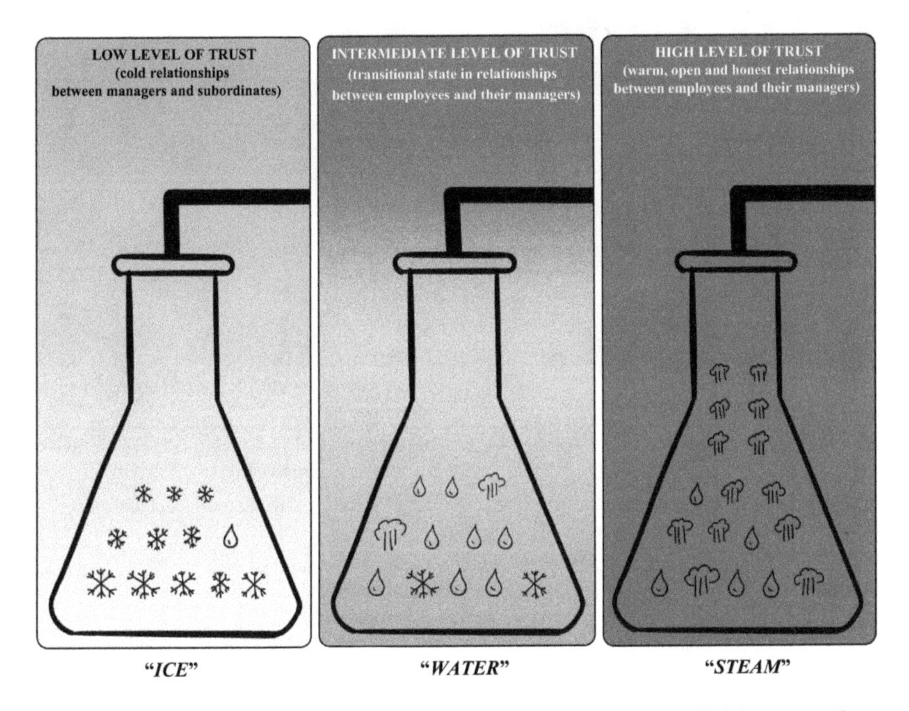

"*ICE*"—LOW LEVEL OF TRUST—COLD RELATIONSHIPS BETWEEN MANAGERS AND SUBORDINATES

In the absence of trust of subordinates in their managers, information about risks will be frozen like "*ice*", and unmoving at all levels of the corporate hierarchy. Information is not transmitted to higher managers because of the cold relations between managers and subordinates. Subordinates fear that they, their colleagues, and their immediate supervisors will be punished for disclosing this information to senior management, and the problems they raise may well be left unresolved. At some hierarchical levels, sites or specific sections within an organization, there may well be much better relations between leaders and subordinates ("*a drop of water inside an icy vessel*"—see corresponding image). If these proactive employees and their leaders begin to escalate bad news about the risks of their unit up an organization's "*frozen*" hierarchy, then they are likely to face "*icy*" attitudes from their superiors. That information will not pass beyond this level of management but remain "*frozen*" and inaccessible to senior management. As a result, a company's top brass will remain in the dark about the development of possibly serious events and will not be able to take adequate measures to prevent a catastrophic scenario unfolding. As a rule, the low level of trust of subordinates towards their managers is only one of the dynamics that indicates trouble within an organization. There are several other indicators that characterize relationships in a "*frozen*" organization.

Organizational model	Multi-level rigid hierarchy
Power distance between managers and subordinates	Long
Model of communication between managers and subordinates	Monologue (one-way from the top to the bottom)
Relationship between managers and subordinates	Lack of mutual respect and trust
Managers' trust in subordinates	Low
Opportunities for subordinates to challenge management orders	Limited
Involvement of employees in solving strategic tasks of an organization	Low
Level of trust of subordinates towards managers	Low
Relationship between subordinates	Collectivism at shop floor level, but at the same time competition for the attention of superiors
Risk management approach	Reactive model (risks only addressed after serious accidents occur)
Willingness of managers and employees to investigate and analyze the systemic weaknesses of an organization that led to previous accidents	Low—tendency to blame individuals
Penalties	Frequent/rigid/punitive/no objective investigation
Willingness of managers and employees to analyze organizational shortcomings and discuss existing problems and risks	Low
Level of concealment of information about employee errors, work-related injuries, risks and problems, etc	High
Quality of risk-related information transmitting across corporate hierarchy	Low

"*WATER*"—INTERMEDIATE LEVEL OF TRUST—TRANSITIONAL STATE IN RELATIONSHIPS BETWEEN EMPLOYEES AND THEIR MANAGERS

Senior management can take several steps to increase employee confidence in management, with the goal of improving the quality and speed of communication about risks: publicly stating that they actively wish to receive information about risks, that no one will be disciplined for disclosing this information, and that the highlighted problems will be solved. Some employees may respond positively to this call and pass on risk information, but senior management should not be under any illusion that, even if some of the most confident employees respond positively, this does not mean every worker will from then on communicate openly and honestly with their superiors. It is likely to take several months or even years to "*unfreeze*" the corporate hierarchy, especially at the lower levels of an organization. For most employees to feel confident to speak out, they need to witness first-hand that management's promises to anybody highlighting a problem are genuine: those bringing the news were not penalized, and the issues they raised were indeed resolved in a way that benefited everyone.

Organizational model	Multi-level flexible hierarchy
Power distance between managers and subordinates	Medium
Model of communication between managers and subordinates	Monologue prevails, but some examples of dialogue between managers and employees
Relationship between managers and subordinates	Respectful
Managers' trust in subordinates	Average
Opportunities for subordinates to challenge management orders	Some examples of active listening to subordinates
Involvement of employees in solving strategic tasks of an organization	Medium
Level of trust of subordinates towards managers	Average
Relationship between subordinates	Cooperative
Risk management approach	Variable—examples of both reactive and proactive approaches
Willingness of managers and employees to investigate and analyze the systemic weaknesses of an organization that led to previous accidents	Medium—more serious incidents usually investigated in detail
Penalties	Occasional, reasonable and fair/objective investigation of incidents
Willingness of managers and employees to analyze organizational shortcomings and discuss existing problems and risks	Average
Level of concealment of information about employee errors, work-related injuries, risks and problems, etc.	Average
Quality of risk-related information transmitting across corporate hierarchy	Average

"*STEAM*"—HIGH LEVEL OF TRUST IN RELATIONSHIPS BETWEEN EMPLOYEES AND THEIR MANAGERS (WARM, OPEN AND HONEST)

Friendly, open, and trusting relationships between managers and employees lead to information about risks being quickly and efficiently transmitted up the hierarchy of an organization ("*evaporated out of the vessel*"). Employees can use a variety of communication channels, including informal ones based on personal relationships between subordinates and managers, to quickly convey critical issues up to senior management level. Workers are confident that risk information will be positively perceived at all levels within the hierarchy. Senior management recognize the value of this information, as it enables them to proactively initiate risk management at appropriate corporate levels so that the issues are promptly resolved. The valuable role of the employees who raised the problem in the first place will be recognized by public praise and suitable rewards.

Organizational model	Flattened hierarchy—fewer management levels
Power distance between managers and subordinates	Short
Model of communication between managers and subordinates within an organization	Two-way dialogue
Relationship between managers and subordinates	Trusting
Managers' trust in subordinates	High
Opportunities for subordinates to challenge management orders	High—feedback welcomed
Involvement of employees in solving strategic tasks of an organization	High
Level of trust of subordinates towards managers	High
Relationship between subordinates	Mutual support and assistance
Risk management approach	Proactive (preventive measures taken to address risks before they become a problem)
Willingness of managers and employees to investigate and analyze systemic weaknesses of an organization that led to past accidents	High (detailed investigations undertaken for even minor and near miss accidents)
Penalties	Rare: look to find systemic causes—poor management decisions and flawed corporate process—rather than blame individuals
Willingness of managers and employees to analyze organizational shortcomings and discuss existing problems and risks	High
Level of concealment of information about employee errors, work-related injuries, risks and problems, etc.	Low
Quality of risk-related information transmitting across corporate hierarchy	High

This model was presented to the participants of the pilot project. Below is the feedback received from employees from every corporate level of this critical infrastructure company.

Results of responses to anonymous surveys within the framework of the pilot project: *Do you understand the thermodynamic model of risk information transmission that has been presented to you?*

	Fully understand	Mostly understand	Don't understand very well	Don't understand at all	Difficult to answer	Number of respondents
All survey participants	54.9%	40.2%	4.2%	0.0%	0.8%	264
Senior management, heads of departments and directors of sites (middle managers)	77.8%	22.2%	0.0%	0.0%	0.0%	36
Lower managers: deputy directors of sites, chief engineers of sites, heads of workshops, heads and representatives of HSE services at sites	47.1%	47.1%	4.8%	0.0%	1.0%	104
Engineers, foremen, and ordinary employees who operate critical infrastructure at sites	54.8%	39.5%	4.8%	0.0%	0.8%	124

Results of responses to anonymous surveys within the framework of the pilot project: *Do you agree with the principles of this model?*

	Strongly agree	Rather agree	Rather disagree	Strongly disagree	Difficult to answer	Number of respondents
All survey participants	45.8%	45.5%	3.4%	0.8%	4.5%	264
Senior management, heads of departments and directors of sites (middle managers)	50.0%	47.2%	0.0%	2.8%	0.0%	36
Lower managers: deputy directors of sites, chief engineers of sites, heads of workshops, heads and representatives of HSE services at sites	47.1%	42.3%	3.8%	1.0%	5.8%	104
Engineers, foremen, and ordinary employees who operate critical infrastructure at sites	43.5%	47.6%	4.0%	0.0%	4.8%	124

Results of responses to anonymous surveys within the framework of the pilot project: *What stage do you feel your organization is at in the framework* of *the "Thermodynamic Model of Risk Information Transmission"?*

	Steam (warm relationship)	Water (transition state)	Ice (cold relationship)	Difficult to answer	Number of respondents
All survey participants	8.7%	61.7%	20.5%	9.1%	264
Senior management, heads of departments and directors of sites (middle managers)	5.6%	66.7%	22.2%	5.6%	36
Lower managers: deputy directors of sites, chief engineers of sites, heads of workshops, heads and representatives of HSE services at sites	14.4%	61.5%	17.3%	6.7%	104
Engineers, foremen, and ordinary employees who operate critical infrastructure at sites	22.6%	60.5%	4.8%	12.1%	124

Interpretation of responses: according to the majority of participants in the pilot project, the relationship between managers and subordinates in this company is in a transitional state (neither "*cold*" nor "*warm*"). A one-way, top to bottom communication channel between managers and subordinates prevails; the level of trust of subordinates in their managers is average; the level of trust of managers in their subordinates is average; and the quality of information about risks transmitted up the corporate hierarchy is often average (in some cases low). It is interesting to note how the responses of senior managers and ordinary employees differ regarding the assessments of "*warm*" and "*cold*" relations. Among ordinary employees, four times more (22.6%) believe that the relationship between subordinates and their management is warm, honest, and trusting when compared to senior managers (5.6%); five times fewer ordinary employees (4.8%) believe that the relationship between managers and their subordinates are cold as compared to senior managers (20.5%). This disparity may be due to the fact that ordinary employees frequently work in small, tight-knit units (regular work teams) under the direction of a single line manager. Such working practices tend to engender a very powerful team ethos and a correspondingly strong bond with the immediate line manager, with unwritten codes of behavior such as "*united we stand, divided we fall*"; and "*all for one and one for all*". That is why some employees could perceive their relationship with their immediate line manager as trusting.

The results of the surveys indicate that this theoretical model is easy to understand and is shared by most respondents within the pilot project. Based on this model, a set of practical solutions to improve risk information transmission was developed at several of this company's sites selected for a pilot project (Chap. 4 of this handbook).

3.6 Recommendation No. 6: Middle Management Are Allies of Senior Management in Building an Organization Where Active Dialogue Between Superiors and Subordinates Is Welcomed

Middle managers have a better understanding of the critical risks of an organization than lower-level employees

According to the HSE director of an oil company, the middle managers in charge of production facilities know more about the situation at an organization than shop floor employees or lower-level managers. If a company only asks for the opinion of shop floor employees about the critical risks of an organization, they may not always get an accurate assessment of the situation. Most lower-level employees are only familiar with a small proportion of the total work of a business, and only aware of problems with the equipment on which they directly work. Failure of a given piece of equipment (which a shop floor employee can report) can potentially threaten to disrupt the whole production process. However, in most cases, the middle management level will have backup solutions to cover specific equipment failure without disrupting production; a lower-level employee may not even know about these backup options. This means that a temporary shutdown of most individual pieces of production equipment will not affect the efficiency of the entire production process. As a result, getting information about critical risks only from lower-level employees may just lead to an increase in information noise, making it more difficult for senior management to understand the true picture of safety at a site.

The HSE head of a mining and metallurgy company also believes that it is a mistake to rely mainly on shop floor workers for information on critical risks. In his opinion, 95% of the critical risks in a large industrial company are recognized at lower and middle management level. The most serious concealment of risks does not happen at the bottom of the corporate pyramid—between employees and lower or middle managers—but at the top, when middle management hide critical problems from senior management, or at the very highest level, when top managers hide them from shareholders. Here concealment can have even more serious consequences. Therefore, once honest dialogue has been established between senior management and owners about critical risks and how to handle them (see Recommendation No. 1), the next step is to establish the same honest dialogue between the leadership and the middle management level.

The middle management problem

Some interviewees consider that in the traditional hierarchical model, the middle management level is the key blocker of risk information coming up from shop floor employees and lower managers.

The vice president of an electricity company believes that middle managers are often the main culprits in distorting or editing reports from their subordinates when they refer them up the corporate ladder. He has seen this occur many times, when middle managers have comprehensive information about problems within the production unit they are running, but often decide not to say anything to headquarters.

An HSE consultant working mainly for oil and gas as well as in air traffic control offers this example from his own experience. In one company, senior managers requested to receive information about problems from ordinary employees, but this did not happen. They told the respondent that this was because of an impenetrable "*clay layer*" of middle managers who did not want to report any serious problems, fearing this would reflect badly on their effectiveness as managers. It would suggest that they were failing to cope with their duties, that they were incompetent in solving problems, that they expected senior management to deal with these issues.

In the experience of a global director of safety for the oil and gas industry, the attitude of senior management towards occupational health and safety is usually very positive, showing a commitment to creating a corporate culture in which bad news can be conveyed seamlessly from the bottom to the top. However, the middle management level (the so-called "*permafrost layer*") often creates an obstacle to this free flow of risk information. According to the respondent, most critical infrastructure companies pay a lot of attention to the leadership skills of senior management—but in truth, communication and problem solving would be improved more effectively if greater attention was devoted to looking at how middle management are functioning.

According to some respondents, most of the problems with risk communication from middle management are the result of owners and senior management setting overly ambitious goals for production, profit and costs. Middle managers cannot then freely discuss the problems that inevitably arise when they have to try and meet these targets in the imperfect reality of the production site. They are torn between the often unfeasibly ambitious corporate plans imposed by senior management, and the reality of operating outdated infrastructure. In addition, they are under the pressure of a company's cost reduction strategy, which restricts both the resources available to modernize critical infrastructure and the wages payable to the workers operating the equipment. All too often, strategic goals and objectives are not discussed with middle managers: they come down to them as a diktat from owners and senior management. Many companies have created a culture of "*no bad news*", where subordinates can only report achievements and successes. In this culture, it is very difficult for middle managers to pass risk information up the hierarchy to senior management and owners. Clearly then, there will be far less concealment of risk information if senior managers show that they actively want to hear the opinion of their subordinates, and then give them the resources to solve the problems they have identified.

The vice president of a gas pipeline construction and repair company believes that risk concealment at the middle level will effectively disappear if senior management are as eager to receive bad news as they are for good news. Middle managers conceal problems because senior management make it clear they do not want to hear about them. The problems of concealment at middle management level often arose because of the penalties that senior managers used to impose on production sites managers for the "*offence*" of giving an honest and objective report on local problems.

The head of an oil production facility explained why middle management seldom disclose the problems at their sites to their seniors: they are afraid of negative feedback from senior management, and of being ignored or misunderstood by the top brass. The middle level is boxed in by a strict framework of business targets. For the most part, senior managers want to hear about achievements and successes, not about problems. Middle managers are compelled to hide information about risks because of the priorities of an organization and the attitude of senior management. It is not that mid-level managers have an inherent desire to hide information. It is more that their bosses simply refuse to listen when they say that it is impossible to follow the plan and hit all the targets that have come down to them from headquarters. Mid-level managers know what their bosses want to hear—"*everything is fine, it's all under control*"—and thus site problems are simply not mentioned.

The HSE manager of a metallurgy company echoes what other respondents have said. He attributes this "*economy with the truth*" to the corporate culture of a company, formed over decades, in which there is an unwritten taboo on bringing bad news to the boss.

The HSE head of a chemical company believes that it is senior management who set up the systems that make it difficult for middle managers to openly discuss production problems. Middle managers cannot challenge the CEO's decisions on production and financial plans at the sites without consequences for their own careers. The ambitious plans launched from headquarters are often impossible to implement without serious violations of safety regulations.

Mid-level managers are interested in reporting problems: they want to share responsibility with senior management for solving them, and they need headquarters to allocate the financial resources. If executives show middle management that they are ready to listen, discuss and find resources to solve problems, they will be pushing at an open door.

The HSE director of an oil company does not consider that the problem lies with mid-level management: they only act as their bosses direct them. Therefore, middle management will only change if executives demonstrate that they want to hear about problems, that no one will be punished, and that resources will be allocated for them to tackle the problems they are facing. Senior managers should demonstrate a positive response to signals coming in from the field, in a way that all production site managers can appreciate. They need to prove that they will listen attentively to their site managers, explore problems with them, allocate resources, show appreciation to those who take the initiative, and so on. Middle managers will then know exactly the kind of response they will receive when they do go ahead and inform bosses about a serious problem at their site. If colleagues have had a previous positive experience when they warned headquarters about a risk, then all site managers will be encouraged to do the same, knowing they can expect a sympathetic and constructive reception.

Summing it up, the main reason for middle management concealing information about risks is because senior management are unwilling to hear about problems. Subordinates withhold this information from superiors, not in general because they are dishonest or malicious, but because senior management have made it clear one way or another that they do not want to hear about the problems that their subordinates are facing. Middle managers have got that message, and will only send reassurances and good news to the top. If senior management demand their subordinates fulfill production plans at any cost, without giving adequate resources, then they will respond by keeping quiet about the real picture on the ground. If instead senior management establish a dialogue with subordinates, and genuinely look to find a balance between achieving production indicators and maintaining occupation health and industrial safety, then middle managers will no longer hide risk information. They will quickly adjust their own practices, and in turn encourage their own subordinates to share problems with them. Risk information will only be communicated effectively throughout a company hierarchy if senior management: (I) genuinely want to receive information about problems; (II) do not penalize middle managers; (III) reward employees who make honest, objective reports; (IV) provide the necessary resources to solve production problems.

If penalties for bringing bad news are removed and subordinates guaranteed job protection whatever "*awkward truths*" they reveal, then most employees will willingly reveal site risks to their superiors. Not least because they recognize that the control and reduction of these—be it faulty machinery or inadequate procedures—is in their own self-interest. Not only does it reduce the likelihood of injury to workers operating the equipment, but it also protects their income and job security by reducing the chances of a serious accident that could mean a lengthy plant shutdown or even collapse of the whole business.

Attempts to bypass mid-level management

If senior managers believe that middle managers are hiding the real state of affairs in a company, then they may decide to bypass them entirely, and communicate directly with shop floor employees at the sites concerned.

The HSE head of a gold mining company has an example of this. On visits to production sites, the CEO would give some site workers his personal email address so that they could inform him directly about problems, thus bypassing the traditional managerial hierarchy. One day, he had an email from an employee at a remote field complaining that the toilets at the site had not been working well for over a year. As none of the field managers had paid any attention to requests from the workforce to solve the problem, the employee had decided to write directly to the CEO. It might be a small thing in terms of the overall business, but it mattered to the day-to-day comfort of employees at the site. The boss immediately forwarded the message to the head of the field, hiding the sender's details. Unsurprisingly, the toilet problem was solved within a couple of days. A week later, a second message arrived from the employee, thanking the boss for sorting out the repair of toilets, even though they were located several thousand kilometers from headquarters. The CEO admitted, in personal communication with the respondent, that he had been ashamed to hear that some managers in the field could not solve such minor problems without involving their superiors, and requested that he intervened. The respondent was irritated by the situation, not least because it suggested that he was not managing his field teams effectively.

This case raises a major potential headache for senior executives. If shop floor employees across a company are taking their problems directly to the leadership, bosses will drown in messages requesting that they intervene in what are essentially local site issues, thus undermining the entire hierarchical structure. Lower and middle managers would become largely redundant if every local problem landed in a CEO's inbox. A better solution would be to address the quality of work of the entire management team, so that non-critical site and workforce problems are satisfactorily resolved at the appropriate management level. Employees should only be going to the next level of management if they cannot solve a problem after raising it with their immediate supervisors.

One of the key questions that the authors of the handbook asked the first 30 interviewees was concerned with the creation of additional specific channels for communicating risk information directly from field level workers to senior management, bypassing the traditional hierarchy of lower and middle managers. At first glance, this seems a logical approach if executives remain ignorant about serious risks because managers all the way up the line are failing to pass critical information upwards. The respondents identified advantages and disadvantages of such an approach.

Advantages of introducing an entirely new alternative communication channel between shop floor employees and senior management—and why it is reasonable to question the wisdom of relying on a traditional corporate hierarchy as the only channel for obtaining critical risk information:

- **The presence of an alternative communication channel reduces the likelihood of risk concealment by middle and lower managers, and shop floor employees.** If senior managers have an independent system for getting information directly from the field, their subordinates will quickly realize that it is useless to try and hide risk information, because it will eventually reach senior management anyway and they might then be reprimanded for concealing it. An alternative channel can reduce the temptation for middle managers to distort or conceal the real situation, a subterfuge that is in any case a laborious and risky process. If they know that information about problems in their area of responsibility will inevitably reach senior management directly from shop floor employees, then that time, effort and risk will not be worth it—better to take the initiative and speak out first, sending accurate reports up to senior management via the traditional hierarchy of an organization.

  It is worth noting that immediately after the Chernobyl accident—when the plant management failed to communicate the gravity of the radiation situation and the state of the reactor to Moscow—a system was put in place to automatically transmit information about the operating parameters and radiation situation from every nuclear power plant across the USSR, effectively bypassing the individual site operators and directors in the flow of risk information. Some of the respondents have given other examples of the effectiveness of such alternative channels from their own experience. For example, one of the contractors of a mining company was involved in the provision of services for lifting goods. Employees there committed a safety violation during cargo handling. Under their contract with the mining company, the contractor should have paid a large fine for allowing such dangerous conduct. Instead of paying the fine, the contractor proposed equipping all their cranes with video cameras, set up to send a direct live stream both to the headquarters of the organization and the contractor's office. Crane workers, realizing that anything dodgy they did during loading operations would now be recorded, began to avoid any behavior that could be later analyzed and lead to a penalty or their dismissal. After the introduction of this monitoring system, safety violations during cargo handling fell by 70–80% from the level before the cameras were installed.

  Another example comes from the oil industry. In one exploitation region, video cameras were installed at oil rigs. A team of specialists with extensive practical drilling experience was assembled at headquarters, and the team began to analyze drilling operations to identify any hazardous actions. This expert real-time analysis of drillers working in a remote region allowed immediate feedback to the drilling rig supervisor on anything that they had observed that required flagging up, most importantly any dangerous activity. Videos were edited at headquarters to exemplify driller errors, and then played in front of the drillers involved. This proved to be very effective feedback in preventing them repeating their mistakes. In the first month of operating the system, about 600 violations were detected. By the third month, violations had dropped by 89% to only 50–60. Here then was empirical proof: if workers realize that their superiors are monitoring them and analyzing their mistakes, they will be more vigilant about observing safety rules. The disadvantage of such a system is that nobody within the company can simultaneously observe and analyze dangerous conduct 24/7 on the dozens of cameras installed on every rig across multiple sites. This company is now actively working on introducing automatic monitoring of the driller's activities. This project, called "*computer vision*", aims to "*train*" artificial intelligence to automatically identify dangerous conduct.

Based on this example, it is reasonable to conclude that the introduction of a comprehensive system of direct feedback on the situation on the ground leads to a reduction in operational safety violations. This alternative communication channel delivers much better information than previously received via the traditional models that rely on lower and middle management. Any alternative risk information channel acts as a security camera. Aware of the camera—whether literal or metaphorical—employees realize it is impossible to conceal their actions. It makes more sense for a worker to report dangerous actions, errors and risks to the management proactively, rather than waiting for management to come to them with the recordings that prove their wrongdoing. This kind of approach shows that while the traditional hierarchy can still perform many useful functions, it is worth considering a range of alternative channels to collect accurate information on the situation on the ground. The more channels providing information, the clearer the picture that senior managers can form of the true state of affairs, and the better their decision-making can be.

When senior managers receive evidence of unsafe practice on the ground or risks being ignored at the bottom of the hierarchy, it is important that they do not automatically seek to blame their subordinates. They need to do all they can to minimize the tendency for lower and middle managers to be defensive when discussing risks identified by their subordinates, as this can have a devastating effect on an entire company. When they disclose risks identified by ground level workers, it is good practice to reassure managers at all levels that the leadership have no intention of penalizing them. On the contrary, they want the whole company to work together to prevent major incidents by maintaining and modernizing equipment. In this context, information about critical risks from employees or from automatic risk transmission systems is valuable to all levels of management. The best approach is for a senior manager to discuss the information received with the middle manager in a calm, non-threatening manner, elicit their views of the risk situation and together discuss ways by which the problem can be solved.

- **Having an alternative channel motivates middle and lower managers to actively seek information from down the corporate hierarchy.** An alternative communication channel will help the leadership to understand how well the middle and lower levels of management are communicating with shop floor employees about operational problems. This will motivate those managers to be more deeply immersed in the problems and challenges facing their workforce. That extra motivation could well be needed, as local managers can sometimes ignore what is happening on the ground. The fact that employees can go straight to senior management if they feel issues are not being addressed by their line managers will be perceived as a threat—"*the sword of Damocles*"—for lower and middle managers. At the same time, it will encourage employees to become de facto internal auditors for an organization, and to proactively identify risks in their area of responsibility.

- **Other pluses.** An alternative channel can transmit risk information very quickly from employees to senior management as it bypasses other levels of management. This could be crucial in a rapidly escalating emergency and make the difference between a full-on disaster and a close shave. Workers also see that senior management actively want to hear from them and that their opinions and observations are highly valued by the leadership.

Disadvantages of introducing a completely new alternative communication channel between shop floor employees and senior management—and why the traditional hierarchy should not be compromised for the sake of direct communication between senior management and shop floor employees:

- **An alternative channel undermines the sustainability of the traditional hierarchy.** The presence of alternative information channels can undermine the work of management. After all, if senior managers hear about risks directly from shop floor employees and make their own decisions, it is not clear exactly why middle and lower managers are still needed. If they are excluded from the decision-making process, then the reported issues will be decided at the top level. This will have the inevitable consequence of information overload for bosses. If this form of communication continues for any length of time, a company will just grind to a halt. All the interviewees who raised this issue advised against invalidating the traditional hierarchy in this fashion, and recommended working to strengthen it: trusting middle and junior managers, listening to them, seeing them as allies in a shared endeavor, and delegating responsibility and resources to them to identify and manage critical risks.

  If the middle managers already in place continue to hide information despite these initiatives, senior managers will then need to recruit new qualified and loyal middle and lower managers who understand and accept the importance of getting accurate feedback up the corporate hierarchy. Some leaders try to control every aspect and layer of their organizations. But with a large technology company, it is physically impossible for senior management even to know about, let alone actively manage, everything that happens. Employees on the ground will understand the situation more fully, and thus make better informed decisions on risk management, than senior managers based at headquarters thousands of kilometers from the production site. This is another reason that most of the respondents were against replacing or weakening the traditional hierarchy by establishing an alternative risk communication system: it takes local decision-making out of the hands of managers at a site. There was unanimous agreement that leaders should support their lower and middle managers to improve the quality of communication up and down the hierarchy.

- **When middle managers fear for their careers, they may take reprisals against employees who, by speaking out, negatively affect the organizational climate of a company.** If mid-level managers are liable to be penalized when their subordinates disclose information about risks, then they are likely to fear shop floor workers going over their heads to senior management. They will try to identify any employee who seems to have a strong sense of civic duty and a high degree of confidence and endeavour to dissuade him/her from speaking out to senior management. They may even take reprisals against such proactive employees. This will silence most of the workforce, or at least ensure that any information reaching senior management from the shop floor will not be objective, but "*edited*" as another "*good news story*" to avoid such reprisals. At this point, senior managers can end up in a lose-lose situation. If they fail to protect employees who have had the courage to speak out and contact them, those employees will lose confidence in senior management. However, if they publicly support the whistleblowers and make an example of a middle manager by publicly punishing them for taking reprisals against their subordinates, they will alienate and de-motivate their middle and lower managers. In general, neither scenario will help create an atmosphere of trust within a company. Middle managers will be left trusting neither their bosses—who have created a system that undermines their authority and scrutinizes their work—nor the workforce, any of whom may go over their heads by reporting directly to headquarters about the real situation on the ground. Ultimately, this whole approach will damage confidence and trust throughout an organization. Instead, leaders need to empower and reassure their middle managers, and encourage them to discuss their problems openly. The declared policy of a company, and the systems it puts in place, should enable both managers and ground level employees to report information about risks and problems without fear. This is more productive than undermining the traditional hierarchy by creating alternative information channels.
- **Employees can use an alternative channel as a tool to challenge decisions of middle and lower managers.** If a channel is set up for employees to take problems directly to senior management level, they are being given a tool that allows them to bypass lower and middle managers and effectively challenge their decisions. In companies with a reactive culture, this could create internecine conflicts and increase mistrust between different levels of the hierarchy. Such alternative channels should only be implemented in companies that already have an open corporate culture, where managers are ready to hear about problems and solve them, and where there is no penalty system for mistakes. If the workforce is well trained and competent and feels secure, then such a system

can be effective and function as an additional communication channel. It can help all levels of management to better understand where the problems are on the ground, supported by the professional assessments of the people who are operating the equipment on a daily basis. In a healthy corporate culture, middle managers will not feel that this means senior management do not trust them: on the contrary, they will welcome having an additional channel available to facilitate the communication of important risk information, which may not have been visible to even the best site manager.

- **Shop floor employees do not have detailed critical risk information.** The value of the risk information they can provide is lower than that received from middle management. Lower-level employees have narrow competencies and limited information about the work of the rest of the organization. Shop floor employees often fix problems that could affect their own personal safety or might lead to failure of the specific equipment they are working with, which may not always be critical. They also may not be aware of the technological condition of the equipment they are operating or have the training to make such evaluations. Many critical infrastructure companies have specially trained employees who evaluate equipment according to clear criteria and objective assessment—the ratings of such experts can be relied upon. In addition, shop floor workers are usually not thinking in relation to the whole organization or even their workshop. Thus they do not see the wider picture of risks across a company. Therefore, if there are direct channels for employees to transmit risk information to senior management, then headquarters risks being flooded with information noise about insignificant risks or fragmentary information about production sites. In addition, as some employees prefer to work on new equipment, they will do their best to discredit older equipment even if it is still perfectly serviceable. As mid-level managers receive information from many subordinates, they have a greater breadth of understanding of the situation. An issue that seems critical to the employee affected by it may seem less significant to a manager making risk assessments across a whole industrial site.
  The head of a power plant cites an example. In his plant, there are 400 workers, but only about ten shift supervisors and shop managers actually understand the critical risks of all the equipment in use across the whole plant. Senior management do not need information about all risks—only those that could have serious implications for their company as a whole, including a major accident. These critical risks are not usually well understood by shop floor employees, but only by middle and lower level managers who have an overall view of the entire site. In addition, middle managers have the industry experience to be able to rank risks, and determine the likelihood of the occurrence of adverse events. There has been a trend in recent years for people with a good business background but no industry experience to become senior managers. However, most middle managers have worked their way up from the shop floor to being head of a production site. It is these people who best understand the intricacies of a production process, and can make professional assessments of the state of critical infrastructure.

For all these reasons, it is more important for a company leadership to build good communication with middle managers than lower-level employees. Senior management can usefully visit sites and meet with the limited number of shop floor employees who monitor critical risks on a daily basis, but this is only a small proportion of an industrial company's workforce.
It does not follow that the information about other less serious risks that shop floor employees provide should simply be ignored. The task for senior leaders here is to ensure that their middle and lower level managers have the resources they need to act independently. Then managers at site level can deal with the secondary problems their workers have informed them about, without requiring the attention of senior management.

- **If you ask for risk information from every employee, you will need huge resources to reduce and control all the risks.** If a company starts collecting information about all risks from every employee throughout an entire organization, so many are likely to come to light that a company simply does not have the resources—organizational and financial—to tackle them all. But if employees see that the managers have paid little attention to their reports, fewer will take the initiative in the future: why waste your time and effort raising an issue if you will probably just be ignored? If managers ask employees to report risks, they must be prepared to give feedback to everyone who takes the initiative to do so.
- **Maintaining an alternative channel requires significant resources with no guarantee of genuine, practical improvements.** Responding to all the data coming in via an alternative communication channel is challenging and expensive in terms of time and organizational resources. A company can invest hugely in investigating every negative comment made by employees, to find that only a few percent of these messages were relevant—everything else was no more than information noise. The HSE manager of a metallurgy company has an example from his practice. Two years before the interview, 600,000 safety audits were conducted in his company, and the results were entered into a data system. A year before the interview, 400,000 audits were conducted, and the assessment methodology was slightly adjusted to reduce information noise. HSE units spent a lot of time entering data into the system. However, the practical impact in terms of improved safety has been modest. The respondent had to go to the CEO for permission to reduce the amount of data being collected, since most of it has proved of little real-life value to the company. The safety audit system made it possible to monitor the managers who had to respond to the problems identified, but a huge amount of time and money was spent conducting these audits. In thc respondent's view, the work of any alternative communication channel for risk information may have similarly modest results.
An analysis of the pros and cons suggests that the creation of a completely new channel to communicate risk information, as an alternative to the traditional hierarchal systems, will only be effective in organizations where senior management (I) actively want to receive information about critical issues; (II) do not

penalize any employees for highlighting problems; and (III) support their subordinates to solve any problems.

However, almost all the respondents asked about introducing alternative channels advocate strengthening the traditional hierarchy—especially the relationship between senior and middle managers—as an essential first step to improving the identification and control of critical risks. A majority also recommend that an alternative channel should only be introduced as an addition to the more conventional chain of command systems, and not as a replacement.

Development of automatic electronic channels to transmit information about the operational situation, bypassing both the operators and managers of industrial sites, is seen by many respondents as a promising and useful development for critical infrastructure companies. One respondent from the oil industry makes the following powerful argument in its support. It is difficult for workers to feel offended or insulted if it is a machine that is collecting information about their performance and sending this on to their superiors. However, they may well resent a flesh-and-blood colleague who reveals their shortcomings and mistakes by '*shopping*' them to the management. Such grievances build up and have a negative impact on the entire production workforce. For this reason, alternative transmission channels will generally operate better if they are based on automation rather than on the reports of employees.

Middle management and senior management can work as allies

Senior management can only build an effective system to obtain accurate information about risks, and change the safety culture in a company, by working with the middle managerial level. Therefore, the best strategy must be to make middle managers allies and not enemies.

In a proactive risk management model, senior managers see their subordinates as colleagues and valued partners, who can be entrusted with identifying risks and solving difficult issues. In this model, the organizational hierarchy of a company becomes compressed and broader. With fewer levels involved, information flows more quickly up and down between different levels of management.

In contrast, a reactive model casts junior managers more as mere subordinates and assumes they will always look to avoid responsibility for problem-solving, cannot be trusted, and need constant, close supervision. Communication is a one-way flow of directives from above, and the organizational structure looks like a tall narrow pyramid with multiple levels, in which information is transmitted slowly step by step, through successive layers of management.

Building an atmosphere of trust between senior and middle management

The HSE manager of a metallurgy company believes that to avoid middle managers feeling under pressure to conceal risks, the leadership should create an atmosphere of trust in which middle managers are encouraged to share urgent problems at their sites with their seniors. They should create a non-judgmental and supportive environment, and always provide the necessary resources from headquarters to solve the problems middle managers have raised.

To achieve this, several respondents recommend that senior management should:

- emphasize that they trust middle management;
- ensure that middle managers disclosing risks and problems are not penalized or dismissed;
- be honest and transparent in their intentions and decision-making;
- appreciate and reward subordinates who provide accurate information;
- show that they want to work together with middle managers to solve problems, and not leave them to tackle local issues alone;
- either demonstrate competence and an understanding of production matters or be straight about the limits to their experience and listen to the professional expertise of middle managers;
- assure middle managers that their position is not under threat if site employees bypass them to communicate risk information to the leadership. Senior management are simply collecting the information they need to make the best possible decisions. Middle managers who understand this are not afraid of alternative communication channels and welcome their more proactive employees as contributors rather than seeing them as a threat to be rooted out;
- identify and empower the employees who are most supportive of a more open approach to risk transmission and best appreciate the importance of the leadership receiving objective information;
- identify and neutralize "*blockers*" (employees resistant to change and open communication) among middle managers, rather than assume all managers at this level are the same and cannot be trusted.

Dialogue between senior and middle managers

Many of the interviewees emphasize that dangerous operational practice, and the concealment of risks, often stem from middle management struggling to negotiate and adjust the over-ambitious production plans set for their sites by headquarters. This leaves them no choice but to try and implement the plans—even if this means increasing operational risks, or falsifying results. Respondents from one oil company recommend that as part of the business planning process, the leadership invite mid-level managers from all production divisions to headquarters to discuss the feasibility of the plans and assess the risks associated with implementing them. In this way, senior and middle managers can thrash out realistic and achievable production plans, identify the main risks involved, and agree on the allocation of resources to address them. A helpful exercise during these sessions is for participants to swap roles, with middle managers becoming seniors, and vice versa. This gives both groups an opportunity to "*stand in each others shoes*" and view the problems from their colleagues' point of view.

The head of HSE of a metallurgy company cites the example of an operating committee that meets regularly at headquarters to analyze the work of all the company's industrial sites. The committee is attended by the CEO and key senior managers. Each session is conducted as a video conference with the management team of one of the sites, during which the site director (middle management) reports to the top brass about the work of their facility. Before the session, a report with production statistics is sent to headquarters. The site director reports on progress towards production indicators, and on any problems or incidents that have arisen. Senior managers then establish a dialogue to discuss the reports. The quality of the committee's feedback to the on-site team is crucial: avoiding reproach and blame and focusing instead on practical solutions to any difficulty raised. Middle managers appreciate this collective approach to problem solving—especially if the necessary resources are forthcoming—and thus become motivated to provide accurate and objective site reports, including the bad news as well as the good.

The HSE director of a gold mining company believes that senior management should regularly ask production site managers what serious risks they have identified over the past week, what measures they propose to address them and what is already being done. This demonstrates the new priorities adopted by the leadership. In response, mid-level managers and their subordinates begin to pay much closer attention to the ongoing management of critical risks. According to the respondent, if this approach is consistently implemented, discussion of critical risks and how to reduce them soon becomes a corporate habit across a company.

The HSE head of an electricity company explains how their leadership include middle managers in risk-related discussions. The first step for this company was to set up a leadership committee for the protection of health, safety, and the environment. The committee meets once a quarter chaired by the CEO, and includes deputies, functional directors from headquarters, and the directors of all the industrial sites. They discuss progress towards the implementation of the annual safety plan and identify any key issues across all the sites. All mid-level managers provide a short report on the general situation and any current problems at their facilities. This allows leaders to gain a clear picture of what is happening across the whole company, share experiences and discuss ways to resolve existing issues. Upon returning to their production sites, middle managers hold meetings of the site HSE working groups, which include lower managers, to discuss the goals and objectives set by the committee.

Many interviewees emphasize that the most effective way for senior management to get an objective picture of current risks—in addition to communicating with their middle managers—is to make regular visits to a company's production facilities. Just as it is important for them to get information from middle management through the traditional hierarchy, leaders also need to hold meetings with the workforce on the spot and discuss both problems and solutions. These site visits provide another vital way of obtaining objective information about the situation on the ground across a whole company's operations, in a way that does not challenge or undermine the stability of the traditional hierarchy (the issue of senior management conducting industrial site visits is discussed in more detail in Recommendation No. 7).

Delegate the authority to stop the operation of faulty equipment to middle and lower-level management

The managing director of an electricity company believes it is very important that employees at production sites have the authority to stop critical equipment that is not working as it should. To make this possible, there must be clear assessment criteria whenever equipment is showing signs of malfunctioning, and employees need to be clear who has the authority to order a shutdown. It is also important that there are no penalties for taking such action, even if it turns out after investigation that the shutdown was not justified. If a complex piece of industrial machinery appears not to be running normally, a swift and urgent assessment of the situation may well conclude that immediate shutdown is the safest course of action. However, it is also possible that the shutdown may be shown after investigation to have been unnecessary. It is important that management show they are committed to never penalizing employees acting in good faith with a "*better safe than sorry*" approach. On the contrary, whatever the result of any subsequent investigation, employees should be praised for acting promptly and courageously to prevent a possible serious escalation of the situation.

Results of responses to anonymous surveys within the framework of the pilot project: *The middle and lower managers have the power to stop the work of a workshop; and even the entire operation of a plant if critical risks are identified. Employees are given the right to refuse to perform unsafe work. In your experience, how often do managers and employees exercise these rights?*

	Very often	Often	Occasionally	Rarely	Never	Don't know	Number of respondents
All survey participants	2.8%	13.2%	24.2%	40.2%	10.4%	9.2%	326
Senior management, heads of departments and directors of sites (middle managers)	0.0%	17.1%	29.3%	39.0%	7.3%	7.3%	41
Lower managers: deputy directors of sites, chief engineers of sites, heads of workshops, heads and representatives of HSE services at sites	2.8%	18.5%	24.1%	42.6%	5.6%	6.5%	108
Engineers, foremen, and ordinary employees who operate critical infrastructure at sites	3.4%	9.0%	23.2%	39.0%	14.1%	11.3%	177

Interpretation of responses: managers and employees rarely exercise the right to stop production or refuse to perform unsafe work. Most staff tend to continue, despite the presence of concerning and potentially dangerous technological problems.

The HSE head of an oil company poses a rhetorical question—why do employees constantly make calls to headquarters about even minor issues? His answer: because they are not sure of their authority to make certain decisions without attracting negative consequences. In fact, these continuous appeals to headquarters arise because of fear in the opposite direction: senior managers are afraid to delegate authority and responsibility for risk assessment and decision-making to site production units. Consider the case of a field manager who independently makes the decision to halt work for an entire oil facility because of some operating issue. Such decisions are inevitably based on incomplete information—at the time, all that the manager can be certain of is that some of the equipment is running abnormally, and that if it fails there is a significant risk of an accident. The stoppage inevitably leads to a drop in the daily oil production rate, which affects the production and financial indicators of the entire company. After the stoppage an investigation is launched, which concludes that the equipment fault was not in fact critical. However, this conclusion is based on detailed information which has only become available to investigators after full examination of the faulty equipment.

Senior managers at headquarters, having seen the investigation report, consider that there were insufficient grounds for the stoppage. They conclude (unfairly) that the field manager was too hasty in deciding to shut down the entire facility, incurring huge costs to the company. In their view, the field manager at this level should have collected more information before making such a major decision. But from the perspective of the field manager on duty, the information available to him at that moment indicated possible serious equipment faults, and every minute's delay made a serious incident more likely. He did not have the luxury of time to gather more operational information by conducting a more detailed analysis of the equipment.

However, leadership teams worry that if other field and site managers start making similar shutdown decisions, then business will suffer. They must be seen to identify those responsible for such a sudden drop in production and hold them to account. The asymmetry of risk information here is hardly fair. The decision to punish middle managers for the "*reckless*" shutdown of the facility is taken at headquarters based on information that only became available after several weeks of investigation; while the field manager's decision was in reality a rapid judgment call made under great stress, based on the inevitably incomplete operational information available at that moment.

News of the discontent of senior management—and the penalty imposed on the hapless field manager—duly reaches the head of the field. Understandably, he decides that if a similar risk occurs in future, he will behave differently: to avoid making an independent decision to shut down the oil field, he will simply report the situation to headquarters. From his point of view, senior management seem to want these decisions to be their sole preserve, so he will obediently pass all responsibility in such critical matters to them. This is likely to have a very negative impact on accident prevention in future—not least because valuable time can be lost waiting for permission from headquarters to stop equipment.

When leaders pull decision-making back to their own level in this fashion, depriving the middle and lower levels of authority, they are taking on the entire burden of responsibility. They are effectively reducing field managers to statisticians, with no power to make rapid, independent decisions in response to changing operational situations. The official policy of this oil company states that there will be no penalties for site managers who act independently to shut down facilities—but in reality, they know this is not true. They will be enormously reluctant to take major decisions, even if a full-blown emergency is developing before their eyes.

Once they have been criticized and punished for taking independent action, field managers will make sure they stay out of trouble. They will never again risk being reprimanded and accused of alarmism or causing an avoidable negative impact on the company's performance. This creates a very dangerous situation:

- they will choose to convey only positive information to headquarters;
- they may not inform headquarters of observed risks;
- they may report a significant risk but play down its potential consequences, so the chance of early preventative action is missed;
- even when faced with potentially critical risks that demand an immediate response, they will wait for decisions from headquarters.

Conversely, if leaders want to reduce the likelihood of risk concealment at production sites, they must take the following action:

- encourage subordinates not to be afraid of disclosing observed risks;
- jointly discuss ways to reduce risk and delegate authority and resources to deal with them;
- delegate real authority to managers on the ground to stop all site operations if they consider it necessary;
- avoid punishing site managers for equipment shutdowns, even when they later prove to have been unnecessary—on the contrary they should be acknowledged and praised;
- demonstrate that such positive feedback is evidence of their trust in their site managers, and that they have confidence in their ability to do whatever is required to ensure safe production in the long-term.

The HSE director of an oil company cites an episode that occurred in his own organization three years prior to the interview. Site managers (middle management) were given the power to halt field operations to prevent emergencies. One day, a manager exercised this authority—and as a result, the company lost several million dollars in revenue. Later, it turned out that he had overestimated the catastrophic potential of the situation. Fortunately, company leaders had the wisdom not to penalize him or accuse him of reckless action. Nevertheless, there were negative notes in senior management's feedback, particularly questioning the competence of the field manager to assess the criticality of the situation. Therefore, instead of gratitude for the preventive shutdown of the facility, he was left with a sense of shame that he had made a mistake and incurred the disapproval of the leadership. This treatment remained in the field manager's memory, and from then on he became more cautious about independent decision-making. For instance, the respondent suggested that after this incident not all information about observed risks at the field was reported to headquarters. The managers of other fields were aware of how senior management had reacted to this case, and they too became more cautious about imposing preventive stoppages. This shows how even a single mixed response by senior management towards a single site manager can inadvertently create unwritten rules of conduct for middle managers. After criticism from headquarters, field leaders may start hiding or downplaying risks, and will become hesitant to take the initiative even when faced with a catastrophic escalation of events.

The leadership of the company did learn from this episode. In similar situations since then, senior management have always thanked site managers for taking the initiative to stop operations, whether or not the action later proved necessary. Nevertheless, it took several years for the entire company to overcome the consequences of this single piece of negative feedback after an unjustified stoppage. Clearly senior managers need to be very careful about how feedback is given. They should consider how their appraisals or comments might affect the willingness of employees throughout the company to report risks and make decisions on preventive risk mitigation.

The company has also ranked the possible risks that the fields might face and determined appropriate levels of on-site responsibility. Decision-making on minor and more routine issues is delegated down the hierarchy: people at the field itself have the authority to make operational decisions, without waiting to hear back from headquarters. With any technological problem that could develop into an emergency, or with certain complex work, field managers and headquarters take joint responsibility. The field provides headquarters with all the information available about the issue in question. Headquarters, in turn, provides the field with technical expertise through a dedicated scientific and technological crisis committee. This brings together highly qualified specialists in both drilling and field operations. In practical terms, it works like this. A designated technical expert is on duty around the clock at headquarters. In the event of a call from the field, they quickly assemble the specialists' team to obtain a comprehensive analysis of the situation. It is beneficial for managers at the field to consult experts at headquarters, because this leads to the sharing of responsibility and of the expertise to reach the best solutions to complex technological problems. The leadership also benefit as they have access to detailed site information, making it less likely that a serious incident develops that could damage the whole organization. Decisions are made through constructive dialogue. Responsibility is shared: it falls neither on a single site manager, nor on a senior manager at headquarters with limited knowledge of the situation in the field.

Headquarters are there to serve production sites, not vice versa

In recent years, this same oil company has adopted the principle that headquarters are there principally to support the functioning of the fields—not as an overseer and controller, but to assist the fields to operate more efficiently. To this end, headquarters provide scientific and technical assistance to the fields, and allocate resources to them to reduce the likelihood of technological risks. They also advise on how to manage field operations more efficiently. The company leaders have declared that they do not want field performance to be simply reported up the hierarchy: this responsibility is a shared function. The message disseminated throughout the company is: "*Headquarters are there to help employees in the field*". All this represents a move away from a top-down monologue model—where headquarters issue orders down a clear vertical hierarchy, and local units carry them out—to a dialogue model where headquarters and the fields discuss risks, problems, and challenges cooperatively and on an equal basis. They take shared responsibility for achieving industrial targets, tackling problems and maintaining a good safety record. This respondent was asked what he thought an ideal model for transmitting risk information from the shop floor to senior management should look like. His answer was that headquarters should be on a par with production sites as a partner, rather than hanging over them as an overseer and censor. The transition from traditional vertical subordination to horizontal partnership is not a step change—it is an evolutionary process of organizational development.

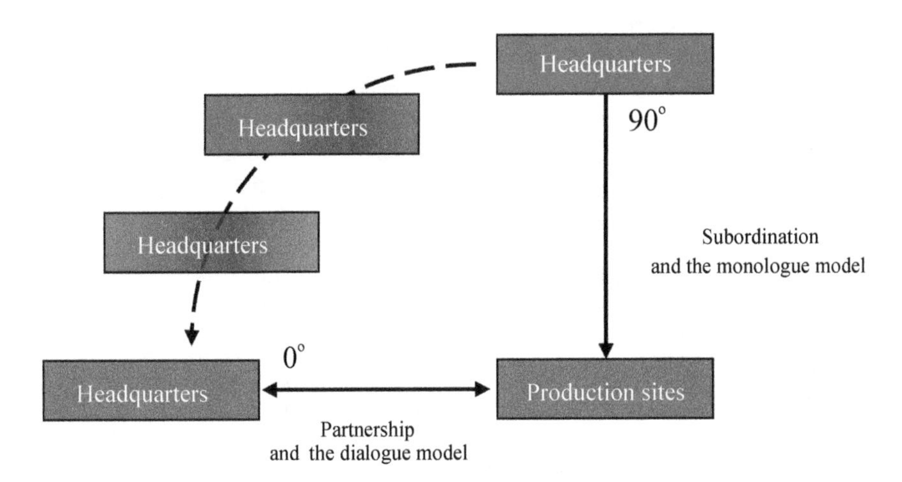

This company is gradually moving from a rigid hierarchy model (90°) to a model where relations between headquarters and fields are on an equal footing (0°). It cannot yet claim that this has been fully achieved and that all communication is horizontal. The current situation is this: (I) headquarters set goals for the field;

(II) headquarters set policies; (III) headquarters set plans for the fields for oil production, etc. However, headquarters tell the sites that if they are unable to fulfill any of the goals and directions set, then they should bring this straight back to senior management, without fear of reprimand. When this happens, both sides are committed to working together to adapt the plans so that the targets are acceptable to both parties, and to find an optimal solution that will address the problem at the site, but without leading to an increase in accident rates. In other words, the goals initially set by headquarters are adjusted through dynamic dialogue to reflect the reality of field operations. In the organizational development of an industrial company, achieving "*0 degrees*"—horizontal communication in the dialogue between headquarters and the field, or full partnership—can be seen as an ideal goal. The same applies to the aspirational "*goal 0*", or zero tolerance to injuries, in the field of industrial safety.

The head of HSE in another oil company shares a similar experience from the beginning of her career in the industry in the mid-1980s. The headquarters of this company was in an oil region, next to a very large field. Over the following decades, the CEO would become a multi-billionaire and one of the most respected oil executives in the world. Once a month, he would gather the administrative staff at headquarters and remind them: "*It is not us at headquarters who make the money —it is the people working at the rigs in harsh conditions!*". This brought it home to these staff that they were there to help production units, and not vice versa. The CEO espoused this mantra for many years and influenced the views of most of his employees. He also made it clear to employees that when managing critical infrastructure, a short-term business philosophy is suicidal. For example, he fought to change the attitude among some novice employees that they were there just as temporary workers: many came to the field on a three-year contract. The CEO encouraged them instead to treat their jobs as if they had come to work in this oil region for 20 years. Everything possible was put in place to assist employees to relocate with their families to the local area. A huge social infrastructure was constructed, offering affordable apartments and generous facilities for families, so that people would see their work for this company as a central part of their lives. Salaries were very generous compared to the national average in the country. Very fast career growth was possible, partly because new fields were constantly coming into operation and the organization was rapidly expanding. Though the climate in the region was extreme, everything was done to make employees' lives as comfortable as possible. This strategy shows a deep insight into the value of mutual support, shared interests, and common values. When employees realize that they will be working in a company for 20 years or maybe even a lifetime, they have a different attitude to how they go about their job, including how they operate equipment: with more care, a proactive disclosure of risks, and preventive repairs. This approach clearly had a positive impact on the workforce: employees looked forward to a long-term, personal relationship with the company, to forming close social and family ties and making the company and the region their home. From the senior leaders of the company, the consistent messages to each employee were: "*you are needed and welcome here for many years*"; "*everything you desire—work, family, colleagues, an apartment, social infrastructure, good salary—is yours for the taking*"; and "*take pride that you are part of a new industrial project that will benefit the whole country*". All company employees lived in small single-industry towns near the oil fields. Many site managers lived next to regular workers in the same street. Any employee could easily make an appointment with the senior managers of the company and discuss any problems or risks that were troubling them. Moreover, there was no strict hierarchy or stern subordination. To make the best possible decisions, managers recognized that they needed to hear the opinions of every employee who was working with critical infrastructure on the ground. This approach means that employees will generally do everything they can to give them comprehensive information, as they feel valued and honored that managers are turning to them for advice. News of the company's open attitude and the management's willingness to listen spreads quickly throughout the company, especially in a small, single-industry town. People naturally begin to reach out to their superiors. Eager to gain their trust, they will work hard and with integrity to prove to their leaders that they are all in this together.

Move company headquarters closer to key production sites

Headquarters and production sites should be in constant communication so that there is no barrier to discussing risks.

The HSE head of a fertilizer mining company explains that their senior managers are located close to production workers. Their headquarters are in the city where the mineral salt is mined, and from headquarters to the main mine is only a 15-min drive. By contrast, most extractive companies have headquarters in capital cities closer to governmental and financial centers, hundreds or even thousands of kilometers away from their production units. Locating headquarters near the production sites makes it easier for different levels of management to communicate, and for senior management to keep track of the situation at the mines, because all production meetings are held at headquarters. In this mining company for example, many production meetings are attended by both senior and middle managers. Mineral harvester team leaders are also invited to some meetings to discuss key production issues. A typical example: the CEO of the company is very familiar with the design of the harvesters used for mineral salt extraction, including operational strengths and weaknesses. Even though the company still operates a traditional hierarchical structure, senior managers personally know all the key lower-level managers responsible for controlling critical risks, and employees are free to request a personal appointment with the CEO. This physical proximity builds cooperative working values and assists the leadership in making faster and more balanced production decisions, even on complex issues, than they could if the headquarters were far away from production.

The managing director of a company supplying heat and electricity to a city of more than 12 million people gives another example. Over several decades, the company has developed a culture of regular reporting from subordinates to managers about operational problems within its infrastructure. This culture has probably emerged for several reasons. Firstly, the company is managing critical infrastructure, the failure of which can immediately affect the quality of life of millions of people. Everyone working there understands the enormous consequences of major accidents, and they do not need to be convinced of how important trouble-free operations are for their fellow citizens. Secondly, because the company infrastructure is all within the city, the physical distance between employees and managers is small: any employee could drive to any of the company's managers within an hour or so and discuss a problem in person. Shop floor personnel who have the experience to assess the consequences of specific risks are required to immediately report any deviations in the operation of the equipment to their supervisors. If resources are not available on the spot to solve the problem, supervisors will report to senior management to quickly get them authorized. In short, the company has created a management system that allows them to respond very rapidly to solve production problems. Employees at this company willingly report any risks they encounter in their area of responsibility, in large part because they know they can quickly gain the attention of senior management, and the problems will quickly be resolved. The respondent said that in his whole time as the managing director, every single occasion that an employee noticed a dangerous development they put the situation under surveillance and reported it. Shop floor workers always inform their superiors about significant risks, and they in turn report to senior management.

The head of risk management at a nuclear power plant admits that one of the reasons why risk communication systems were effectively implemented in his organization was that the managers responsible for implementing the proposals were physically located at the nuclear power plant during the entire period of project implementation. The respondent was on this team and attended all meetings where identified risks were discussed, and his colleagues were also members of the emergency team: "*we managers worked together with everyone; we were part of the family; we were in a group. If we said something, then the workers of the nuclear power plant believed what was said*".

Reduce the number of managerial levels

The HSE vice president of an oilfield services company believes that even in a large industrial company operating over a wide geographical area, there should be no more than five or six management levels, with no more than six deputies under each manager at any level. In his experience, this is the optimum structure. With fewer levels of control and a flatter corporate structure, it is easier to manage the company, and the transmission of information up and down the hierarchy is more efficient. He cites the example of one of the world's leading gold mining companies, where the field managers (middle management) personally know all the workers at their respective sites.

However, the respondent considers that senior managers should take great care in making decisions about the delegation of authority and responsibility down the corporate hierarchy. In projects associated with oil production—for example, the commissioning of a new field in an unexplored geological region—the transfer of broad authority to people on the spot is justified. In this case, experienced local production managers and workers understand the specifics of working "*in the field*" better than the staff at headquarters, likely located several thousand kilometers away. Nevertheless, in the operation of a sophisticated fixed facility like an oil-refining complex or a nuclear power plant, lower-level employees should be required to follow clear instructions. Any initiative to modify these instructions or operate equipment beyond design specifications should have the full approval of higher managers, informed by qualified experts and by the suppliers of the equipment.

Agents of change

The HSE manager of an oil company believes that nowadays, many companies are trying to move away from traditional hierarchical management models. Special training is available to teach managers how to communicate with employees. For example, his company has a very innovative production site where they are beginning to level out the structures of top-down communication and introduce new principles of organizational behavior. At other production sites where traditional management practices persist, meetings of the drilling teams are held every day, attended by employees and their immediate supervisors. As a rule, the style and agenda are very prescribed: pressure to implement the drilling plan, penalties for errors and target shortfalls, blaming and shaming of subordinates. The same applies throughout the chain of command. By contrast, at the innovative production site, weekly employee tea-parties are held with middle and lower management present. With this more informal environment, discussions about issues and problems faced by specific rigs are much more open and productive. Sometimes there will be a barbecue during working hours to "*break the ice*" and "*oil the wheels*" of communication between leaders at different levels, so that it becomes much easier to discuss problems. Sometimes mid-level managers visit the rigs in person, and make time to speak with the drilling teams while they work.

But back at headquarters, many of the "*old school*" managers do not welcome these innovations. Their main criticism, often unspoken, is that this approach erodes the traditional hierarchy within which they have worked through their entire careers. There is institutional inertia and a resistance to change. They believe it is fine for employees to form friendly working relationships at the same organizational level but they are not willing to establish more friendly relations with subordinates. These "*old school*" leaders are fearful of closer, more trusting ties as they would make it harder for them to shout and berate the workers, to manipulate and control them through threats and authoritarian attitudes as they have done for decades. If their employees begin to see them as friends—and therefore to some extent equals, rather than as people to fear—it will no longer be the leaders in charge, but employees controlling and manipulating them. For many long-standing managers, a new company instruction to build more trusting and friendly relations with employees can seem unfamiliar and threatening: they simply do not know where to even begin. Therefore, there is much skepticism and unofficial criticism of this approach from outside the innovative production site, including many senior leaders at headquarters. It is ironic that the "*old school*" managers' position contradicts the stated new corporate goals, which were created and actively promoted by senior management. Nevertheless, their resistance is understandable: for 15–20 years they delivered acceptable results for the company through their tough management approach and practices. Suddenly the leadership is expecting them to abandon the way they have always treated and managed their subordinates—which though repressive, has proven to be effective in achieving results. Leaders are now encouraging them to make friends with their subordinates, whom they have always tightly controlled and never considered as partners in improving the efficiency of the company. Those adhering to the "*old school*" are doubtful that this new dialogue model will allow them to achieve the set production targets—conveniently forgetting that the tough monologue communication that used to deliver successful results required them to ignore many critical risks.

With less rigid hierarchical relationships, some individual middle and lower managers can have a very good influence on the production process, so senior management do not need to supervise them so closely to achieve improvements in the field of industrial safety and risk management. If senior management can identify these more proactive employees amongst the critical production workforce, they can use them to build a better system for delivering feedback on technological risks. In this respondent's company, one of the senior managers likes to say that "*the leaders are like preachers in the religion of safe industrial production and their task is to gather disciples who will then go preach the same principles*". These positive influencers need to be recognized, promoted up the organization and given more authority, so that the reluctant majority of conservative employees see the success of their example and are encouraged to follow in their footsteps.

The HSE head of a mining and metallurgy company emphasizes that many employees are initially skeptical of many senior management initiatives, believing that as soon as there is a change in leadership these new ideas will be quickly canceled or abandoned. Many employees think: "*I have already survived many company leaders, and no doubt this one will leave soon enough after receiving a huge bonus. Why follow the ideas of this latest one, when everything will change all over again as soon as the next one arrives?*". It is vital that within the first three years of the launch of a new system there are plenty of illustrative examples of positive outcomes from the proactive disclosure of risks, alongside penalties for concealing uncomfortable truths. Only then will employees see the evidence that this is not a temporary gimmick: that rewards for disclosure and penalties for concealment are inevitable, that accurate risk reporting really is important to the leadership, and that they are expected to change their familiar behavior and start working in the new way. The full implementation of a new risk information transmission system is never a quick process. Even with a totally committed leadership team, it will take a large industrial company between three to five years to establish effective operation. It will always require constant feedback, analysis and fine-tuning—alongside a steadfast commitment to invest the necessary resources, both time and money—to create meaningful, lasting change.

Another HSE director in the mining industry believes that when a company's leadership are trying to bring in new attitudes and practices, it generates a broad spectrum of reactions. Some employees will immediately accept the innovations and change their behavior. Others, though willing, will take longer to change their work attitudes, while some may be secretly skeptical, but will make a show of going along with things to hide within the rest of the pack. Finally—and inevitably—some employees will be highly resistant, and will never accept the change either to their attitudes or their behavior. It is better for managers to identify this last group—the "*hard blockers*"—and issue them with an ultimatum to comply, which if not met will lead to their dismissal. If they hang around, they are likely to remain intransigent, spreading unrest and resistance in the workforce, which will make the implementation of change that much slower and harder. A clearly laid-out motivational scheme should be created to encourage the rest of the workforce to embrace the change and to secure their long-term future as employees in the company. From the respondent's experience when his company introduced these changes to working practice, the employees broadly divided into three camps: 10% who fully believed in the process and rapidly changed their behavior; 80% who were cautious about the innovations; and 10% who clearly opposed them. Leaders need to enlist the 10% who fully believe in the changes as active allies, and promote them so they can act as agents of innovation on the ground, and help convince the remaining 80% of the benefits of the new developments. All employees need to be given reasonable time and opportunity to adapt to the changes, and encouraged along on their journey by proper training, allocating them meaningful tasks, and supporting and monitoring them appropriately. Well managed, most of the cautious 80% will slowly come round to the new way of operating, especially when they see the benefits for their everyday work life. Most people are amenable to change once they see it working in practice and will join the ranks of the believers, and even start to persuade others to follow their example. But inevitably, as with all things, there will be a small number who remain resistant and obstructive to change. These entrenched opponents will never be convinced—they should leave the company.

The CEO of a consulting company in human performance and human factors, with extensive experience in power generation, recommends that if leaders are looking to initiate major organizational change they should concentrate initially on the shop floor workers who act as "*opinion leaders*" within their teams. In a team of say eight workers, there is usually one who takes on the role of informal group leader, who the others instinctively look up to. This may be by dint of personal characteristics—confidence, maturity, intelligence—or because of their experience—high competence, longest track record, being a union activist etc. These employees set their team's behavior: when to start and stop work, professional standards, teamwork and so on. Though designated as 'ordinary' workers, they are the de facto leaders. Involving these "*opinion leaders*" as cheerleaders for change from the outset—perhaps giving them additional training or introductions to more senior managers—will bring better results. Senior managers often view these employees unfavorably: "*That guy is a pain in the bum, and I don't want to talk to him*". But instead, managers need to be persuaded to build bridges with them and establish effective two-way communication channels. Their efforts will be more than repaid by the real boost it will give to the chances of organizational changes being swiftly accepted by the wider workforce.

The head of the HSE department at an electricity company also believes that when introducing changes, it is important to focus primarily on those that are supporters from the outset. Gradually, these "*believers*" will swing the bulk of the vacillating workers behind them, and eventually the number of adherents will reach a critical mass and overwhelm the naysayers. Management do not need to wage war against the new ideology's opponents—it is easier and more effective to give support and recognition to the supporters. In this respondent's experience, it takes at least 2–3 years to successfully introduce the initial changes in a company's values and set the ball rolling towards a revolution in corporate culture.

The HSE manager of a metallurgy company agrees that whenever senior leaders are looking to introduce major changes, it is essential to identify employees who, from the outset, have supported the new ideology and been willing to completely re-configure their behavior. Indeed, these workers can become a living embodiment of the new way of doing things, and help persuade the hesitant majority. There is a big difference between dismissing staff after an accident investigation and dismissing those who refused to take on board the new principles of a company. In the first case, one person may be a sacrificial scapegoat, while others who were guilty of the same or even worse safety violations are left alone to continue with business as usual. This culture of "*don't get caught*" is pervasive in many companies. In the second case, though most employees are willing to change their behavior, a "*resistant rump*" will continue to disregard and even undermine the new rules. These people need to go, because they have shown that they refuse to change.

The head of a power plant says that he would always be looking to remove any employee he does not trust. Lack of trust between a leader and a subordinate will never end well. A manager cannot afford the time to double-check every bit of information that subordinates provide. If an employee is discovered (often after the disastrous event) to be lying or withholding information about important safety issues, then dismissal is the best course of action.

According to the HSE vice president of an oilfield services company, no more than 15% of the workforce should leave a company each year. Rapid staff turnover can be extremely dangerous for an organization, as skill levels may fall, experience on the shop floor disappear, and management become chaotic. Fundamentally changing the culture of a large critical infrastructure company will take at least a decade, because although shop floor workers can be relatively easily replaced, the same is not true of the senior management. Most managers have been employed by a company for longer than the workers they are supervising, have grown up in a reactive corporate culture and know nothing else—reshaping their core values and beliefs cannot happen overnight! Nevertheless, if the leaders of change focus consistently on high values and long-term goals, the results can be amazing. If even 20% of the workforce of a company are, or are willing to become, proactive employees who fully share the new values and goals, given time they be the force that will win over the remaining 80% who are more doubtful.

The HSE manager of a metallurgy company insists that there are no bad people: they are just different. Employees are malleable—they can learn and adapt, though it takes time and perseverance from the leadership to change the character of a company. This is never a quick process—it will take at least three or four years of very consistent work to change the corporate culture of an organization, for workers at all levels to embrace a more open and honest dialogue about issues of risk.

The HSE director of a gold mining company maintains that if a company has a multi-level hierarchy, then major changes initiated by the senior management will take time to filter down to the very lowest level. Senior management need to be prepared for it to take several years before they see the first fruits of the ongoing changes. Leaders need to understand that while some people will quickly recognize the new rules of the game, some will be resistant, at least initially, to accepting them and will slow down the changes.

The HSE head of a mining company believes that there is no need to attract "*new blood*" from other companies—instead, you need to develop and invest in the employees you already have. The most actively supportive employees should be recruited to train their colleagues, along the lines of a production development school. In this framework, it is not specialist teachers who conduct the training, but the most motivated employees who pass on their work experience to colleagues in the enterprise.

The HSE manager of an oil company believes that it is very important to recruit students and recent graduates and put them directly into training in the advanced skills that a company needs. Otherwise—since most companies have at least some leaders who still cling to the ideas of the traditional hierarchy—there is a risk that new employees, coming straight from college, will be unduly influenced by "*old school*" managers, and encouraged to follow their now obsolete values and practice. The introduction of a more active dialogue across a company is a process that requires constant modeling and reinforcement from senior management, and in some cases, even a new generation of leaders. Factors such as age, experience, and ingrained hierarchical ways of thinking all tend to foster inertia and slow down the restructuring of communication from a directive monologue to a shared dialogue. The choice of mentors for young professional recruits is extremely important—from the very first day of employment, they need to be witnessing how dialogue communication can improve an organization. Years later, when they become senior leaders, they will pass on this early experience to their own newly recruited professionals. If a skilled mentor invests time and effort in a trainee specialist, that trainee may well eventually go on to become a leader and will in turn pass on the accumulated expertise. When young employees come to a company without much experience, a company should invest in the best possible in-house training they can provide. This investment will be repaid many times over in the future, and help cultivate a new generation of leaders with far more progressive and productive business attitudes.

Training employees in new skills and a new ideology of behavior

According to the HSE vice president of an oilfield services company, senior management need to realize that whatever ingenious solutions they come up with, nothing will work unless shop floor employees are brought onboard. Trying to introduce new systems when people do not understand them is just a recipe for anarchy. People need to be taught, given the opportunity to practice their new skills in various safe situations, and rewarded for the successful delivery of the new principles. In this way, innovations become part of their everyday working landscape: consistently followed, they are completely integrated into people's value systems.

The HSE head of a chemical company believes that it is necessary to train managers and employees to both give and receive feedback based on the analysis of critical risks, and then to make decisions based on the information received. Employees need to practice this process of information exchange and subsequent decision-making for it to become effective.

According to the HSE head of an oil company, managers need to understand that changing the entire corporate culture must start with the training of both managers and employees in a dialogue model of behavior and the acquisition of new skills. Employees need to be taught how to identify risks; how and through which channels to report them; and how to devise and deliver solutions to reduce them. If a company adopts new corporate standards, then people need to be taught how to apply these standards in practice to their real work environment. Without this training, the ideas and processes in the new strategy will remain theoretical, and will not be translated into meaningful change across all areas of a large industrial company. The leadership must create a team of internal trainers to deliver this knowledge to employees at all levels and all sites. The leading trainers from each site should regularly attend update sessions at headquarters, and discuss progress and next steps with senior management. The most outstanding trainers could be selected to do an internship at other companies to gain additional experience or undergo additional professional training. To inspire employees even more, it is recommended that board members sometimes visit production sites to train shop floor employees. This will motivate employees and managers at the site to comply with the new corporate standards, as it demonstrates that senior management are clearly taking these issues very seriously. The importance of training is that it allows employees to build up familiarity and experience with the new corporate regulations in a safe space, before taking it back to their workplace. Experienced employees can become the key "*agents of change*", nurturing the new corporate values among their colleagues. From the respondent's experience in this company, the most effective training programs consist of 30% theory and 70% practice. This is a useful guideline for those responsible for devising new employee training programs.

3.7 Recommendation No. 7: Use Different Upward Risk Transmission Channels

According to some respondents, it is not worth trying to implement every conceivable tool or new channel for reporting risks in an organization at once. It is better to launch pilot projects in various departments and look at the results. Then the most effective solutions can be rolled out to the whole company.

1. **SENIOR MANAGERS SHOULD VISIT PRODUCTION SITES AND COMMUNICATE WITH EMPLOYEES WHO WORK WITH THE CRITICAL RISKS OF AN ORGANIZATION**

Employees will only trust senior management claims that safety is their top corporate priority if their superiors demonstrate a genuine commitment to solving risk problems on the ground. One of the most effective ways for senior managers to do this is through making personal visits to the production sites. A face-to-face exchange between an employee and a manager is the most effective way of communicating information about operational risks. According to some respondents, other methods of transmission like anonymous mailboxes or telephone hotlines are far less effective. During these trips, management can obtain first-hand, detailed information on the real state of production processes. This is vital in order to identify and prioritize key risks and guide decision-making when implementing solutions. If employees see that the problems they have raised are being solved, their trust in senior managers inevitably grows, which will naturally increase their willingness to report future problems quickly and truthfully.

The HSE head of an oilfield services company cites the following example. Throughout history, successful military generals often made visits to the front line, to assess the situation, communicate with officers and soldiers, determine the best tactics, and raise morale with their presence on the battlefield. This is rare in corporate practice: senior management usually make their decisions without leaving their "*ivory towers*" or obtaining a realistic assessment of the real situation at production sites. These central decisions go down the hierarchy and put employees in a difficult and demoralizing position. On the one hand, they do not feel confident enough to complain to management or criticize their decisions. On the other hand, they know that they cannot fully comply with what they have been ordered to do without violating safety precautions, because the problems and limitations on the ground simply make it impossible to attain the imposed targets. According to the respondent, the solution is to narrow the information "*reality gap*" between the workforce at the site and senior management at headquarters. This can only be achieved if senior managers are willing to leave the comfort of their headquarters and reach down the hierarchy to communicate with low-ranking employees. Employees will have more confidence to raise critical risk information when they see that managers genuinely want to hear their feedback and will use that information in making their decisions.

Why some senior managers are reluctant to visit production sites.

A significant number of respondents note that the main obstacle to senior managers regularly visiting industrial sites is a lack of time due to their busy schedules, but there are other obstacles too.

The HSE head of a mining and metallurgy company shares his experience. On one occasion, he stipulated that a KPI for the management was that they must each visit four different industrial sites every year. This did not work out well in practice. Most of the managers kept postponing their trips until the last possible moment, and then in November and December started calling the respondent asking him to identify the plants closest to their headquarters. It was clear that they simply saw this as a box-ticking exercise: get the visits done as quickly as possible and do the absolute minimum to meet the target. This obviously defeated the intended purpose of the visits. The following year, the respondent changed the KPI: each senior manager had to visit an industrial site at least once a quarter, and this forced them to travel regularly throughout the year. However, the following year the HR director of the company was replaced. The new director was opposed to including this requirement in the list of indicators for senior management and removed it. As results, the site visits by senior managers soon stopped. Many of the managers had noticed that being more familiar with the production sites brought benefits for the entire company. Nevertheless, as soon as it was no longer mandatory, the managers found numerous excuses why they were needed at headquarters and had no time to visit the sites. Trips to production facilities do not appeal to many managers for several reasons: the journey, especially to remote sites, can be lengthy and uncomfortable; it takes them away into an unfamiliar and possibly challenging environment; it requires focus and immersion in listening to staff about problems at the site; and some senior staff are simply uncomfortable talking to shop floor employees. Moreover, they must take responsibility for any new problems that are brought to light. This may require making difficult decisions and taking immediate action. If they make a mistake, they may be left very exposed, and new resources may need to be found to reduce the risks identified. If the main priority for a company is to maximize profit by constantly pushing the equipment to its limits, then any production problems the visiting manager identifies will be seen primarily as an unwelcome distraction and a barrier to achieving corporate goals. Should an accident occur as a result of risks previously disclosed to them but not acted upon, managers will no longer be able to claim ignorance about the risks. They can no longer shift responsibility onto their subordinates for running equipment unsafely or not maintaining it properly, or claim that their supervisors neglected their duties, or failed to pass on vital information. It is now the managers who will be in the firing line!

An HSE manager of a manufacturing company shares a similar experience. On one occasion, the respondent managed to convince the CEO to visit a production site and conduct a safety audit. Fifteen minutes before the scheduled meeting, the CEO rang to inform the respondent that his car had broken down and he could not make it to the meeting. The respondent replied: "*No problem. Where are you? I will come and pick you up*". Surprisingly, after this incident, the CEO's car never "*broke down*" again—in future, he always made it to the sites for safety audits. The respondent had to be very insistent and inventive to persuade other senior managers to visit sites: "*The site's really not that far away; the local team are really friendly; I've booked a top hotel for you; the food's great...*". It was only after senior managers visited the production facility that they realized the value of such trips.

According to the HSE head of an oil company, an average industrial company does not have channels for direct communication between senior managers, line managers, and shop floor workers. Many large critical infrastructure companies have a multi-level management hierarchy, so there is an information disparity between what the leadership at headquarters understand about operational issues and the actual situation at the production sites. In most organizations, communication between senior management and shop floor employees is little more than an ostentatious formality: rallies and large assemblies, production managers touring a site on important dates, meetings with specially selected employees. A shop floor worker can hardly pull a senior manager aside during events like these to tell them about a critical risk. And even in organizations where there is a stated policy commitment for leadership to communicate with regular workers, information about problems passing up through the hierarchy is rarely useful: almost inevitably, messages are significantly distorted along their way and arrive at headquarters mostly stripped of relevance.

The HSE manager of a gold mining company maintains that this practice arises as a consequence of the monologue model of communication between managers and employees. Many managers do not want to receive feedback from subordinates, considering them just as "*worker units*" who should carry out their orders without asking questions. If bosses are only interested in whether their orders have been carried out by their subordinates, then they will only send reassuring reports back to headquarters—it is more than their job's worth to say anything else.

The head of HSE in an electricity company gives several reasons why managers do not want to speak directly with employees at production sites. Firstly, they want to avoid discomfort. They assume that, given half a chance, employees will start venting their dissatisfaction with the situation in the company, and bombard them with all kinds of complaints. Most of these will not even be about production or safety issues: wages too low, holidays too short, working conditions uncomfortable, etc. Secondly, managers will usually not have access to the resources to solve any of the problems raised by their subordinates, so responding positively is almost impossible, making them look both uncaring and ineffective—not a good look in front of the whole workforce. For all the power and status senior managers appear to have, they are likely to have all kinds of concerns about opening conversations with shop floor employees: "*What can I say to them?*", "*I can't do anything*", "*It'll only leave me feeling guilty*", "*These conversations are so awkward and disagreeable*", "*Why put myself through all this aggravation?*".

The HSE director of an oil company also makes the point that senior managers are people too: like anyone else, they can feel awkward and embarrassed talking to strangers, and anxious that a conversation with an employee will take a difficult turn and end badly.

A former chief mining regulator gives the following example of why some managers are reluctant to visit production facilities. One day, a senior manager of a large company who ran a mining division went to visit one of the mines. It turned out that this manager suffered from claustrophobia. He experienced a panic attack while underground and had to be urgently returned to the surface. Some managers are not mentally prepared for the harsh and difficult conditions underground. Add to that the challenge of facing miners who will not hesitate to express their opinion about what they dislike about their work and one can understand why senior managers may not be so eager to visit the workers at the coalface. According to the respondent, the best senior managers in the field of mining are those with enough confidence to fully engage with the workforce: listen carefully to what they have to say, empathize with and understand the conditions and challenges of their daily work, and bring the professional expertise to engage with the problems they raise.

According to a lower-level manager in an oil company, site visits by senior management are a rarity in most companies. In the unlikely event of a senior manager visiting a regional site and asking the workforce about their problems, their answer will usually be a deafening silence. For decades, the culture in most companies has been "*only bring good news to the boss*". In a company where this is standard practice, employees will just keep their heads down when senior managers are around: to coin an English phrase, "*ask us no questions and we will tell you no lies*". At the same time, middle managers will do their best to focus attention entirely on achievements and successes at the production site. Lulled into a false sense of security by these bland assurances, the boss will return to headquarters under the cozy illusion that everything at the site is going well.

The managing director of a gas distribution company says that some leaders believe that their valuable time should only be spent on general management and finance issues, not on analyzing production problems at facilities far from headquarters. Some of them genuinely seem to believe the reassurances that they receive from production sites that everything is fine, and so see no reason not to continue with "*business as usual*"—nothing bad has happened so far, so why should anything bad happen in the future?

A risk expert in the oil and gas industry has a similar opinion. Many senior managers believe that the main task of senior managers is to control cash flow, maintain share prices and provide a profitable return for shareholders. Some executives mistakenly believe that an MBA (with a focus on finance) is a kind of driver's license for running a successful organization, and hold that the first principle of business is to provide dividends to shareholders. This respondent disagrees. In his opinion, the first principle of business is to stay in business! To properly manage a company and secure the long-term future of its assets, senior managers must be willing to reach down the corporate hierarchy and "*get their hands dirty*". Only then will they gain a proper understanding of what is really going on in a company and be able to make balanced, prudent decisions affecting the long-term success of an enterprise.

An HSE consultant specializing in oil and gas, but also in air traffic control, quotes workers and site managers commenting on the technical competence of some of their senior managers: "*Well, they parachute these guys with MBAs and so on into these positions and they haven't a clue about what happens on the frontline. They've got an MBA and some kind of financial and economics degree but they don't understand the process, they don't understand the business: the sharp end of things*". Without this knowledge and experience, it is impossible for senior managers to ask the important questions and raise issues with shop floor employees and site managers—they just end up showing their ignorance. No wonder they are reluctant to visit sites and engage with production teams to gather essential front-line information.

A risk management consultant specializing in oil and gas once attended a seminar in a rich oil-producing country. A couple of guys dressed in work clothes sat next to him and they had a friendly chat: one of them was the director of a small drilling company, and the other a drill operative. The respondent learned that in their drilling company there is no communication barrier between senior management and shop floor employees: they are comfortable sitting down in a bar together to discuss pressing production problems. The respondent expressed regrets that things were different in his own country, where many oil companies have boards of directors without any engineers. Instead, they often consist entirely of professional managers without any experience in production. These top executives might have finance qualifications, but they lack engineering knowledge and have no direct contact with people on the front line. According to the respondent, CEOs should be people with lived experience—an engineering education and self-assured familiarity with production processes—if they are to adequately understand the principles of a critical infrastructure company, and be confident enough to visit facilities on a regular basis to engage directly with the production teams.

The HSE head of a chemical company echoes many of the same reasons why he believes senior managers are uncomfortable with site trips—but perhaps puts it more bluntly:

- they prefer to sit in their comfortable offices and fly to meetings on private jets rather than communicating with shop floor employees in hazardous and remote workplaces;
- they do not want to be dragged into local problems. Many managers believe that as soon as they hear about a problem, they share the responsibility for solving it. If an incident occurs because of a problem which has already been reported to a manager but not resolved, it is the manager who will be blamed, not the staff on the ground;
- they do not know how to communicate with shop floor employees. This is especially true for leaders with a purely managerial or financial background. Those who have come up through the ranks generally understand the concerns of shop floor workers much better, and are able to speak with them in a language they understand;
- they lack technical qualifications and do not understand the intricacies of the production process, so cannot accurately gather or appreciate the important technical information that shop floor employees could potentially provide them with.

The safety manager at a nuclear power plant believes that senior managers fear their employees will overwhelm them with a tidal wave of problems they cannot solve. They may lack the expertise to even identify the potentially critical issues that could lead to a serious incident. Being shown up like this in front of a subordinate can be shaming and undermine a manager's authority. Better not to ask in the first place than be faced with things you cannot deal with.

Managers can make better decisions if they have objective information about the problems of production sites

A psychologist and consultant in the field of organizational behavior believes that if senior managers do not visit production sites, they will not have all the information they need for good decision-making. Poor decisions can compound existing problems, hamper development and maintenance of safe operations, and eventually jeopardize the future of an entire organization.

The HSE head of an oil company tells the following story about a senior manager who visited a production site. He managed to gain the confidence of the workers and within 30 min, they had told him all about the main problems they were facing. The meeting was attended by the general director of the site (middle management). While his subordinates were talking, he sat in silent shame, well aware that throughout his entire time in post he had only ever sent good news up to headquarters. After the meeting, the senior manager expressed his shock about all the dire reports he had heard from the workers. "*How could things be this bad without me knowing?*". On his return to headquarters, he began to make completely different decisions and try to solve the critical problems he now knew about: after some truthful feedback, his decisions were more grounded in reality. For years, the senior management team had made decisions based on all the falsely reassuring information received from down the hierarchy. Unless managers visit production sites, they will remain ignorant of what is really happening in the field, and their decisions will be based on entirely false premises. The traditional monologue model of communication, prevalent in many large industrial companies, automatically encourages employees to pass on only good news up the hierarchy; bad news of safety and technological problems is understated, embellished, or not transmitted at all. This is not a fault of lower-ranking employees, but stems from the unwillingness of senior managers to engage properly with the production process and site problems. No wonder that when leadership do at last visit their production sites, the experience can be a shocking eye-opener. Finally, they realize the colossal reality gap that exists between the sanitized information they have been receiving for years from the field, and the reality of what they now hear directly from the workers and can see with their own eyes.

The head of HSE in a petrochemical company says that if senior managers want to understand how things really are on the ground, they must visit production sites and meet with employees. He pointed to the documentary series "*Undercover Boss*" as an example: a project involving more than hundred real companies and their leaders. For the documentary, senior managers were disguised and sent to work as shop floor employees under false names, thus taking top company officials right down to the bottom of the corporate hierarchy, so they could appreciate the situation first-hand. Afterwards, the financial director of one of the companies participating said: "*What I saw in the reports sent to our headquarters from the field was completely different from what was actually happening*". At the end of each series, the senior managers revealed to their "*colleagues*" who they really were. It is interesting to note that when shop floor employees were unexpectedly summoned to headquarters for this final "*reveal*", a significant number assumed that they were going to be dismissed. This reinforces the suspicion that many employees are (possibly subconsciously) afraid of communicating with their superiors, because of their past experiences. They take it for granted that a call from a superior always means bad news: either a "*monologue order*" that they cannot question, or a punishment or reprimand for some "*mistake*" that was probably not even their fault. Most ground level workers could hardly begin to conceive that the call to headquarters might be for the leadership to invite their suggestions about how to improve their company.

The head of a production facility at an oil company (middle management) suggests that senior managers are often under the illusion that everything is fine inside a company. In most companies, reports to superiors are often embellished as they proceed up each level of management, and managers are not in the habit of reaching out directly to employees below their immediate subordinates to obtain objective information. To do this would represent a real gamble for senior managers: they may succeed in gaining first-hand and undistorted information about the risks an organization is running, but this will probably diverge widely from the accepted assessment prevailing at headquarters. According to the respondent, managers should ensure they regularly visit industrial sites to gain a better understanding of the problems shop floor employees are facing. Otherwise, they will be forever "*looking at the world through rose-colored spectacles*", and for a leader this can be catastrophic. The truths they uncover may well be unpalatable, but will greatly diminish the risk of misjudging a risk situation, making inappropriate and possibly calamitous decisions. A leader's visits to production sites also increase the level of trust across an organization: they demonstrate that senior management are genuinely interested in what is happening at the sites, giving shop floor employees the chance to have their say and to provide feedback.

Most of the leaders interviewed believe that managers need to remember that the work that makes a profit for a company is not located at headquarters, but at the sites where the product is being produced. This is where value is created. It is then only logical that they should devote a lot of their time to production issues, including site visits, so they become familiar with the whole production process from top to bottom. Regular management site visits are a relatively simple antidote to the problem of poor-quality feedback in an organization. Better feedback means better management of production processes, which makes a company not just safer —by proactive control of critical risks on the ground—but also more profitable, by improving decision-making and production planning.

Even after visiting industrial sites, senior managers are unlikely to be 100% aware of all that is happening on the ground. But they do not need to know every detail of every operation. What is important for them is to:

- understand how a company is managing critical risks;
- identify key problems and manage their resolution;
- show their employees that they are committed to solving key production issues;
- encourage the views and opinions of employees at every level of a company;
- show employees that management can be trusted and that when a significant problem is identified, it will not be ignored.

The head of the nuclear design department of an international electricity company believes that senior managers must be convinced of the value of site trips to expose a range of potential problems in how an entire organization is operating. Timely identification will allow these issues to be corrected in time, before they develop into significant issues which will inevitably be harder to rectify and require more resources.

It is possible to motivate managers to visit production sites by demonstrating that it will give them a deeper understanding of a company's problems, and thus enable them to make better decisions and significantly reduce the occurrence of critical risks.

The CEO needs more training than anyone else in the company

Careful preparation is important prior to visiting production facilities if maximum benefit is to be derived. Senior management should undergo special training in a range of important leadership skills, including public speaking, facilitating discussion and complex debate, conflict resolution, etc. It is also important that senior managers lead by example. This includes being familiar—and complying at all times—with the requirements and policies of their own company, especially around issues of labor protection, risk reduction, and industrial safety. In addition, before visiting each production site, managers should be provided with detailed information about the specific technological processes at the site, and any negative incidents and problems that have previously occurred there. This should include reports from the production department, HSE and other key departments, so that the CEO and senior leaders have as full a picture as possible of what they are likely to find and the most important areas for them to focus on. Brief CVs (with photographs) of lower-level managers and the workers who operate critical equipment will also assist in their task and help build good relationships with key staff once they are on site. During the entirety of the visit, senior management should have immediate access to the most experienced experts (advisers) in industrial safety and production processes in order to accurately assess the criticality of problems raised by shop floor employees.

Managers must prioritize production sites where critical risks have already been identified

Employees may be aware of risks that exist in their own area of responsibility, but generally do not have the information to make a comparative assessment of the critical risks across the entire company and understand how all these risks have been ranked by headquarters. Therefore, the first port of call for senior management should be those facilities where risks have been identified that are critical to a company's survival. Senior managers should send their critical risk assessments to their subordinates who are directly managing these risks, and ask them to provide headquarters with regular feedback on the dynamic situation on the ground, so that plans can if required be modified and resources supplied without delay. If a company has many industrial facilities, then senior management need to prioritize the most problematic.

The fact that the leaders are there at all to evaluate critical safety and technological risks sends a powerful signal to employees of the importance they attach to this issue for the whole company. Employees will think: "*If these things are so important for the top brass, then it means we should think about them too*". This is especially important for employees who wish to build a career in a company. The regular appearance of senior managers motivates managers and employees alike to be more conscious about their work, maintain equipment in good condition and comply with rules and regulations.

Managers must show employees that their opinion is important to the leadership

According to many managers interviewed, it is very important that on their trips to production sites, leaders demonstrate they actively want to hear the views of shop floor employees on how to control critical risks. This requires a temporary leveling of the hierarchy: workers competent and experienced in operating key equipment on the production floor meet on equal terms with senior managers who have resources for an open discussion on how to control and reduce critical risks.

If senior executives never visit a production facility, staff are bound to think that headquarters are not interested in what is happening at their facility—they do not care about safety and employees' opinions, but only about hitting their production targets and making fat profit margins.

Site visits provide opportunities for senior management to see how clearly employees understand organizational goals and perceive the instructions of senior management

A consultant in nuclear safety, with long experience in nuclear power plant operations, believes that 50% of senior managers' working time should be spent discussing with staff how they interpret a company's values. When visiting production facilities, senior managers should not only observe employees at work, but also make plenty of time to talk with them about the values of an organization.

By visiting production sites, senior managers can see for themselves whether their instructions are reaching down to and being followed by shop floor employees, and whether the workforce understand and share the goals of the company, both in terms of production and in the field of risk management and safety.

Demonstration of approved behavior

Some executives attend production sites to present awards to their best employees. This demonstrates the kind of behavior approved and encouraged by a company and can motivate other employees to follow the example of their most productive and conscientious colleagues.

Senior management trips to industrial production sites should be regular

One interviewee explains: "*Safety is not a sprint, but a marathon. If you don't keep running, everything will slip back to the starting point, no matter how far you have travelled*". To become ingrained into corporate culture, safety issues should be frequently monitored on the ground by senior management site visits. According to several respondents, a CEO should be timetabled to visit every key site at least once a year. Not just the CEO, but deputies and key managers from headquarters should also make regular timetabled visits so that the whole leadership team gain a better understanding of the real situation on the ground. Every site knows it is then under constant attention from above, via a timetable of visits throughout the year, every year. During the first visits to the production sites, the CEO should not expect employees to immediately open up and report every problem. It is natural that employees will at first be wary of attention from senior managers. It is important that they quickly recognize there will be no penalties for expressing their opinions to their superiors, and it is invaluable to show positive examples of problems being solved as soon as possible. If senior management make their site visits regular and gain the trust of employees by acting on what they are told, then employees will gradually open up. Managers also need to be taught how to collect information on risks and problems without causing conflict. None of this can be hurried.

The management of one petrochemical company constantly visit industrial facilities and communicates with shop floor employees. Over five or six years of this focused practice, the employees have gradually come to communicate more freely with managers: they have become secure enough to share the problems they are facing and make suggestions about how things could be improved. This has helped to reduce the discrepancy between the information local employees have about critical risks, and the limited picture that used to filter through to senior management.

The head of risk management at a nuclear power plant recommends that CEOs of critical infrastructure companies create a staffed office at each production facility. All employees and site managers at a facility are then told in advance when a CEO will be in the site's office, specifically to receive anybody who wants to raise an issue with them. The respondent suggests that a CEO's visiting schedule should be set a year ahead so that employees can book their interview well in advance. Executives should make themselves available at each site office for up to fifteen days a year (depending on the size of the company). It is strongly recommended that executives adhere strictly to the schedule they have set, and do not postpone or cancel visits. According to the respondent, the regular availability of a CEO for communication directly with shop floor employees is a decisive factor in obtaining better information about the critical risks of a production site.

Senior management visits should be safe for shop floor employees

Top-level executives should not go to industrial sites to find culprits and reprimand them. They must visit with the purpose of improving cooperation and communication with shop floor employees, and with a longer-term goal of identifying and solving critical risk problems that require the resources and intervention of headquarters.

For employees to feel they can open up about problems, they must trust senior management. This will only happen if they are confident that revealing difficult information will not threaten their job or career—that neither senior management nor the production site leadership will penalize them in any way. One of the respondents maintains that senior management will only get relevant information in person, in a confidential face-to-face conversation with an employee and not in a group setting. For most shop floor employees to start giving real feedback, senior management will need to make regular visits over an extended period, during which real solutions to critical problems at the site start being implemented. The whole workforce can then see that the arrival of the senior management team is not a danger. On the contrary, it is an opportunity to raise serious issues about their site—some of which may have remained unresolved for years and may carry a significant threat to themselves and their colleagues—and witness them finally being addressed and rectified.

If the leadership cannot solve a problem brought to light during a site visit, they must explain to the workforce why that is. Employees must believe that they are being heard, that their voices matter, and that management are determined to solve the problems raised and wishes to continue receiving feedback, even if some issues cannot be resolved immediately. An honest and transparent approach when dealing with subordinates will eventually foster trust between employees and senior managers, and convince workers that they are all on the same side and share the same goals.

Senior management visits should not make production site managers feel threatened

Senior management visits should not be perceived as a threat by the middle managers in charge of production sites. If middle managers feel they are being blamed, they will immediately revert to previous behavior: cease communicating openly and censor their reports to conceal production problems so as not to be the target for criticism and censure.

Most of the respondents recommend that when leaders conduct face-to-face meetings with shop floor employees, local site managers should not be present. However, these meetings must not look to load blame onto the site managers. Finger-pointing and scapegoating must be avoided at all costs, so that shop floor employees are not worrying that their managers will take revenge on them if they disclose problems, and the site managers will not be tempted to try and control the feedback of their subordinates by instructing them on what they should and should not say.

The message must be loud and clear: the leadership are here to support the site managers in tackling site problems that come to light, not to blame them or load all the responsibility for finding solutions onto them. Instead, they should be consulted as equal partners on what they think should be done, and what additional resources will be required. The role of leadership here are to evaluate the solutions that the site managers suggest, fine-tuning if necessary and coordinating the allocation of resources to implement the plan. This approach will motivate site managers to be open about problems, as they know they will be taken seriously and provided with extra resources as necessary. This will encourage them to become more proactive in identifying critical risks as they arise, and dealing with them promptly, as they can see this is what the leaders want them to do.

If senior management return to a site to find that an agreed action plan is not in motion, then the site managers need to be pulled up immediately and told in no uncertain terms that they risk losing their jobs if progress continues to stall. If this inaction persists, then dismissal is justified. The message from leadership must be clear to everyone—the site manager has not lost their job because of a production problem being identified, but because they have failed to follow an agreed plan and utilize resources effectively after a problem had been identified.

Middle managers generally adopt the behavior of their superiors, so if top executives of a company regularly go down the hierarchy to talk directly with workers, this will soon be mirrored by mid-level managers. When leadership show their subordinates how they encourage open discussions about critical risks, middle managers will be motivated to do the same and maintain regular dialogue with both lower management and shop floor employees.

How senior management should conduct visits to production sites

Site inspections can be carried out with or without warning. With a pre-announced visit, production site managers will have time to prepare for the audit. Naturally they will do all they can to have the processes and employees working to optimum efficiency and safety, in line with company policy and regulations. They want to demonstrate the best possible version of the site that they possibly can, according to their understanding of current accepted operational standards. Senior managers should be aware of this and not be deceived into believing that they are necessarily seeing a true and complete picture of how the site normally operates.

One of the executives interviewed cites cases from their practice where site managers had given all the workers a list of questions that senior managers might ask during their upcoming visit. Sometimes recommended answers were even attached to the list—and of course, these were all "*good news*" answers about how well everything was going. Another clear indicator that middle managers are trying to "*butter up*" the visiting leadership is when every single employee at the factory seems miraculously to be wearing spotless overalls. This is obviously window dressing—in heavy industry, few workers manage a shift without their clothes getting dirty—and should raise suspicions that some other awkward truths are being concealed from the visiting top brass.

When senior managers visit production sites without warning, they get a much clearer picture of the real situation at the facility. It is still advisable to hold meetings with randomly selected site employees to make sure it is not a group specially selected and tutored by the site managers to respond to likely questions in a particular way. Site visits on the night shift can also be very informative as there are usually no middle and lower-level managers around. In the absence of their superiors, shop floor employees and team leaders on shift will be more talkative about both production problems and more general concerns. Even unannounced daytime visits by senior management are usually monitored by site managers. In order to avoid their visits becoming "*the special director's tour*", senior management should include visits at night or on public holidays.

Leaders can also hold factory meetings with the entire workforce. Admittedly, not all senior managers are good at communicating with a large audience, but many should be reasonably comfortable with public speaking after many years of internal corporate events—and there is always extra training available.

To make the information gathering as effective as possible, senior managers must meet face-to-face with the employees who are monitoring the critical risks of the enterprise on a daily basis. Suitable employees can be selected but should not include middle and lower managers, so that workers are free to voice their concerns and opinions without fear. The site managers can be told: "*Thank you, I have heard your opinion—but I need this meeting to hear the opinion of other employees*". Alternatively, the visiting leaders can simply walk around the site and engage shop floor employees in conversation along the way.

In one-on-one meetings, the workers must be reassured that whatever they say will remain anonymous and that their opinions matter. The leaders must make the purpose of their visit and conversations with employees absolutely clear:

- "*No one will be penalized for helping us solve safety and technological problems. Conversations are confidential and employees will not be in trouble for giving their honest opinion. On the contrary, we welcome true and objective 'warts and all' assessments, however negative*".
- "*We want to hear about the real situation on the ground and discuss the risks and problems that you face in your daily work. This is so we can make work safer and more efficient for everybody and make the whole organization stronger*".
- "*We are all partners here, and we have come from headquarters to help you solve safety and technological problems, but we can only do this if we all work together to first identify them, second decide what to do, and finally put those solutions into action*".
- "*You and your colleagues are the best people to help us leaders to figure out the situation on the ground. You are the people who work on the front line of production every day and you know better than anybody the realities of what is working well and what is not working well*".

Leading questions should be avoided. For example, if a leader asks a worker, "*Is everything going OK here?*" they are likely to answer with a yes. Open questions are far more productive: "*Tell me about what you're doing here*", "*What do you see as the main risks in this process?*", "*What do you suggest we do to reduce these risks?*", "*Where do you think we should invest money at this site?*", "*What are your main safety concerns in your day-to-day work?*", and so on.

After hearing from shop floor employees, visiting executives can begin to form a clear picture of how the site operations really are in respect of health and safety. Rather than asking "*head on*" about specific problems, it is better to approach from a more general point of view, and then gradually move to the thornier issues. If senior management have technical backgrounds and are familiar with the operation of complex equipment, they can get a head start by studying manuals, and incident and safety logs. They are then fully briefed to be able to ask highly specialized questions about the operation of the equipment. Employees will quickly realize that they are dealing with a professional who knows their stuff and respond accordingly.

After dialogues with shop floor employees, the visiting leadership team should gather with the site managers to present their results, again being mindful to protect their sources and without looking to blame. They should arrive at conclusions about what needs to be fixed at the site and agree what resources will be required and where they are coming from. An implementation plan needs to be drawn up with a clear timetable and designated responsibilities—who will do what, and by when.

This program needs to remain under the supervisory control of senior management, so they can check on progress through reports from site managers and further site visits.

An equal dialogue between senior management and shop floor employees

During conversations about the risks of an organization, the traditional boundaries between senior managers and shop floor employees should be set aside so that everyone feels they are in a proper two-way dialogue, communicating on an equal footing about how to solve problems in which they all have a stake. Some critical infrastructure companies prohibit leaders from visiting industrial sites in business suits, and stipulate they dress instead like shop floor workers to help break down traditional hierarchical suspicions. The managers should shake hands warmly with employees and address them like everyday work colleagues to help create an informal atmosphere that encourages confidence and trust. This requires a degree of skill and confidence from the senior managers if it is not to seem false and patronizing. But it is always worth the effort, as it softens the over-deference and unease that could otherwise inhibit employees and make an open dialogue very difficult to establish.

When senior managers genuinely communicate on equal terms, it can improve employees' job satisfaction, self-esteem, and loyalty towards the company. The HSE manager of a metallurgy company cites an example. In his company, senior management no longer sit at the head of the table during meetings but mix in amongst employees and managers from different levels within the organization, to demonstrate they are fundamentally equal, irrespective of their position in the hierarchy. When there is equality, communication can be built on mutually respectful dialogue. A hierarchy based on dominance and submission inevitably reduces communication to a one-way monologue, from bosses down to subordinates. Asking representatives from the shop floor to create a professional commission can have a very positive effect on workers' motivation, encouraging them to adopt a more responsible attitude to equipment, and take more initiative to report problems to superiors. Having established opportunities for honest dialogue—through their own representatives even if not every individual—helps shop floor workers gain belief that their voice matters, they are listened to and their contributions are valued.

An oil company executive shares his observations. In some production sites, he often meets workers who remember how the former head of the company would come from headquarters to visit the oil fields, engage in conversations with shop floor employees, and seem to genuinely want to understand what was happening on the ground. Every worker felt that the CEO was "*his*" and had their interests at heart, mainly because he was friendly and respectful towards them all. You could ask him anything, raise any production problem, with the very real expectation that it would then be sorted out—if the CEO promised something to workers, he delivered every time. This was in contrast to another oil company where the respondent had worked. Here the CEO also traveled to the production regions but never bothered to talk directly to shop floor workers. Few employees remembered his time in charge or had anything positive to say about him.

The safety manager at a nuclear power plant notes that when a manager views his employees as professionals and experts in their field, meetings and communications with them will be much more productive, because there is an atmosphere of mutual respect and confidence.

A senior HSE manager of an oil company believes that leaders should involve workers who regularly operate critical infrastructure in the process of finding solutions for problems disclosed to senior management. According to the respondent, they should avoid imposing their own solutions: often these will not be appropriate for the specific situation, and will be less likely to find support among the workers who are responsible for implementation. Employees working on the front line of production work know best what needs to be done—get them involved right from the start!

On production trips, the company's top brass must show that they have come specifically to hear the opinions of shop floor employees on critical risks at the facility and how to control them. An employee has a right to have his voice heard if he has competence and experience in controlling critical equipment. The senior management's main function is to exercise their authority to secure resources and influence policy and regulations. This non-hierarchical communication does not undermine the status of senior managers. On the contrary, it shows that they are confident enough to be open and ready to listen to their subordinates. And if employees feel that they can influence the decisions of the company's top officials, their self-esteem, loyalty, and motivation to work will all grow enormously. They are then much less likely to casually turn a blind eye to violations of safety regulations or malfunctioning critical equipment.

Visits to non-production areas

Senior managers should not just visit the production zones of a site to discuss critical risks. They should also enter non-production areas—toilets, dining rooms, showers, locker rooms, etc.—to see with their own eyes the reality of everyday life for shop floor employees. The condition of these facilities says a lot about how a company really regards its workers. If these facilities are poor or badly maintained, this suggests that a company does not care much for employees' well-being. If conditions are good, then the message is also clear: the company wants them to feel safe, valued and looked after. In return for good treatment, workers will generally work harder, value their jobs more, see the company as a valuable part of their lives, treat equipment well, and observe safety rules—so there is a lot to gain. These behaviors fit much better with most employees' natural inclinations. Most do not want to risk their own life and health in an accident at work, or lose their good jobs —whether through disciplinary dismissal for breaking regulations, or through production shutdowns and financial losses after an equipment failure.

One oil industry executive believes that top-level managers should be familiar with the reality of working on the production frontline and understand how shop floor employees are treated. The factory uniform and the safety equipment issued to them, the machinery they operate, the everyday risks and challenges they face when doing their job: this is all valuable additional information when leaders are looking to improve safety and production, or need to deliver a fair response to employees' complaints and requests.

The HSE manager of a metallurgy company believes that senior managers need to identify areas of both comfort and discomfort for workers in critical production. More personal questions of their employees are advisable too—their daily work routine, their commute into work, the food at the canteen, wages, and so on—to show that these issues are also important to their bosses. Any significant or safety-related issues that arise should be dealt with as soon as possible, so that employees see that there is a genuine commitment to solving problems—improving working conditions is an obvious place to start. Seeing long overdue improvements to daily working conditions will encourage employees to start sharing critical operational issues.

One of the respondents, the HSE head of a metallurgy company, had some negative experiences at one enterprise. At this site, there were problems with the quality of the overalls, the food and the prompt payment of wages. Trade unions and workers were constantly raising these issues when the CEO visited their production site. Later it became clear that because these issues had not been dealt with, the workers grew so resentful and fixed on their uncaring treatment that they stopped considering wider operational issues. It was only after these basic (and entirely reasonable) demands were finally met that a constructive dialogue could begin on production problems and how to improve safety.

The head of HSE in a mining company tells the following story. Some of the company leaders arrived at the production site and decided to have lunch. The workers' canteen was not a pretty sight, but they still went ahead and joined the employees for lunch. Mortified, the canteen director ran up to them and begged them to come through to the separate VIP zone where the site managers dined. The executives politely refused, queued up and sat down with the regular workers eating the same food. This gave them a perfect opportunity to experience the workers' everyday reality and chat to them directly about the food and the canteen environment, etc. After this incident, the leaders spoke to the site manager to express, in no uncertain terms, their dissatisfaction with the state of the facilities and the quality of food. Within a few days, the menu had radically improved, and improvements were made to the canteen. Of course, the reputation of senior management among the workers shot up: they had shown in a simple but profound way that they were happy to share a meal and talk with their employees to understand and improve their working lives. The site managers learned from their bosses' example and started regularly dining in the staff canteen, where they would chat informally with their workers.

Other possible formats for senior management trips to production sites

Senior managers can conduct safety audits during production site visits, which might include a tour of all areas to assess safe working practices, discussion with workers on existing safety problems, and drawing up an agreed program of safety improvements. The key task for the senior manager leading the audit is to establish clear standards for what is expected from employees, and what is unacceptable. Safety audits should be cascaded down the production site hierarchy so that all managers and team leaders visit the areas of the site for which they hold any responsibility.

Corporate meetings could also be organized at different production sites, rather than always convening at headquarters. This demonstrates to all employees that senior management are paying attention to every facility, however remote. It also provides an ideal opportunity for managers at the production site hosting the meeting to conduct a safety audit. This will allow visiting colleagues to appreciate the positive and negative experiences of another site management team. If one production site, for example, has exceptional statistics on accidents and injuries, then leaders from headquarters and other site managers are encouraged to go along to the meeting at that facility to learn from their experience. The psychology behind this recommendation is shrewd: if managers from other facilities are told that they must attend in order to learn from the best, they will be eager to find flaws, and compare their own practice with this exemplar. If the safety statistics turn out to be flawed or "*massaged*", then the visiting managers will draw their own unfavorable conclusions about the leaders of that site. With the entire top and middle management teams there to witness this shameful exposure of exaggerated safety claims, every other site manager will in future think very hard before trying to gloss over their own safety performance.

The HSE head of an oilfield services company believes that visiting key production sites should be included in the KPIs of all senior managers, including company CEOs. When traveling to the regions, top executives should take with them not just the vice presidents for production and industrial safety, but also the chief financial officer and the head of HR. This sends a clear message to people on the ground that the whole senior management team have a keen interest in the problems of industrial sites from all operational aspects.

The head of an oil production facility (middle management) believes that to build trust between managers and subordinates there needs to be frequent systematic communication. This should include regular meetings to discuss short, medium, and long-term tasks for the team and to monitor progress towards achieving set goals. In his company, this is a widespread practice at all levels of the hierarchy. The leader also holds personal meetings with each of his deputies, and they in turn hold similar face-to-face meetings with their subordinates. It is vital that the meetings are based on honest communication and undistorted feedback. For example, the respondent will begin a meeting with his deputies by telling them that he is imperfect as a manager and is aware of many shortcomings and asking them to give him honest feedback. This "*breaks the ice*"—if an employee has just been invited to give his opinions on the manager's faults, he/she will more easily accept the manager's criticism in return. This helps build trusting relationships between manager and subordinates, and allows them to share their concerns about problems without avoiding unpleasant truths. This of course must all be in complete confidence and not go beyond the four walls of the meeting room. If an employee later hears from someone that the manager has revealed the details of their conversation, all trust will be lost—probably for good. Just as in any relationship, confidentiality is crucial for building and maintaining trust between managers and their subordinates.

Decisions arising from site visits

The worst possible outcome of a site visit when employees have opened up, is for there to be no subsequent action from the leadership. Workers need to see production, safety and/or well-being improve in response to the issues they raised. If resources for improvements are for some reason not immediately available, then senior management should be very wary before asking employees to be open about existing problems. A lack of remedial action will only lead to disappointment, and workers will be more suspicious about voicing their concerns next time they are asked.

If a CEO or senior manager from headquarters makes a trip to an industrial site, it is essential that they do not leave the site without committing to tackle one or more of the problems that are highlighted during the visit. On return to headquarters, a designated member of the senior leadership team should take responsibility for handling the issue and go back to the enterprise as soon as they have reached a workable solution. This demonstrates that raising problems with senior managers actually works: that their proposals and initiatives will get to the top of the organization, where the authority and resources exist to deliver effective company-wide solutions. The next time the top brass visit their site, workers will be eager to share other possibly more serious problems, in the expectation that these too will be solved.

With several successful site visits under their belts, headquarters can probably afford to reduce the frequency of future appearances, as the site employees will have grown accustomed to their active feedback producing positive results. Any future problems or risks can then be successfully reported through the traditional hierarchy channels without the need for repeated face-to-face meetings.

Summing up, senior management visits should show all employees at the key industrial facilities that it is safe to give critical feedback, without bringing negative impacts down on their own heads. Senior management must gain trust by explaining to their workforces in face-to-face meetings how they will handle critical risk information, how decisions will be made, and what resources will be made available to solve the highlighted problems—and then they must deliver on their promises. Workers will be encouraged by the results of their colleagues' previous risk disclosures, and quickly learn that honesty about site problems works better than concealment and lies.

Executives can use direct communication with shop floor employees to send a signal to middle managers

With the option of visiting industrial sites, senior managers have two distinct channels for obtaining information from the production front line: (I) site tours and personal meetings to hear directly from shop floor workers; (II) the traditional chain of command up through an organizational hierarchy, where they hear reports from the middle management level. If middle managers know that a company leadership will now be gathering information about serious problems directly from the workforce and independently of them, then they too will be motivated to approach low-level employees about problems and solve them using their own initiative, without waiting for the arrival of high-ranking officials. A proactive senior management will motivate middle level leaders lower down to engage with problems on the ground, instead of just sitting in their comfortable offices. However, the arrival of senior management at regional enterprises should not destroy the existing hierarchy of decision-making, so that middle managers feel they are being ignored and bypassed. A dialogue between senior management and shop floor employees is only needed to collect information first-hand, so that senior managers gain a better understanding of the critical risks they are ultimately responsible for. Company leaders should never charge in like super-heroes and try to solve every problem for every employee on the site. Their focus must be critical problems across the organization—other matters should be left to the employees' immediate supervisors to sort out.

Regular tours of an industrial site by mid-level managers

No CEO or senior manager can physically meet face-to-face with all employees of a company. But they can set an example to all managers of how to engage with teams who are managing critical risks, listen to shop floor employees and encourage them to share their opinions, and tackle the problems that are identified rather than "*shooting the messenger*". The principle they want to embed throughout the organization is that "*problems should be solved at the point of their origin by working together*". This can only work if internal communications between employees and their direct managers is working well. By senior management leading the way, this approach can be cascaded down to all departments that manage critical risks, as subordinates will always try to imitate their leaders and follow their example.

One of the interviewees gives an example involving the director general of a metallurgical plant (middle management). This enterprise employs 16,000 people. Every day the director sets aside time to make a tour of one of the production areas, talking to individual workshop managers, highlighting their achievements and areas where they need to improve. After several weeks of these daily rounds, he walks around the entire plant, visiting all areas and rating every workshop, with the highest score winning an award. Any employee can use the company intranet to access the assessments received by their own workshop. This rating system encourages all the lower-level site managers to improve their position. To move up, they will need to tackle any problems within their area of responsibility so they score higher on the boss's next visit. This reward system also encourages line managers to visit other workshops and sites to learn lessons from those coming out on top.

The head of HSE in an electricity company cites a similar case at his company. Every month, the manager of each power plant is obliged to conduct a complete site tour, talking with as many employees as possible. A couple of new employees accompany them on this tour, so the manager can show them the key risks and the most dangerous areas of the plant and share the company policies on occupational safety and labor protection, including what the company expects from them. Advice from the production site leader is usually remembered by these employees years later, especially coming as it does at the beginning of their career. A personal chat about safety from the head of the facility can really inspire them to do their best, and this early focus on safety tells newcomers that it is one of the company's core values. Every third Wednesday of the month, each power plant holds a "*Safety Day*". In the morning after an introductory meeting, middle and lower managers disperse throughout the enterprise and spend the day identifying shortcomings and achievements in the field of HSE. In the evening, they gather to "*debrief*" and agree on a corrective action plan based on the urgency of the issues they have identified.

The head of a power plant gives a similar example. At his plant, there are regular meetings of the whole workforce to enable employees to bring any questions to the plant's management. 25% of the questions are collected from employees in advance so that the management team have time to prepare detailed answers: operational questions about highly sophisticated technology often require careful analysis before a satisfactory response can be formulated. The other 75% of the questions come live, directly from the audience of employees. If the session is led by a mid-level manager, this creates a great opportunity for employees to raise all kinds of issues.

2. FAULT LOGS/RISK REGISTER/RISK DATABASE

Companies should create a database of risk reports from their employees, so that information can be accumulated, analyzed, and acted on. The data must be carefully processed and stored so that it is not lost, and the employee who provided the report must be given direct feedback. A statistical analysis should be made of all incidents, to identify which operational issues are most frequently implicated in causing incidents.

The vice president of an electric utility company supports the view that there should be parallel channels for communicating risk information. One example of an additional channel over and above the traditional hierarchical chain of command is a fault log. He recognized the value of such a system when he was a middle manager running a power plant at which an electronic fault logging system was introduced.

All entries in the fault log were available for viewing across the whole organization. Once a risk was recorded, it could not then be deleted but only archived when the risk was eliminated. Recording faults was a mandatory duty for equipment operators and a requirement of their job description, and management demanded strict adherence to the system across all areas of the plant. Once an employee had recorded a problem in the log, they automatically shared responsibility for solving it—with the support of their superiors, who were notified immediately whenever a new entry was made. Sharing responsibility with their superiors encouraged employees to become more proactive in identifying and tackling risks.

The data collected in the log was analyzed by the management of the plant and the risks graded. From this, a detailed action plan was devised and made available for review by all employees of the company. First, they prioritized funding to address critical key equipment problems. Interestingly, once details of the organization's production problems were made easily available, many lower-level managers and team leaders found they had sufficient capacity to eliminate many of the risks identified. They proposed these solutions to their superiors, requiring only permission to reallocate existing resources.

It is essential that once risks are logged they are not ignored, so that employees can see that the system is not an empty exercise, but is being actively used by the management to manage problems. This then encourages employees to continue to identify problems, and in time to become the de facto internal auditors of an organization. Based on the data in the system, site managers create a targeted program to eliminate the most pressing problems. With these decisions made at an early stage, they have an empirical basis for requesting the additional funds they need from their superiors and justifying actions and costs to regulators.

Some technical issues raised were difficult for middle managers to fully understand, so they had to request the assistance of the shop floor employees who had first logged the problem. Only in this way could they make a full assessment and devise appropriate solutions. This encouraged open dialogue on equal terms between managers and proactive employees—a productive partnership, with workers bringing their practical experience and a detailed understanding of the problem, and managers contributing their experience in the industry, access to resources and the authority to impose solutions.

The respondent recalls several occasions when the fault log flagged up problems that could have led to serious accidents. The system was instrumental in enabling prompt identification, cooperative decision-making, and effective action, so that a catastrophic development was averted.

The HSE director of an oil company believes that to minimize human error, a critical infrastructure company's register of identified risks should be digitized as a database and maintained with appropriate software. Every month, a meeting of risk managers at every site should be convened to discuss progress towards addressing the problems. The respondent also suggests that any issues that have not been fully mitigated at production site level should be escalated up to the next level of the managerial hierarchy, so that additional expertise and resources can be made available. In turn, any issue that cannot be tackled effectively at that level should again be escalated to managers based at headquarters, so that senior management are aware of these serious issues and can rapidly employ the necessary resources to bring the situation under full control.

The HSE manager of an oil company believes that only the most critical risks entered in the database should be transmitted up to senior management, so that they can focus their attention on the most significant problems across an organization. All risks entered in the database should be accompanied by full technical details, but a summary of the most serious risks helps busy senior managers to assess their criticality more quickly.

The safety manager at a nuclear power plant maintains that data about serious risks referred to senior management should not be too detailed or time-consuming for them to analyze. It needs to be concise, clear, and detailed enough to establish the current state of play at the production facility. Ease of perception will increase the willingness of senior management to engage in finding solutions, and thus the quality and speed of their decision-making in tackling a critical risk.

The HSE director of a gold mining company gives an example of the escalation of critical risk information through an organization's hierarchy. The respondent believes that at each production site a special committee—including six to ten of the most qualified and experienced employees at the site—should analyze the information in the risk database every week. The committee's main job is to identify the ten most significant risks at the site and record the measures being taken to manage them. Once a month, the site director brings the heads of workshops together and the same process is repeated to create a top-ten risk list for the whole site, responsibility for which then transfers to the site director. Once a quarter, another more senior risk committee convenes, with directors from every site of the entire company agreeing on the top ten risks across the whole company. These ten risks are then brought to the attention of senior management, who now assume the responsibility for them. This filtering and prioritizing system, operating from the bottom up, works to bring production risks to the most appropriate level of the hierarchy. Simpler problems requiring fewer resources are dealt with at the bottom, and the most serious risks requiring major resources are referred to the top. Safety thus becomes a shared responsibility throughout the organization.

A safety consultant with managerial experience in oil and gas, chemicals and mining believes that every time a company encounters a new risk, it should be classified in the risk register. It will be assessed according to its prevalence and frequency, the severity of its potential consequences, the probability of it getting worse, and finally the steps and costs required for its mitigation. This ensures that all the key information is there to help managers decide what to prioritize. However, the respondent noted that a company's lawyers may advise against recording estimates of the likelihood of serious events occurring and the potential cost of their mitigation, considering it legally prudent to avoid calculating the costs and benefits of reducing certain risks.[5]

[5] The authors of the handbook recommend that readers familiarize themselves with the cost–benefit analysis in the case involving the weakness of the Ford Pinto fuel system to rear-end collisions in the 1970s [Dmitry Chernov, Didier Sornette, Critical Risks of Different Economic Sectors. Based on the Analysis of more than 500 Incidents, Accidents and Disasters, Springer, 2020, pp. 61–62, https://link.springer.com/book/10.1007/978-3-030-25034-8].

The HSE head of a chemical company shares his company's experience. Their chemical plants all maintain operational and business risk registers. The former collect information about technological problems that may have a negative impact on production. The latter contain information about problems that could harm the business: financial pressures, public relations disasters, environmental accidents and so on. Different groups of leaders regularly review each register and take prompt corrective measures to reduce risks and manage existing problems across the whole organization.

3. **STOP CARDS**

Some companies have introduced STOP [Safety Training Observation Program] cards, which document hazardous conditions or unsafe actions that employees have become aware of while performing their duties. Field staff are trained and encouraged to complete these cards. They can choose either to write their name on the card or remain anonymous. The cards are not intended to identify a specific employee as a violator, nor indeed as a whistleblower: their purpose is to identify hazards and dangerous actions in the workplace, in order to make systemic changes and prevent recurrence. STOP cards may be limited to those facilities where critical risks are most likely to occur. Every STOP card is entered into a single company database so that decisions and solutions can be tracked, and no details are lost. Details from each STOP card are collated, and summaries sent regularly from production sites to supervising managers at headquarters. Included is a ranking of the problems and suggestions on possible solutions. Senior management routinely review these reports, and make decisions to eliminate the problems or reduce the risk of their escalation.

In the first year of the system's operation, employees can be rewarded simply for the *quantity* of risk information they provide, i.e. for the number of STOP cards filled in. The following year, rewards will be offered only according to the *quality* of information provided and the significance of the risks identified. Automation of the data allows swift analysis and easy transmission throughout an organization. The respondents agreed that a reward for each identified risk is unnecessary—better to reward employees for the best card of the month or year through a competition. In the early days, rewards for the best card can be material—for example, a bonus or a medal; but they become less material in subsequent years—for example, gratitude and public praise from senior management.

In one large oil company that runs a STOP card system, there is an issue with feedback, as it is difficult to acknowledge all the named employees who have filled out a card. 95% of the personnel at the company's fields are contract workers, and they tend to have a high staff turnover. Often, by the time the company has taken measures to address the problems received via a STOP card, the employee who filled it in has left the company. However, as far as possible, it is good practice to thank employees who submit useful STOP cards and keep them up-to-date with details of what has been done about the issues they raised.

4. ELECTRONIC FORMS AND SMARTPHONE APPS

The head of the well construction and repair department of an oil company gives an example from his personal practice. He used to work for an oilfield services company, where a senior manager was implementing a proactive approach to risk management. If an employee saw a problem, they could fill out a special electronic form on their personal computers—with or without their personal details. The system would automatically forward the messages received to different levels of the corporate hierarchy. Employees were aware of this, and were confident that if they reported a risk on this form, they would see a solution in time. To motivate employees to take an active part, a competition was held regularly for the best messages. Employees who highlighted the most pressing problems were rewarded with a small cash prize. After a few rounds, hundreds of employees were reporting problems. Each message generated instructions to the appropriate division of the company to resolve the problems outlined. On the corporate intranet, each department could view its own list of tasks, and the senior managers continuously monitored their progress. Every data entry stayed active on the system until a successful solution was in place, and the employee who had first sent the message confirmed that the problem had indeed been rectified. There were times when employees logging a problem suffered as a result of their initiative, when the work to solve the problem fell to the line manager of the worker initiating the report. The line manager might then retaliate by refusing to assist that employee in the future. However, these occasional cases did not detract from the overall success of the system in resolving problems raised by production site employees.

To ensure success in introducing this kind of system in a critical infrastructure company, senior managers should:

- encourage employees to take the initiative in making reports and ensure that they are not penalized for raising problems;
- take part in discussions about the most difficult problems logged through the system;
- deal promptly with any crisis situations that arise when solving complex problems;
- personally thank the most active employees who have identified serious issues;
- remain actively involved in solving the problems being reported, especially in the early stages. This shows employees that the company is invested in making the system a success, and they will begin to fill out forms on a regular basis without fear.

The HSE head of a mining and metallurgy company believes that for employees who operate critical infrastructure, there should be clear and concise instructions on how to identify potential operational risks. It is important that the reporting of serious risks is so quick and simple that it can be done on the spot. If the employee has to fill out a lengthy form to report risks, they will simply not bother—and the project, however well intentioned, is doomed to fail. The report form must be automatically delivered to line managers, middle managers, senior managers, and employees of the independent production control systems, who report to the board of directors.

Completing electronic forms, fault logs, and STOP cards on computers is of course already outdated. In one oil company, they are now creating smartphone apps so that any employee who sees a safety violation at work can photograph or video the transgression and send it with a comment directly to the HSE department. The message automatically provides geographical coordinates, so the location of the incident is logged. As soon as the message is sent, it is automatically emailed to the managers of the relevant production facility, senior executives and the HSE department responsible for that site. The site managers are then required to report to headquarters on the measures they are taking to address the violation and prevent recurrences.

The chief risk officer of a national power grid shares his experience of introducing a company-wide smartphone app that allows shop floor employees and contractors to quickly alert management about safety issues they have observed in the course of their work. The app was made available to contractors too because the company outsources much of its work for the maintenance of substations, transformers, and pylons, and the installation of new electric grid assets. The aim of the project was to identify risks and problems as early as possible and resolve them before they could become major problems. To launch the project, an anonymous survey was given to 450 employees and contractors. This survey was itself set up through an app, as it would otherwise have been too time-consuming to organize. The results indicated that senior management, much to their surprise, were simply not aware of many of the problems on the ground that were flagged up in the survey by lower-level managers and workers. This was additional confirmation that the project could fulfill a valuable role for the company. Risk seminars were held for each production unit, senior management team and the board of directors. Participants were asked to review the risks identified in the survey and add any others they knew about. Each risk was then assessed and ranked according to urgency and severity, and a prioritized action plan drawn up to tackle them, including the resources required.

Data entered in the app goes to a special team of a dozen company managers, so that no issue is left unnoticed, unanalyzed, or unresolved. This team is in direct contact with senior management, and can promptly inform them of any serious issues and get rapid decisions on what action should be taken. Many workers gave very positive feedback to the project team: "*Look, it's fantastic, I've had this issue and was not able to resolve it, and now I've reported it to you and within a couple of weeks it has been resolved*". Seeing a quick resolution of problems they had identified really motivated employees to buy into the new scheme and use it as often as necessary. The project team convinced senior management to hold a bimonthly general meeting, where the CEO gave employees some success stories: instances when employee feedback through the system had worked to solve a serious problem. According to this respondent, employees are generally distrustful of perfect-looking corporate codes and rules, and only trust in what they see with their own eyes. If they witness an employee being penalized for revealing a problem, they know that whatever is written in the codes and rules is simply not true in practice. On the other hand, if they see employees receiving praise from management for flagging up problems, then this motivates them to embrace a new risk alert system. In this example, senior management promised all employees that there would be no reprimands even if the problems were the result of worker errors (with the obvious exception of deliberate violation of regulations). The CEO publicly praised employees who reported serious problems. He added: "*I'm not angry or worried if you report something that has gone wrong. Nobody will be punished if, for example, they report an accident in a substation which has been caused by somebody's stupid behavior. But I will be very harsh and punish everybody who is not communicating these sorts of issues, because these people are depriving the organization of the chance of learning from the mistake. And I will not tolerate people depriving us from learning from... mistakes or from near misses that we have*".

5. **INDEPENDENT PRODUCTION MONITORING SYSTEMS**

According to many of the interviewees, production monitoring systems should be independent of production management. They should operate in parallel to the production system, with feedback directly to senior management or the board of directors, so they can monitor the actions of the entire company.

Production monitoring personnel need to make regular unannounced site visits to conduct independent audits. They do not need to inspect everything—which would be an impossible task—but instead work on the basis that a detailed analysis of 10% of operations can inform accurate and valuable conclusions about the state of an entire organization. Similarly, auditors cannot possibly talk to every single employee—but they can learn a great deal by interviewing a random sample of workers operating critical equipment.

With effective production monitoring, lower and middle management will quickly realize that it is no good trying to hide the situation in the field, because any problems will inevitably come to the attention of senior management. This will encourage line and middle managers to proactively report problems at production sites to headquarters, and take the initiative to propose solutions.

Experience suggests that a production monitoring service needs to recruit professionally qualified analysts who can assess all incoming risk data without bias. The quality of their performance—from diagnosing problems to the effectiveness of their recommended remedial actions—needs to be constantly reviewed.

The experience of one industrial company offers a good example. The production monitoring service in operation here is subordinate to the board of directors, but is audited by a professional external assessor, also reporting to the board. Project documentation and cost estimates are also independently reviewed.

Some respondents also suggest giving some control of safety decision-making to shop floor employees, who clearly have a strong personal interest in making the environment where they work as safe as possible.

6. **PROCESS IMPROVEMENT PROPOSALS/RATIONALE PROPOSALS**

One of the interviewees comments on a highly experienced leader in the coal industry. He held a strong conviction that the best way to get feedback from employees on problems within an organization is through what are known as process improvement proposals or rationale proposals (the terminology varies between different regions). Essentially, these are ways for employees to share their ideas—changes in design, technology, equipment or working practice—that could improve operations. Many companies invite employees to regularly submit their proposals, often by making a "*best idea*" competition out of it. This can provide management with a great deal of valuable information about the existing problems within a company. Employees tend not to propose ideas that involve a minor tweak or concerning things that are essentially working satisfactorily—they are suggesting significant improvements and overhauls, that could help to highlight major problems and safety issues. Some of these may previously have been unidentified. It is also a safe way for employees to share their concerns and ideas, as they have been explicitly invited to do so by the management. In essence, a process improvement proposal system encourages employees to both identify a problem and propose a solution—the management benefit from both. The system provides a positive and safe "*wrapper*" for all employees to flag up potentially serious flaws in an organization. Unsolicited bad news requires urgent action from leadership—whereas process improvement proposals can be solicited and analyzed, but do not demand immediate implementation or remedial action.

The HSE manager of a mining company believes that operational development goals, including those for safety issues, are better anchored in the workforce by including shop floor employees and low-level managers in both setting and achieving them. In most industrial companies, managing process improvement proposals is a rather bureaucratic business. To make a proposal, a heap of papers and justifications need to be submitted, which can be time-consuming for employees. This respondent recommends submitting these proposals in A3 format, breaking the page into six equal parts: (I) the name of the project, (II) the essence of the project, (III) the problem that needs to be solved, (IV) the desired outcome, (V) actions to be taken, and (VI) the project budget. Workers devise improvement projects together with lower-level managers and present them personally to the site manager. The simplicity of the submission form allows the manager to quickly evaluate a suggestion and make an operational decision on its implementation. Analyzing a range of these project summaries, each covering specific issues and improvements, gives the manager a good overview of the risks at the site.

Employees are not stupid—if the process improvement proposal system produces good results, then they will continue to use it. Workers become familiar with it as an effective channel for conveying their ideas and concerns up the hierarchy, and take pride in their contribution to the wider organization. This is especially true if they see their suggestions taken up and resources provided to implement their ideas, something that would be impossible without support from higher management. Such experiences provide enormous motivation for employees to become increasingly invested in improving an organization's operations, by identifying risks and suggesting solutions. Employees see that they are needed, that their opinion is valued, and that they can influence decision-making. This increases job satisfaction and improves performance, motivation, and company loyalty. If some improvement proposals are implemented at one site, then employees working at other sites will want to do even better. Internal competition can boost employee participation and bring even more useful risk information to the management's attention.

However, it is probably misguided to make it compulsory for employees to submit improvement proposals. The head of a power plant gives the following example. He was on a business trip to a large production site, where a mandatory system for submitting proposals had been introduced. The site manager said that in the first few years after the system was introduced, employees contributed some really interesting and relevant suggestions. But gradually, the submissions became less helpful, with a lot more information noise and very few workable proposals. It appeared that employees were just plucking any old idea out of the air just to meet their compulsory quota. It reached a point where the company had amassed 3000 different proposals from employees, which would have required ten years of intensive work and huge resources to implement—it was clearly impossible to do this, even for a small proportion of these suggestions. This is bound to disappoint employees and make them less likely to contribute in the future. If managers are not to be swamped with ideas that they do not have the resources to implement, it may be prudent to limit process improvement proposals to ideas involving critical equipment, operational safety, and overall organizational efficiency.

7. **APPLYING A COMPANY'S EXPERIENCE OF SUCCESSFUL SAFETY PRACTICES**

Sometimes, it can be difficult to start a dialogue based around the problems of an organization. Many employees and managers are not immediately willing to start discussing their past mistakes. It is easier to begin introducing dialogue by talking about the points of strength of the company concerning safety and risk management, to help disseminate the best practices throughout an organization.

The HSE head of an oilfield services company shares the following experience. When his company reached a plateau in safety performance, senior management tried to figure out how to continue to drive up standards. To this end, they opened a dialogue with employees and lower-level managers and soon realized that it is vital not only to study accidents and things that have gone wrong, but also to identify areas where the company has got safety right. Across all production sites, they began to count the days that passed without an incident. They also collected feedback from employees about what they were doing on a day-to-day basis to minimize the risk of accidents, and examined the behavior of production teams that had the best safety statistics. These best practices were then implemented across the entire company. An effective bottom-up communication channel was established, based on the willing participation of all shop floor employees and a management who actively welcomed information about the steps employees were already taking to improve safety. Neither employees nor managers were threatened by this information as it did not focus on mistakes but on successes—everybody likes sharing good news! Senior management supported this initiative by introducing additional information technology, so that all employees could easily communicate and discuss these successful experiences across all departments. This shift in emphasis from negative to positive was so powerful that it changed the entire corporate communication culture. Two-way dialogue up and down the hierarchy has now become the corporate norm. Such profound changes were achieved by senior managers showing their commitment to constantly communicating with shop floor employees on an equal footing. Once workforce is engaged in discussing the good news, they can gradually be encouraged to extend the dialogue to include safety problems and negative issues, and identifying the root causes of safety incidents and poor work practices.

8. **PROBLEM SOLVING BOARDS**

The HSE manager of a mining company recommends setting up a problem-solving notice board at every workshop across all company sites, with the aim of gaining a better understanding of the concerns of shop floor workers. Anyone who sees a problem can enter it up on the board using the following table headings: (I) date, (II) essence of the problem, (III) author of the entry, (IV) action to be taken, (V) who is responsible for the solution, (VI) deadlines. Line managers study the board at the end of every day and coordinate solutions for all the issues based on who is assigned responsibility. This system has obvious similarities with a fault log, where managers take responsibility for fixing any fault that they have been made aware of. The boards are photographed daily by the site HSE representatives, who send details of any new entries up to headquarters. They also periodically collate all the data to generate a report, especially relevant prior to senior management visiting that site. Leaders can easily identify the issues of most concern to shop floor employees and raise these with the entire workforce, including senior and middle managers.

9. ***"RISK HUNTING"***

The head of a thermal power plant believes that "*risk hunting*"—where small teams of employees and managers walk around a facility in search of risks—has proven to be a very effective tool in his experience. It allows all identified risks to be gathered in one operation and presented to the site director who, in turn, analyzes the risks and designates who is responsible for tackling them. Then once a month, the facility holds a meeting of all managers, during which progress on mitigating all the problems is assessed. This ensures that the focus is on effective problem solving and the provision of suitable resources to wrap up issues promptly. Visual documentation—such as "*before/during/after*" photo reports—is a valuable tool to enable progress to be closely monitored.

One large metallurgical plant launched a similar project. 12,000 different risks and problems were identified. However, the project had to be eventually closed. The problem was that when the risks were "*hunted*", they were not assessed and ranked according to their criticality and urgency. The vast array of 12,000 risk records baffled managers of the plant: with no initial assessment it was very difficult to understand which were the most critical and required urgent solutions. The vast majority were minor issues that did not have a serious impact on the safety of production, and just distracted managers from solving the most dangerous and urgent operational problems. This experience suggests that however information about risks is collected, it is important to rank the risks initially so that managers can understand where to direct their immediate efforts to prevent any critical developments.

10. **MESSAGING APPS AND GROUP CHATS**

A low-level manager at an oil company cited the use of modern messaging platforms to provide rapid communication between shop floor employees and their line managers. His production team has a group on the Viber platform, which has proved its value again and again in helping assist information transfer about operational issues. The group chat creates a forum for drillers, foremen, and lower-level managers where they can share information daily, and operational problems are freely discussed. Many lower and middle managers at the site also subscribe to the group, though employees generally remain unaware of this, as most only appear on the group under nicknames, allowing them to observe conversations incognito. Leaders can also message the group, though in the respondent's opinion this is unwise, as it may spook employees, who then stop posting on the group. With hundreds of employees on the chat, it is easy for a couple of leaders to remain in the background, using the information to tackle the problems that come up, while allowing the group discussion to continue without interference.

11. **VOLUNTEER RESCUE WORKERS AS INTERNAL AUDITORS OF COMPANY RISKS**

The head of the HSE department in a company that uses hazardous chemical processes shares the following experience. When they began training field staff running a critical chemical facility on how to respond to emergencies, they identified a need to train volunteer rescue workers. This was a return to the practices of the past, when chemical plants had their own reserve fire teams that could provide on-site assistance to the professional fire brigade when dealing with an emergency. When shop floor employees are trained in tackling emergencies, they gain a better understanding of the existing production risks and how well the site is prepared if those risks escalate. When they witness equipment malfunctioning, they have a better understanding of the potentially serious consequences of its failure—so they will take urgent measures to control risk, even initiating a shutdown if need be.

Initially, the senior management of the company thought that staff would be reluctant to volunteer without a financial bonus. But it turned out that the opposite was true: employees rushed to sign up to the emergency teams. Like most people, employees want to show that they care, that they are brave, decent people. They are willing to take responsibility for providing an emergency response to prevent disasters—not for financial reward, but to save the lives of their colleagues or residents near a production site. Site managers should also task these volunteer teams with the job of proactively identifying health and safety risks, and shutting down production if they see worrisome equipment faults or serious infringements of safety protocols. Some employees even get tattoos to show they are members of a volunteer emergency team and, at some sites, there is a waiting list for volunteer fire training. The teams have monthly goals—for example, to find a certain number of safety violations, which they will report via a dedicated channel to factory directors and on to headquarters. Management then rank all the issues received by criticality and probability using a special risk database, and make decisions based on this analysis. The volunteers are encouraged to attend further emergency trainings with the regular fire crews.

Every site manager also has a KPI that requires them to reward these volunteer teams when they flag up safety violations. The teams soon become de facto site safety auditors with broad risk control powers, and are highly motivated to uncover more violations. Their immediate supervisors are similarly eager to prevent further infringements, while management's role at all levels is to make sure the resources are in place to implement remedial measures. This process inevitably creates a degree of information noise—in their eagerness, some volunteers will report issues too minor to bother with. Nevertheless, all the data is entered into a single risk database. Experience shows that there is no need to restrict the teams' activities: they soon learn to regulate themselves, letting insignificant issues pass and focusing instead on serious matters, independently ranking them according to criticality and probability of occurrence. The most important learning from this example is that, if company leaders create the right conditions—specifically that risk information is actively welcomed and rewarded—then employees at all levels will willingly play a role in identifying and managing critical risks, including offering their own solutions.

12. ANONYMOUS MAILBOXES AND HELPLINES

All the respondents who mention anonymous communication channels are skeptical about their effectiveness compared to more "*open*" risk communication channels. The HSE head of an oil company maintains that anonymous postboxes do not work in most organizations, because people must have trust before they will willingly share information about problems in their area of activity. If there is trust, then posted messages do not need to be anonymous—if there is none, then people will not risk posting anything critical, even anonymously.

For companies with a reactive approach to risk management, anonymity can reduce the threat to employees who disclose information to senior management. If the identity of the employee who reported a problem remains unknown to their colleagues and immediate supervisor, then work relationships should not be damaged. The best safety guarantee is for the problem to be promptly solved. It is vital that senior managers do not penalize the employee's colleagues or supervisor. On the other hand, companies with a more proactive approach to risk management generally do not need anonymous risk communication channels.

The HSE manager of another oil company gives some statistics on an anonymous hotline that went directly to the head of the HSE function for directional drilling. More than 11,000 people work in this oil company, but over the year the hotline was in operation, there were only three calls. One caller reported that there was no drinking water on the drilling rig—a problem quickly resolved through the regional manager. The other two cases were even less serious. Either the company was virtually perfect… or the hotline was not working!

The head of the HSE department in a mining company cites similar statistics from his work at various coal and mining companies. In his experience, 80–90% of anonymous messages transmitted through drop-boxes and hotlines were about personal grievances: employees trying to take revenge on their superiors or their work colleagues. Most of the rest were small local issues, leaving only a very small number that actually raised significant safety concerns. In other words, most of these anonymous messages were no more than information noise.

To address this problem, the CEO of one company instructed the internal safety managers to monitor the hotline, and only inform him of a problem if it was mentioned three times. To work effectively over the long-term, senior management must respond promptly and publicly to hotline reports that are significant. More cautious employees, seeing that management seem really committed to reducing risk, may then decide to use the hotline or mailbox to inform them of a problem they have noticed. It is important that there are no rewards for disclosing risks through anonymous mailboxes and call lines—employees should have no interest in using these channels beyond the fact that they want the problem to be solved.

13. **DISCUSSION ABOUT CREATING A NETWORK OF SENIOR MANAGEMENT REPRESENTATIVES**

Several respondents recommended creating a network of special senior management representatives covering all production sites, whose role is to receive risk information directly from employees who, for whatever reason, do not want to communicate with middle and lower managers. Employees must recognize that this team is acting in effect like the ears of the CEO/owners/board of directors. Information shared with them should be of sufficient importance to justify bypassing normal hierarchical channels and going straight to the top. However, other respondents disagreed with this model, believing the possible benefits are outweighed by the disadvantages, such as undermining trust, especially the efforts of leadership to build good relationships with site managers.

3.8 Recommendation No. 8: The Words of Leaders Should Be Supported by Their Actions: Problems Once Identified Need to Be Solved

If leaders want their employees to believe them when they say they want to improve the quality and speed of risk communication, then they must consistently address the issues that their subordinates bring to their attention. If identified problems are not satisfactorily solved, then employees will inevitably lose faith, and will not bother to disclose risks to their superiors anymore.

When asked how to increase employee trust in a leadership, 45 out of 100 respondents said that the most important factor was that leaders should never say one thing and then do another—their words must be matched by their deeds. If not, employees will in the future simply not believe what leaders say. This is especially relevant if senior managers call for risk disclosure and then do nothing to solve the problems that their employees have dutifully reported. Trust will be lost, and once lost it is a hard thing to recover.

Results of responses to anonymous surveys within the framework of the pilot project: *In order for employees to trust managers, their actions should match their words.*

	Strongly agree	Rather agree	Rather disagree	Strongly disagree	Difficult to answer	Number of respondents
All survey participants	75.4%	22.9%	1.4%	0.0%	0.0%	280
Senior management, heads of departments and directors of sites (middle managers)	86.1%	13.9%	0.0%	0.0%	0.0%	36
Lower managers: deputy directors of sites, chief engineers of sites, heads of workshops, heads and representatives of HSE services at sites	76.0%	22.1%	1.0%	0.0%	1.0%	104
Engineers, foremen, and shop floor employees who operate critical infrastructure at sites	72.1%	25.7%	2.1%	0.0%	0.0%	140

The HSE head of an oil company cites the following example. The CEO of a company may state that the company's priority is "*goal zero*"—in other words, a declaration that the only acceptable target is zero accidents, injuries, or oil spills. He can start walking up the stairs holding onto the handrail, using his seat belt in corporate cars, wearing a helmet when he arrives at a production site, and so on. These are easily achieved at no real personal cost. The real test will come when discussing things like production plans. If the boss has boldly declared that "*safety trumps everything*", but at production meetings only wants to focus on achieving the latest ambitious production plans and profit margins—and shuts down any discussion about the safety issues and production risks that those targets will entail, and their impact on his "*goal zero*"—then no one will believe that safety is really a priority for the company. In other words, a leader's eye-catching declarations must be reflected in the real decisions he makes when planning future production and financial goals.

The HSE head of a mining and metallurgy company also emphasized that the words of managers should be matched by their deeds. He always tells his fellow managers "*If we're not yet ready to launch a new initiative, let us wait until we have all our soldiers in a line before we make any announcements*". Managers who successfully build trusting relationships with their employees never make promises that they cannot fulfill. If they are not yet ready to make safety issues paramount instead of profit, it is foolish to make any heroic claims about prioritizing safety.

The head of risk management of a renewable energy company believes that he can only gain the trust of his employees by following through on his promises. If he does what he says, then he becomes a role model for subordinates. Leaders should try to show fairness, honesty, and openness in all their actions, so that subordinates can witness the standards that a company expects them to follow in their own area of work.

Issues highlighted by employees should be addressed

Employees will only learn to trust senior management when they see that top managers will act to solve their problems. When employees report risks or problems, they do so in the belief that managers will make the right decisions to solve the problem or at least reduce the risk. A company may well not have enough resources to solve all the problems identified at any given time. If this is the case, then managers must be sure to feed back to the employee who reported an issue and assure them that they will tackle the problem when they can.

The HSE head of an oil company gave the example of a company boss who visits production sites and invites employees to have a "*heart to heart talk*". After such an open conversation, employees expect that any problem they brought to light will be resolved. To meet these expectations, the boss must work with the site manager to resolve the problem, but without punishing them for having allowed the situation to develop. If successful, employees will be ready to disclose additional information on the boss's next visit: they would rather have safety problems sorted out, so their work becomes safer. After a few visits have led to solutions, the boss will have a good reputation for getting things done, and a high level of trust among subordinates. On the other hand, if these site visits lead to criticism and no positive action, people will stop revealing problems altogether and stick to bland reassurances that everything is fine.

The HSE manager of another oil company believes that when it comes to encouraging employees to openly share their concerns about technological risks, the moral qualities of the boss can be more important than their technical knowledge. Employees do not need their leader to understand every minor aspect of their job and the equipment they work with; but they do need to be confident that the manager will take responsibility for solving a safety problem that they have reported, and get it remedied as quickly as possible. Whether the boss has the competence and resources himself, or needs to go to colleagues for technical advice or to the owners for additional resources, is immaterial as long as the problem gets resolved.

The CEO and chief nuclear officer of a nuclear power generation company gives an example from his company, in which an effective process of issuing corrective actions for risk mitigation has been established. Any employee can identify a problem, incorporate it into a streamlined remedial action business process, and then track progress towards its resolution. Each problem identified is analyzed to understand why it has arisen, and what the organization needs to change to make sure it does not recur on another site.

Senior management must have the resources to solve the problems their employees report

When executives say that they want to reduce the likelihood of major accidents, they must have the financial resources to deal with the problems that come up. They must be aware in advance of their capacity to implement remedial actions. For example, they may not be able to build a new facility on an outdated production site —but with careful management they may well be able to maintain existing equipment that is decades old in acceptably safe working condition. Senior managers cannot tell employees that they want to hear about anything and everything that might lead to an accident, and then turn round and say that they do not have the means to rectify all these problems at once. Carrying out extensive operational repairs, making comprehensive modernizations or installing completely new machinery is likely to require a huge investment. Before raising expectations, senior management must ensure that they have the unconditional support of owners, shareholders, and the state to tackle critical problems, however expensive the solution may be. Not all critical risks will cost a fortune to mitigate—some may simply require minor adjustment of equipment or staff reallocation.

Once a list of the critical problems facing an organization has been drawn up, senior management should turn to owners/shareholders for the resources they need to solve these problems. Generally, the resources requested from shareholders exceed their capabilities—or at least their willingness to invest further. Managers should understand this, and be prepared to re-analyze their wish list and set priorities on the most urgent issues. Therefore, when requesting risk information from employees, they are advised to limit this to the most urgent and hazardous problems.

The HSE director of a mining company cites one such example. As a consultant safety expert, he attended a strategic session for a large manufacturing company. The CEO told the mid-level managers present that he wanted to hear about the problems they were facing at the production sites, insisting he was willing to solve all of them. The session was promoted as an open, honest forum but, during the discussion, it became apparent that the subordinates did not consider the CEO to be very sincere and were skeptical about whether he was genuinely looking to solve site production problems. In their experience: (I) it is always difficult to get a meeting with the CEO to discuss outstanding issues; (II) few of the senior managers of the company ever give clear responses (if any) to messages from subordinates about problems at the sites; (III) if a subordinate brings a problem to the CEO, they inevitably end up having to deal with it themselves, because senior managers always look for someone convenient to blame; (IV) there is no point in informing senior managers about problems because they will not approve the allocation of necessary resources, always using the excuse "*there's no money*". The CEO was very surprised to find these negative opinions so widespread among the middle managers of his company, as there was a large fund available specifically for solving urgent problems, which could be accessed without long bureaucratic procedures. He invited mid-level managers to contact him directly in case of future critical risks, so they could be promptly dealt with. He ended by clarifying that this fund was there only to address critical risks. The mid-level managers agreed that they would work to divide their site risks into critical and non-critical, and in future would ask the CEO for his help in solving critical issues.

Frustrated expectations destroy trust

To promise the world and then do nothing is the quickest way to destroy trust in leaders. When managers ask their subordinates for information on equipment problems, they should not raise false expectations. They must demonstrate with their actions that they will tackle all the most hazardous problems that come to light through a planned program of works. In real life, this may take some time—it is impossible to do everything at once. But it should start as soon as practicable, so workers quickly see the truth of a manager's words.

One leader noted that it is critical to be straight with your employees: act as you want others to act. If senior management depart even once from the publicly declared risk transmission procedures, then their credibility will be permanently damaged, and no one will have any part in disclosing risks in future.

The HSE manager of a metallurgy company insists that, if the boss promises to solve something, then it must happen. The worst scenario is a leader who promises everything and delivers on nothing. This automatically destroys employees' trust in senior management initiatives. This respondent believes that, if managers cannot realistically do anything about a given issue, it is better not to make any promises. If the problem is not immediately fixable, then managers should study the situation until they can bring a workable solution back to their employees, and not just put it on the back burner and hope everyone forgets about it. Employees need to see clear communication and resolve from senior management if leaders are to gain their respect.

An important basis of trust is transparency. Subordinates can see a mile off when their superiors are lying to them. If they hear lies, they will no longer be willing to listen to senior management and will just disregard all their promises. If management do not have time to solve a problem, or complications arise, then they need to be straight with the workforce and tell them that everything is not working out as planned. It is also important to manage expectations. Bosses must make it clear in advance that it is impossible to fund every proposal or fix every problem that the employees raise. However, they will all be looked at, their risks assessed and ranked, and the most pressing issues given a funded solution. Any risk that comes to a manager's attention should be actioned and solved—or if not, the employee who raised the issue should be told why this is not currently possible. This involves making clear: (I) the obstacles that are preventing a resolution; (II) why the problem is not a priority for the company; (III) why there is a delay in dealing with the problem. Leaders should acknowledge that they are not omnipotent: "*Even the boss can fail and make mistakes*". Such an honest and transparent approach will win greater respect from employees and help develop a sense of shared responsibility for solving problems.

Justice and honesty when discussing difficult issues

Employees should understand how a manager makes decisions. If leaders can be seen to make fair judgments, then their actions are far more likely to inspire the respect and confidence of their workforce.

If an industrial facility is in an atrocious state, then the manager must be honest about the severity of the situation, and that the only way to pull through is if everyone works hard together to bring things back from the brink. Penalties for negligent work or for ignoring problems may be necessary, but the leader must always be fair, consistent, and make sure to supply the resources required to rectify the situation. This approach will help gain the trust of the workforce.

An interviewee from the metallurgical industry shared the observation that, in some companies, corporate slogans are written to express an idealized image of a company. The corporate media and public relations department will always seek to portray an organization in the best possible light and focus only on the good side of things. However, nobody knows better than employees about the real state of affairs and all the knotty problems that these glossy publications are choosing to airbrush out of existence. Indeed, the employees' opinion of their bosses emerges from the way the leadership manage these issues.

Senior management should have an action plan and make sure it is implemented

One of the strongest factors influencing how much a workforce trust their leadership is if senior managers share ambitions they have for a company, and then successfully reach these goals. Keeping promises about major company developments encourages employees to believe that senior managers can be trusted to keep other assurances around improving safety systems and communication within an organization.

Some respondents recommend that senior management should form an action plan that is shared with all employees, so that they know where a company is heading, what goals it sets for itself, what the leader requires from them, and so on. It is very important to show employees what they can expect from a company in the future: what their place might be in the workforce, what skills they should develop, what the workplace will be like. Employees are less stressed when they understand the leader's plans for the future of the company and recognize the opportunities and limitations that may arise. Establishing clear strategic goals and working tirelessly to achieve them has a very positive effect on the trust the workforce feels towards senior management. It helps boost motivation.

In this respect, the example of cleaners working in a spacecraft assembly workshop is illustrative. Based on their duties, we could dismiss them as "*just floor polishers*". But we would be forgetting the context. In fact, the cleaners help to launch the spacecraft: keeping the final assembly shop spotless will have a direct effect on the operation of the equipment in space. Seeing constant improvements in a workplace, and an exciting picture of their future in a company, will encourage employees to dedicate their careers to that organization. Employees will be willing to "*go that extra mile*" in their duties: perhaps they will stay late to get something finished, or willingly work extra "*on call*" shifts, because they want to be part of something they believe is worthwhile and is destined for success. The strategic goals of a company should be "*packaged*" into a convincing and attractive vision that all employees can buy into and be motivated by. If employees are to believe that a paradigm shift has occurred with risk reporting and communication, then they need to see frequent real-life evidence of this in operation, both in their area of work and across the whole company.

The professional competence of managers and the adequacy of their decisions

Several interviewees emphasize that the level of professional competence of managers and the adequacy of their decisions has a huge impact on the level of trust among employees. Leaders must be professionals in the field in which a company operates, so that their employees are confident that management know what they are talking about.

Several senior managers from the electric power industry express this view. For employees to trust senior management, they must be competent and experienced. Trust is based in part on what employees know about a manager's performance at previous workplaces; this competency assessment is then modified as they begin to see how they are shaping up as the new boss. Employees generally respect leaders who have worked their way up within their industry.

The HSE director of a metallurgy company believes that the head of a critical infrastructure company should be competent in both technical and production issues. They should show employees that they have a good grasp of equipment problems, understand the complexity of any given problem in the context of a company's industrial policy, and have a wider knowledge of the state of affairs across the whole company. If subordinates feel the boss is not competent, they will not respect them. If they do not respect them, they will not trust them, and this is likely to be mutual.

Some respondents expressed concern that these days, many companies operating critical infrastructure facilities are appointing senior people with little or no industry experience. Some managers do not have specialized technical training and their education and career have been more in the world of finance and business economics, so they know little or nothing about the key operations of an industrial facility. Unfortunately, this is a global trend. For such leaders, production is a very grey area. Without previous industry experience, they have only a very sketchy idea of what they are managing, and this unfortunately can lead to unhelpful bombastic behavior. They want to appear competent in the eyes of their subordinates—so they "*compensate*" for their lack of knowledge by being overly assertive, inflexible and categorical. With a dogmatic mindset like this, there is only one right opinion and that is the manager's: the opinions of their subordinates are totally ignored. This tends to result in monologue communication from manager down to worker. If dialogue fails within a company, then there is less shared professional communication about risks.

Defensive and ignorant managers like this try to avoid visiting production sites. They are worried, at least subconsciously, that their lack of knowledge will be exposed in front of subordinates, and they will look foolish. As the respondents point out, a significant proportion of senior managers in critical infrastructure companies now come from the world of finance and economics. Mid-level managers running production sites, on the other hand, will often have begun their careers as shop floor workers and risen through the ranks through merit and hard work. These two levels of management often have conflicting goals. Senior managers tend to be oriented towards getting a financial result—they measure everything in money. Middle management are more focused on the implementation of the production plan, on the production process itself—and not least, on the safe operation of the facility. This can lead to misunderstanding between managers at various levels, and foster mistrust between headquarters and industrial sites. This, of course, has a negative impact on the quality of information about risks and problems being passed between the sites and headquarters. A lack of "*coalface*" experience at the top of a company can affect the level of employee confidence and trust. Shop floor employees start to doubt that managers are focused on the efficient and safe operation of equipment and suspect that they just want to maximize profits —even if that means disregarding the safety parameters of equipment and concern for employee wellbeing.

The best solution to offset this tendency is for any manager who arrives in post with little appropriate industry experience to be given supplementary training. By accepting such training, leaders are honestly acknowledging their lack of industry knowledge and showing a willingness to learn from professionals in technical fields, even though they may be their subordinates. Actively involving competent employees from different levels of a company in developing solutions helps create a respectful atmosphere, and allows dialogue to flourish around the optimum functioning and development of a company.

3.9 Recommendation No. 9: Do not Penalize Specific Employees: Look for Systemic Defects Within the Organization

No penalties for disclosing information about risks, errors and incidents

According to the HSE head of a gold mining company, managers should not penalize individual employees for incidents, but instead look for the systemic shortcomings in a company's operations that forced the employees to commit safety breaches. He gives an example from his practice. In the past, he knew one leader who received an incident report that placed the blame on specific workers. The manager concerned put this report aside and, instead, initiated a real investigation of the incident asking: "*What were these employees ordered to do by the leadership?*"; "*Were adequate resources provided?*"; "*Was employee training fit for purpose?*"; "*Had the workplace been set up for safe work?*". As soon as employees realized that there was not going to be a witch-hunt to find individuals to blame and punish, they felt free to reveal the true motives behind their behavior, enabling management to identify the systemic causes that were the real reason why the safety violation had happened in the first place. Working to eliminate systemic shortcomings by analyzing specific incidents in this way will prevent the recurrence of similar incidents in the future. If systemic problems are solved, employees will no longer have any reason to violate safety regulations.

The head of the HSE department at a chemical company believes that employees need to be shown that disclosure of risk-related information is in their own interests. It should be made clear that preventive identification of risks avoids: (I) the recurrence of emergencies, thereby reducing the likelihood of serious injuries and deaths; (II) staff and wage cutbacks due to production shutdown after accidents or equipment breakdowns. Employee motivation may differ from site to site and country to country. Managers should identify the issues most relevant to their employees and tackle these when devising new models for transmitting risk information.

It is worth noting that for the most part, employees who violate safety precautions are not deliberately trying to do something wrong. Generally, people want to do the right thing and are more likely to be acting to try and save a company's expenses, or cut corners to increase productivity, and so on. Despite such good intentions, they still find themselves being penalized for breaking safety regulations. When investigating incidents, the key task for management is to look beyond individual workers and identify the systemic shortcomings of a company, so that these can be addressed, and the workplace made safer. As well as being a far more effective approach than blame and recrimination, this reassures employees throughout the corporate hierarchy that they need not worry—letting the company know there is a problem will not lead to them or their colleagues being hauled up in front of everyone and publicly reprimanded. The honest disclosure of risk information must come to be seen throughout a company as a positive action and one that any conscientious and loyal employee will carry out without hesitation. Clearly, this will not be the view if employees continue to be penalized for being the bearers of "*bad news*".

The head of an oil production site believes that penalizing employees for incidents is appropriate only if they have deliberately disregarded safety rules. In his experience of incident investigations, he has never once identified deliberate negligence or malicious violation of safety regulations. Generally, employees fail to follow safety measures not out of malice, but because they want to maintain or increase production output. Unfortunately, through error or overload, they find instead that they have caused an accident.

If senior management wish to radically improve the transmission of risk information throughout a company, they must give employees a very clear message: "*When we learn that an accident has been caused by an employee being over-ambitious or go-getting, we recognize the truth that the whole company is to blame, that we allowed mistakes in various aspects of our operation, or demanded production targets that meant the employee felt forced into committing the safety infringement which led to the emergency. We are not trying to blame a specific employee, but to work together to learn from any negative incidents that occur so that we can improve the whole system*".

The purpose of removing recrimination or punishment for accidents is that employees are no longer afraid to analyze their own mistakes and start discussing them with colleagues and management. It is only when fear of retribution is removed that employees will begin to send more honest and objective information about problems to their managers. According to the respondent, we should remember that employees will always feel some trepidation when it comes to reporting bad news to their superiors. The challenge for managers is to reduce that fear, even if it cannot be completely removed. To achieve this, they must openly declare: (I) there will be no penalties for disclosing negative information about the situation on the ground; (II) senior management want to be actively involved in the discussion of the problems facing production units; (III) middle managers are eager to hear from their subordinates; (IV) senior management have resources, and are happy to invest them in solving critical problems; (V) senior managers are more than willing to share the responsibility for solving problems with their subordinates —"*We are in this together. We are all on the same team*". In time, this approach will create a shift in culture, and instead of being afraid to admit there is a problem, employees will be afraid to conceal it. Why would they try and tackle the problem alone or risk hiding it, when they have been encouraged to work together with senior management, to share the responsibility and find effective solutions?

Results of responses to anonymous surveys within the pilot project: *When employees report a problem to a manager, they then share the responsibility for solving it. If employees keep a problem to themselves, they are taking full responsibility for whatever happens.*

	Strongly agree	Rather agree	Rather disagree	Strongly disagree	Difficult to answer	Number of respondents
All survey participants	35.4%	45.4%	14.6%	2.5%	2.1%	280
Senior management, heads of departments and directors of sites (middle managers)	44.4%	44.4%	5.6%	2.8%	2.8%	36
Lower managers: deputy directors of sites, chief engineers of sites, heads of workshops, heads and representatives of HSE services at sites	37.5%	50.0%	8.7%	2.9%	1.0%	104
Engineers, foremen, and shop floor employees who operate critical infrastructure at sites	35.0%	45.7%	14.3%	1.4%	3.6%	140

Several interviewees share their experience within one specific oil company. The policy there is not to penalize employees after conducting incident investigations. The company denies interest in penalizing particular employees, but instead seeks to understand the flawed organizational and production processes that led to the accident and to work out how they can be changed to prevent future accidents. When investigating incidents, the priority is to identify systemic flaws, not the mistakes of specific employees. The company has informed employees of the principles it will follow when investigating incidents: (I) the most important thing for the company is to ensure the investigation is transparent and that all employees actively participate to help identify the root causes of the incident so that it does not happen again; (II) once identified, these causes will be comprehensively dealt with; (III) the company has no desire to penalize individuals.

Based on the results of each investigation, a report is written on what changes the company needs to make in order that employees in future are not forced to contravene regulations or take unjustified risks. Applying these principles to their incident investigations has had a positive impact: employees have become more open, and cooperate willingly in investigations aimed at identifying the causes of incidents. It took time to embed the principle of "*no penalties for industrial safety violations*" but, once accepted, the corporate culture around the discussion of safety issues radically improved. Employees are no longer fearful of penalties: they proactively identify risks in the early stages before incidents occur and are willing to discuss how to eliminate them with colleagues and managers. When incidents had led to penalties, employees had—subconsciously or not—avoided analyzing the causes of incidents. When this threat has been removed, almost all employees showed a real willingness to join dialogues on how to increase production safety.

According to the respondents, harsh penalties based on the results of an incident investigation should be entirely abolished: penalties are only appropriate for cases where serious risk information has been concealed, or employees have blocked efforts to rectify production problems or follow safety regulations. Employees need clear guidelines for what is approved and what is prohibited. Management must instill a proactive approach in all employees to maximize occupational safety, rather than simply looking to blame individual employees for their perceived infringements and disobedience.

The head of the HSE department at an oilfield services company believes that reports to management should only contain information about dangerous working conditions and risks, rather than the actions of specific workers. If a risk report does contain information about individuals' faults, then this should be removed and discouraged in the future, so that employees do not try and use risk communication channels to advance their own interests or pursue vendettas. If a specific employee's name appears, it raises the question of who benefits and who suffers from its inclusion. Managers must focus on analyzing the risk inherent in a situation, not the actions of specific employees.

Most companies have designated HR channels for reporting employees for perceived unacceptable behavior—even then, such information must be handled with confidentiality and care to avoid any suggestions of a witch-hunt or purge. Senior management need to send the following message to the whole company: "*We want to identify risks and improve processes to make them safer, not to penalize people*". Rather than reprimanding those involved in dangerous incidents, managers must endeavor to remove the conditions that put pressure on employees to act unsafely in the first place. Then employees will begin to welcome investigations, since they can see these are now focused on improving safety and working conditions, and not on inflicting punishment.

The vice president of an international oil company believes that if, after an incident, senior management automatically ask the question, "*Who did it, who did something wrong?*", then employees will be afraid to report risks and own up to their mistakes. They will instead try to hide the truth about incidents. The respondent recommends that senior management ask the following question after an incident: "*What lesson can we learn from this incident to prevent a similar thing from happening again?*". In the respondent's company, the Incident Review Team has been renamed the Learning Team. This immediately changed the attitude of the employees towards the team's activities. The learning team does not blame employees involved in the incident, but simply asks the questions: "*What systems failed here? What can we learn from this? What do we need to do differently or change to avoid repeating the same mistakes in the future?*". In response, the employees involved in the incident turn from defensiveness to cooperation and become willing to help analyze the shortcomings in the company's operations, even if this means admitting personal mistakes. As a result, the company produces objective information about the incident, which helps identify root causes and the changes needed to achieve meaningful and lasting risk reduction.

The head of a power plant gives the example of one energy company where owners and senior management decided to work together towards preventing major accidents involving the company's infrastructure. The CEO told the entire workforce that he wanted to hear about every problem, equipment failure and safety incident: "*We will never penalize you for incidents that occur. Tell us about all the problems and safety incidents and we will use that information to ensure that the accident rate at the factory really is reduced*". Confident that they would not be penalized, employees then began to take the initiative in disclosing risks. From this huge influx of new information, much of it about relatively minor issues, the HSE department are isolating the causes of incidents and taking measures to prevent critical developments. According to the respondent, this company has one of the highest equipment failure rates in the industry, but this is because employees are no longer concealing this negative information, and any failure is duly reported. Elsewhere across the industry, the opposite is true. At senior management level, the goal is to keep the official accident rate down, whatever that takes. Employees are tacitly encouraged to hide incidents at all costs, manipulating statistics and reports to prevent an increase in registered accidents. Ultimately, this approach is against everybody's best interests: it just leads to a situation where senior managers no longer have any real idea of the true risks and accident rates prevalent in the infrastructure that they are responsible for. They are only deceiving themselves, and their illusory "*rose-tinted*" accident-free world may at any moment be shattered by a major accident. This may come as a complete shock, even though in reality the danger signs had been there for years, in plain sight for anyone who had been willing to look properly. The plain truth is that some managers just do not want to know about organizational problems.

Even for a manager who has the courage to take real responsibility, there is a balance to be struck here. The goal of minimizing equipment breakdown inevitably brings increased costs—major repairs, modernization, or a total refit. Safety and reliability costs money. According to the respondent, it is not economically feasible to try and achieve zero-accident conditions—it costs too much! Such an extreme goal is only appropriate in manned space exploration. Elsewhere in industry, if equipment failures are not catastrophic or life-threatening, then it is reasonable to expect and accept their occurrence at some point. The only question is the frequency of failure and the amount of damage this will cause. The priority has to be reducing the likelihood of the critical risks that could wipe out a company, result in casualties or precipitate an environmental disaster.

The head of risk management at a renewable energy company also believes that, to improve the transmission of risk information, senior management need to create an atmosphere in which employees know that they will not be blamed or penalized if they reveal problems. Managers should make it clear to subordinates that when they report a problem, they will not automatically be assumed to be responsible for causing it. There is a fine line between making a mistake and being responsible for it. The analysis of many incidents shows that, when employees take risks, it is often because a manager has put them in a position where they felt they had no alternative but to contravene safety procedures, in order to pursue time-saving, cost-cutting, and profitability indicators. Managers should promote the belief that mistakes and bad news can be the very best way for learning hard lessons and for employees not repeating the error—ultimately improving efficiency, safety, and productivity.

A manager at an oil company believes that any mistake can be considered as an opportunity to understand where they, as an organization, have gone wrong. Managers and employees can work together to untangle any incident into its components and understand how to avoid repeating the same scenario. But this can only be achieved when the practice of penalizing specific employees for mistakes is abolished. Managers should not be acting as judge, jury, and executioner, but helping their employees learn from their mistakes so that they do not repeat them. Sometimes employees lack the training or experience to draw the right conclusions from their own errors. The manager's task here is to help them figure that out.

The respondent gives the following example. Several senior managers were walking round a site and saw a gross violation of the safety regulations: the employee in question was doing a great job but did not have the right protection devices to run a particular piece of machinery. The leaders approached the employee to tell him of his transgression, but immediately reassured him that he would not be penalized for this oversight. They invited him to join them in a nearby office, where they gave him a pen and paper and asked him to write down what his mistake was. The worker wrote a detailed account of what he had done, and what would have been the right thing to do. He felt humiliated, but appreciative that there was no official reprimand or penalty. He left with a clear understanding that he must not make the same mistake again, and that if he did his bosses would not be so lenient next time. He was never seen breaking safety regulations again. After this incident, it came to light that, at the time of the visit, the employee's supervisor had left the team to work unsupervised without explaining the safety measures the employees were required to take. The supervisor was spoken to about this lapse, but again without being penalized. After this, the supervisor was careful to always explain to employees at the start of every shift about the possible risks involved with the work. By debriefing everyone involved in this incident, without imposing any penalties, the managers succeeded in permanently changing how the employees were working. This is the important thing—to achieve a permanent change in workforce behavior, not to penalize mistakes.

The head of a power plant believes that there should be an unshakable rule that any employee who reveals a problem will not be penalized for their part in causing it. Repeated demonstration of this policy in action, and praise from superiors for any employee who discloses problems, will convince even more cautious and insecure colleagues to do the same, providing managers with a great deal of valuable but hitherto hidden information.

The HSE head of an oil company believes that the first step in implementing a new "*no penalty*" system should be the declaration of an amnesty: all employees are given the opportunity to report any shortcomings or mistakes in their own work, the times they or their colleagues have taken risks or disregarded safety protocols, and existing problems they are aware of but have not reported—all covered of course by a ironclad guarantee of no penalties. This gives managers information to assess the real situation in their organization. If during this amnesty period, any employee still fails to volunteer information about a problem they knew about that later comes to light, they are liable to be dismissed.

The respondent relates the experience of an attempt to set up an amnesty along these lines in an oil company. The CEO here told the entire workforce that the senior managers were launching an amnesty, that they were poised to hear about all the problems within the organization, that they were committed to solving these, and that nobody would be penalized for disclosing difficult or uncomfortable details. Despite this clear statement, 30–40% of lower and middle managers continued to keep quiet about risks and problems in their area of responsibility. They had worked for the company for years—they thought this was just a new gimmick from the CEO and his inner circle to play about with, soon to be forgotten. These skeptics acted as a damper on the new policy of transparency and the free flow of risk information. In the end, the company had no choice but to dismiss some of them after the amnesty period, when it came to light that they had concealed risks they knew about. They were sacked very publicly to make it crystal clear to the remaining employees that the new system of open transmission of risk information was serious and long-term. It took two years to build trust between the top officials and the workforce, who gradually began to share problems, and enter dialogues about how to solve them. But at this point, two middle managers informed senior management of some serious problems, and were promptly fired. This was a disastrous move by the leadership. As soon as the news of their dismissal spread through the workplace, all disclosure of risks from the bottom up instantly dried up. This is a vivid example of how easily the trust of employees in their senior management can be undermined—one hasty decision which betrayed the CEO's promises was enough to destroy two years of careful cooperative bridge building.

The moral is that, if managers tell employees they will not be penalized for volunteering information about risks, this principle must be upheld at all costs. In the long run, maintaining trust increases the likelihood that an employee who makes a mistake will promptly report it to the authorities, rather than hide it. This patience and good faith will pay huge dividends in the end. The company will find out in advance about emerging problems and be able to tackle them before they become more serious and are still relatively quick and cheap to rectify: far better than being faced with a full scale disaster, and losing a fortune having to deal with the consequences of serious hidden problems about which senior management were ignorant until it was too late.

The head of HSE in a mining and metallurgy company describes how his company decided to make deliberate concealment of safety issues completely unacceptable. An amnesty was announced throughout the company, during which no penalties would be imposed on employees reporting safety problems and violations of the regulations. Middle and lower managers were encouraged to disclose any problems and risks they knew about, whether recent or long concealed, and this produced lots of disclosures. Some of the chief industrial site engineers later commented on the initiative in an informal conversation: "*Well, finally, someone wanted to hear us. At last, we were able to clear our consciences, otherwise it would have been hard to keep working*". At the same time, they made it clear that, in the past, they would never have gone to their superiors to discuss the risks in their area of responsibility. Feedback only became possible when senior management explicitly asked them for it, with the promise there would be no retribution. Voluntary disclosure of risks can potentially damage the careers of many employees. They are admitting their mistakes, so laying themselves open to accusations of professional lapses and failure. Despite the measures they had taken to reassure staff during the amnesty, senior management found out later that some middle managers had continued to hide serious problems. The entire leadership team were shocked: "*Why are you still hiding critical risks, when we have promised that nobody will be penalized for reporting these problems?*". In most cases, the answer was that these middle managers were still focused on achieving their production plans—finally trying to solve problems that they had successfully concealed for ages would put them behind schedule and reduce their bonus.

The HSE vice president of an oilfield services company believes that an amnesty is only appropriate when senior management need quick information about risks within an organization. If there is no hurry, he maintains that patience is the best policy: one should make gradual changes in the corporate culture and seek to create a safe environment for discussing problems. In the end, employees will no longer be afraid to talk about risks, and will be ready to give feedback on systemic shortcomings within a company. But this is only possible if managers have time: changing corporate culture is a slow process and can take years.

The HSE manager of a metallurgy company also believes that the process of obtaining information from employees cannot be hurried, because they need to see plenty of positive examples of how senior management react to bad news. It takes time to build trust and employees must be sure that they are not in danger if they disclose negative information.

Severe penalties for concealing risks, errors, and incidents

The HSE head of a metallurgy company believes that there is no need to penalize employees when they first make a mistake—they need to be given the opportunity to correct this mistake on their own. However, if they repeat the same infringement, immediate dismissal should follow. He recommends that senior managers personally go to production sites to investigate the most serious incidents. By doing this, they can show employees that the incident investigation will not be a search for individuals to blame, but for the right operational improvements and changes to implement for preventing a recurrence. If there is clear evidence that the incident has arisen because risks and failings have been concealed, then site managers must be dismissed—regardless of their popularity, status, and past merits.

The HSE head of a mining company recommends that companies should have no more than three to five key working principles, so that every employee knows, remembers, and follows them in all circumstances. One of the core principles in the field of industrial safety is the inadmissibility of concealment and fraud. Given that many employees distort information when reporting to their seniors, it is advisable to select only one well defined area in which distortion like this is categorically forbidden. For example, you can choose the issue of industrial safety and make concealment of safety violations explicitly unacceptable. If senior manager are found guilty even once of double standards in the application of these principles, the whole system will be discredited—consistency is key. If employees are discovered to have knowingly concealed risk-related information, they should know that they are liable to be dismissed, regardless of previous performance or their status in an organization.

This respondent cites an example from his own experience. A director of a large mine was publicly dismissed for lying, despite many years of loyal service. After that, the flow of reliable information from other production units improved significantly, because middle managers realized that the leadership were serious about wanting a true picture of the situation in the field and was willing to severely penalize employees who concealed the truth. When middle managers grasp the serious consequences of communicating false or distorted risk information, they will begin to insist that their subordinates tell them the truth about any risks they know of.

The HSE head of a production company managing a large number of hazardous chemical processes believes that penalties for communicating bad news should be abolished, and that the reaction of managers to receiving reports about problems needs to change. All employees need to realize that this information, painful and troubling as it may be, can help the company identify critical risks in time to take remedial action—action that could prevent a catastrophic cascade of events. Barriers between subordinates and managers need to be broken down so that employees are not afraid to enter a dialogue about production problems, because they understand that this is a shared threat—they are all in the same boat. In his organization, the management team spent a lot of time reassuring employees that they had no interest in looking for specific workers to pin the blame on, and nobody would be singled out for punishment for disclosing risks. Their aim was to identify the systemic causes of the problems and tackle them. However, even this clear approach does not guarantee that employees will start freely sharing risk information. When investigating incidents, this company still on occasion uncovers deliberate concealment of risks, even though the proactive disclosure of these risks would not have posed any threat to the employees concerned. The senior management have thus started to adopt Al Capone's principle that "*You can get so much farther with a kind word and a gun, than with a kind word alone*". If the owner, board of directors and senior management want to maximize the disclosure of risks, then they should have recourse when necessary to the ultimate sanction for deliberately concealing critical problems: dismissal—irrespective of rank, length of service and skills. The respondent is convinced that having a "*big stick*" like the threat of sacking is more effective in encouraging compliance than avoiding penalties altogether.

The HSE head of a metallurgy company expresses the view that employees need to be clear about the consequences of hiding information. If there is an emergency or accident, and it turns out that they knew about the risks that led to the incident but did not report them, they will be held responsible. In addition to the injuries and deaths that might occur, production may have to be halted, the entire site may even be closed, and they and their colleagues may lose their jobs. If this message is really rammed home, then employees will realize that they have lots to gain by passing on whatever they know about existing critical risks—and lots to lose if they remain silent.

The HSE head of a mining and metallurgy company believes it is important to convey to shop floor employees that a company's policy of transparency around critical risks is not just a management affair, but has as much, if not more, relevance to them. Early preventive shutdown of malfunctioning equipment can save the health and the lives of the operating team, and prevent more widespread incidents that could cost jobs, precipitate environmental disasters, and even threaten the lives of nearby residents. If disruption to the production process can be kept brief and as local as possible, rather than requiring the wholesale shutdown that would likely follow a major incident, then this will be much better for the stability of the operation and the job security and incomes of employees. This respondent also recommends abolishing any penalties for bringing bad news, while severely penalizing any deliberate concealment of critical risk information. Senior management should publicly thank and reward employees who disclose serious risks, and reprimand those who have hidden the truth. However, penalties must always be fairly applied. If the investigation shows that the employee or crew involved concealed the risk because of pressure from their supervisors, then it is the supervisors who must go, while the others might, depending on circumstances, receive a lesser sanction. Two or three such cases should be enough to get the message over, and the company should start to see significantly fewer problems going unreported.

Changing the script: bringing bad news should be seen as positive, not negative

Some of the interviewees express the view that a positive attitude to communicating bad news about risk should be fostered at all levels of an organization. Employees who have the courage to inform managers about progress or regression in controlling critical risks should not be viewed as "*informers*" or "*traitors*". On the contrary, they should be respected by their colleagues for transmitting information that is likely to benefit them all. In the opinion of the respondents, you need to "*change the script*". Rather than risk transmission being seen as a negative action (bringing news about problems), it should be reframed as positive (protecting their colleagues, preventing serious emergencies, saving a company). For most shop floor employees, saving themselves and their colleagues from injury and death is likely to be the strongest motivator.

The head of HSE at a mining company believes that the best career guarantee for an employee who is brave enough to take the initiative and report a serious safety issue is to make them a company hero, whose prompt actions saved equipment, prevented injury to fellow workers, perhaps even prevented a full scale disaster.

Public shows of approval for those who show a strong sense of civic duty are the best protection for employees from supervisors or line managers who would prefer to hide negligent or dangerous practices in their area of responsibility. Publicly acknowledging such employees as heroes shows the whole workforce the qualities senior management wish to see, and courage and integrity like this are to be encouraged across an organization. If people follow their example, over time these altered attitudes will become ingrained company values.

The head of the HSE department at a chemical company also believes that successful examples of risk disclosure should include an assessment of the costs and damage avoided, and be publicized throughout the company. If employees who reported a problem are acclaimed as heroes, their actions will inspire other employees to do the same. At the same time, this very public corporate approval provides protection for the employee, making it virtually impossible for vengeful retribution to be inflicted on them by their colleagues or immediate superiors.

3.10 Recommendation No. 10: Reward Employees for Disclosure of Safety and Technological Risks

Public recognition is a universal reward applicable across an organization

The safety manager of a nuclear power plant believes that the best way to reward employees is to recognize their important contribution to an organization, as everyone derives fulfillment from having their work appreciated and praised. Management should deliver this not just through a private conversation, but also in front of the whole workforce. Expressing gratitude publicly in this fashion provides an opportunity for senior management to highlight the kind of behavior and performance they wish to see from all their employees. Public recognition will motivate the employee to even greater efforts and encourage colleagues to raise their game, including when it comes to communicating risk problems up through the hierarchy.

The head of the nuclear design department of an international electricity company believes that recognition is what employees crave most from their managers—it means a lot to feel appreciated. When an employee receives positive feedback from their manager after taking proactive action around risk transmission or problem solving, then this naturally motivates both them and their colleagues to be more vigilant in identifying future issues in their area of responsibility, and in communicating them to their managers so that corrective measures can be implemented at an early stage.

A consultant in nuclear safety with long experience in nuclear power plants operations suggested that senior management should always be ready and willing to say a genuine "*thank you*" to any employee who has carried out their role within an organization with skill, diligence, or integrity.

An HSE consultant working mainly for oil and gas as well as in air traffic control believes that a heartfelt "*thank you*" is an obvious but excellent way for senior management to show appreciation for that employee's contribution, and helps motivate the whole workforce. The respondent believes that senior managers should do this far more often, along with a warm pat on the shoulder when they want to show their appreciation, rather than handing out food vouchers, iPads, or some other material reward to employees.

A global director of safety in the oil and gas industry also notes that publicly saying "*thank you*" to an employee who has identified a serious risk or prevented a major incident is a powerful signal from senior management to the whole organization. The respondent believes that senior management should make more effort to demonstrate its recognition and appreciation of proactive employees by publishing success stories in corporate magazines and on online channels. These reports should describe in detail how prompt action by shop floor employees has averted major disasters, prevented injuries and loss of life, jobs, money and so on.

The HSE head of an oilfield services company believes that one of the most important non-material reward mechanisms is the public recognition of employees among their colleagues and across the company as a whole. This company has an annual CEO's award for safety, awarded on the basis of specific safety performance metrics that it wants its employees to demonstrate throughout the year. However, this award is focused on recognizing and rewarding teams, rather than individual employees. This company also has an HSE director's award, which is given to the lower or middle manager that the HSE director considers has best exemplified the advanced safety culture the company is striving for. The company also has quarterly safety awards, where employees can nominate colleagues and line managers who they believe have demonstrated outstanding safety performance over that three-month period. Such awards motivate employees to keep constantly improving their safety performance. The details of the polls are not made widely available, so the winners have no idea who has nominated them. However, they feel proud to have been recognized for having made a positive impact on the safety or productivity of their team and the wider organization. These awards are focused on recognition, not material rewards. The HSE department creates media presentations of the award winners and the most outstanding safety performances, and these are distributed across the company's sites around the world. Global recognition like this has motivated the workforce to work harder on safety performance, and many employees now take part in these regular competitions and strive to demonstrate the very best practices in safety and risk control to senior management. Winners have also been offered promotion opportunities.

A risk expert in the oil and gas industry gives the example of a company senior manager and mentor. When he visits the production sites, he would gather all the workers and lower managers together at the end of the day shift and say: "*Guys and girls, thank you for your work today, thank you for keeping us all safe*". He would then stay behind to shake employees by the hand and chat face-to-face.

Remuneration preferences depend on the employee's goals and perception

One regional manager, responsible for the operation and maintenance of turbomachinery in the power industry, insists that it is difficult to rank different types of reward for risk disclosure because their importance depends on the perception of individual employees. Some prefer financial rewards and are not interested in promotions, perhaps because they do not see themselves as having a long-term future with that company. Others prefer non-material rewards, because they believe recognition of their good safety performance will help them achieve promotion within the company or industry.

The head of HSE in an oil company believes that one needs to combine both psychological and material rewards, since employees react differently to different types of motivation depending on what suits their particular needs.

Defining terms: material stimulation vs. non-material motivation

The HSE head of a mining company suggests the need to define terms. Stimulation comes from the Latin word *stimulus*, meaning a sharp metal tip on a pole used to drive a buffalo harnessed to a cart. When employees are stimulated, someone who wishes to motivate or control them exerts external influence in the form of material rewards, so that they do something in the interest of the "*controller*". Stimulation is usually financial—a salary or a bonus. Therefore, any material form of reward could be described as stimulation. In the respondent's opinion, if you offer people material incentives for disclosing risks, many employees will mistrust colleagues who disclose to managers the risks that their team is dealing with. They will react especially sharply if the management then penalizes someone else in the team who they blame for the situation. Workers considering disclosing risks are faced with a dilemma—either to preserve their relationship with their work mates by keeping quiet or disclose to their superiors and collect their "*thirty pieces of silver*". The respondent believes that, to better protect employees who are willing to speak out, it is more effective not to use material incentives to encourage feedback.

Motivation comes from the word *motive*, meaning the reason for an action or deed. Therefore, the most effective way to motivate employees to disclose risks is through non-material psychological rewards: an appeal to a person's better nature, to stand up for high universal values, to act in the greater collective interest rather than their own. These higher motives allow employees to develop as individuals, to step beyond just going to work to earn a living. For most whistleblowers, it is more important to be able to convince the leadership that their ideas and solutions have implications and applications throughout an entire company. This is the highest recognition of an individual's contribution to improving a company. As well as receiving public gratitude and praise, active and effective employees willing to disclose safety issues should be further encouraged by being rewarded with greater work responsibility: perhaps they could be promoted to take a role within areas of the company that are most problematic, so that they can help drive fundamental improvements. If there is no appropriate higher vacancy to reward a worthy employee, then they could be transferred horizontally to a larger work area, with an increase in responsibility and income. If they perform well in this new role, a more senior position can be offered later. Proactive and confident employees are part of a company's talent pool: when implementing new projects, they should be invited to lead from the front.

Non-material motivation is more effective than material stimulation

The head of HSE in a mining and metallurgy company shares his experience. Proposals to start paying employees for disclosing safety problems have been made on the board of directors of every company he has worked with over the past 20 years. This seems to many managers the most obvious way to stimulate employees not to violate safety regulations, especially after a series of incidents. However, if you start paying money for disclosing information about critical risks and triggering preventive shutdowns of equipment, there is a risk that operators will deliberately bring equipment to a pre-critical state so that they can report it and receive their reward. Material rewards are a false way forward and are likely to produce a very distorted picture of risk, because the desire for financial reward will encourage employees to come up with as many problems as possible. Instead, he believes that to the best approach is to identify the most proactive employees and encourage them to focus on relevant information about serious problems. These employees need to be promoted up the career ladder, or given the authority and resources to solve the problems they have helped identify in their existing area of work. In both cases, an increase in wages should be linked to increased responsibility and authority, *not* perceived as a payment for their actions in disclosing risk. The employees concerned should also be publicly acknowledged and thanked for their active contribution in improving the safety of the enterprise.

The managing director of an electricity company agrees that financial incentives for disclosing risks are likely to lead to employees deliberately pushing facilities into a pre-emergency state in order to report the situation and be rewarded. They may then end up intentionally damaging or neglecting equipment in order to earn a premium. It is unwise for senior management to attempt to buy employee loyalty—better to motivate employees to be loyal by non-material recognition. Employees, like other people, want to feel valued and appreciated for what they do. They need to see that the contribution they can make by disclosing risk and safety issues is vital to improving the safety of a critical infrastructure facility. They would love to be acknowledged for doing something that may save the health and the lives of their colleagues and of nearby residents, and protecting the well-being of the whole country which relies on the organization's services. The respondent's main advice is that employees' loyalty to an organization should be encouraged by non-material incentives. If a company tries to buy loyalty with cash, then the entire corporate culture will favor self-serving employees who only care about maximizing their own financial bonuses. If you build loyalty on the values of vocation, duty, responsibility, and teamwork, then over time a company will begin to reflect and embody those very same virtues. A company should nurture those employees who will "*fight for the cause*", and not just for themselves: employees who will not ignore risks, or fabricate new problems, but strive for the greater good and bring serious issues to the immediate attention of their superiors.

The HSE head of an oilfield services company witnessed several occasions where employees tried to bring the equipment to a pre-emergency state, so that drilling could be stopped and they would be rewarded for a preventive shutdown. This is obviously a very dangerous and unwelcome development: to get paid for a preventive shutdown, some employees are willing to gamble with equipment that costs hundreds of millions or even billions of dollars, and may damage the livelihood of everyone involved in an organization and have even wider adverse impacts. This is not a sustainable motivational system.

Several HSE executives from the metallurgy industry agree that it is best to exclude material incentives, since they can trigger a wave of fake messages from workers looking to use the reporting of non-existent or amplified risks to make money. They also shared the same concerns as the three respondents cited above, regarding equipment tampering and neglect as a way of gaining financial rewards.

A psychologist and consultant in the field of organizational behavior also maintains that it is better to focus on non-material rewards. If the rewards are financial, there is a risk that people will disclose information primarily for money rather than from any real sense of duty, and this can have negative consequences for the corporate culture. The best solution is to motivate people to disclose information under the principle of doing the right thing, not for mercenary reasons. People should raise the alarm about dangerous conditions or working practices out of concern for their own and their colleagues' safety, and a desire to safeguard the organization and its operations. We all understand and largely share these inherently good human instincts, but they can be weakened and lost if the community around us—in this case work colleagues and managers—does not foster and constantly reinforce them. Money can act as an incentive to get employees moving—for example, when launching a raft of changes—but in the long run only the more meaningful and enduring psychological reward of recognition, praise and the respect of colleagues and superiors can motivate people to do the right thing.

The HSE director of a gold mining company refers to a related circumstance. At a senior management meeting, they unveiled the motto: "*We do not pay for values*". Everyone agreed that a company should not pay employees extra for working safely: this should be an integral part of the duties of any worker. Similarly, there is no need to reward them financially for disclosing risk information—if there is a reward, it should be non-material. For example, an employee who has identified a critical risk can be thanked in front of the whole workforce, to show their colleagues what senior management expect of them. In such a public acknowledgment, it should be made clear that, by raising the alarm, the employee has helped save the lives—and livelihoods—of their colleagues. Otherwise, the worker concerned is at risk of being labeled a "*snitch*" who has dared to inform against their workmates, and been praised by the boss for doing so. It is crucial that whistleblowers are perceived by other employees as people who took an action to try and protect their workmates, by preventing an accident through the early disclosure of risks. Everything should be done to encourage their colleagues to admire their actions and want to follow their example, rather than resenting them. According to this respondent, in addition to not giving material incentives for conduct that people should see as their duty, avoiding financial rewards will help prevent these kinds of accusations. And if there is money to be made, some less scrupulous workers will create fictitious problems, or even report equipment failures or concerns that they have deliberately created.

The head of HSE in a metallurgy company expresses the same idea by saying that the contract for the transmission of risk information within a company should be social, not material. The most effective social message is "*be your brother s keeper*", i.e. be responsible for the safety of your colleagues and friends. If everyone takes care of each other in the workplace, risks will be identified and reduced much more effectively than when everyone is acting primarily out of self-interest. Safety is not a private matter, but the business of everyone in the team. By working safely, employees are saving their colleagues from injury, and worse, as well protecting themselves. It is a mistake to bring material rewards into what should be the realm of shared duty and comradeship. Many companies have tried to introduce material incentives to solve the problem of risk disclosure, but the respondent did not know of a single successful example in any country where this had been effective in creating permanent change to an organization s operating values.

The head of the HSE department at an oil company also believes that non-material rewards are more effective than concrete ones. The impact of material rewards on motivation, both individually and across the whole company, is debatable; intangibles may take time, but their long-term benefit is undeniable. According to her, business is no different from life: nobody can buy love. The focus must be on psychological rewards, on the pleasure and fulfillment of working for a company where you feel appreciated and part of a team. In the long run, employee loyalty simply cannot be bought with money. Instead, an organization must endeavor to find effective non-material rewards to motivate employees to develop company loyalty. The fact that managers respond positively to employees who take pre-emptive action is far more important than the size or nature of a gift. A personal letter from the leadership to an employee s family to thank them for their contribution to the safety of the enterprise, a photo of the "*distinguished*" worker with the CEO, a dinner with senior management for the team whose prompt action saved the company from disaster—these are all very strong motivators. Employees often remember these honors for the rest of their lives, whereas a bit of extra money is quickly forgotten. This is especially true in single-industry towns, where social bonds are very strong, and news of workers being gratefully acknowledged and lavishly praised by the top brass of a company quickly spreads throughout the city. For example, the CEO of the respondent's current company made a point of publicly congratulating their best contractors for the high safety standards they were maintaining. The entire contractors' workforce were so delighted with this that they posted photos from the award ceremony across their websites and social networks, framed the photos, and hung them prominently in their offices. After this event, the contractors "*grew wings*": they were so inspired to have been recognized and appreciated by the leaders of the oil company that they worked even harder to promote safety and risk disclosure.

The HSE manager of a metallurgy company agrees that it is worth initially focusing more on non-material rewards. As he described it, it seems immoral to pay money to safeguard yourself. You should not pay employees for their safety efforts—you should already be paying them a decent salary for their efforts in sustainable production: "*It may seem crude to say you need to pay for quality and safe production, but you cannot separate out the quality of work from work safety and the implementation of the production plan*". According to this respondent, senior management dinners with outstanding employees from the production sites invited as special guests are a very effective form of non-material reward. At relatively informal face-to-face events like this, the leaders of a company can encourage their most proactive employees to share their opinions on how to improve safety and make immediate changes to implement the best suggestions. The special guests are bound to share their experiences of the evening with their colleagues, their pride for how much their bosses appreciate them, and their willingness to listen and act in response to their suggestions. Naturally, this kind of leadership recognition has a very positive impact on the employees concerned, but also works to motivate the whole workforce.

The HSE head of a fertilizer mining company cites several examples of effective non-material rewards for progress on safety issues. A company ran regular competitions for project proposals, with the first prize an all-expenses-paid dinner with the CEO. These proved enormously popular. One winner, a shop floor IT specialist, took the chance to tell the CEO all about his project idea. As a result, he was promoted and given the opportunity to implement his suggestion across the whole company. Management also ran a competition for best safety practice for all the work teams at the salt mine complex. Here however there was initially very little enthusiasm. The truth is that the salt miners already make very good money, so most of them are not interested in promotion: their current wages are already higher than those of engineers or lower managers. Finally, someone had an idea of how to motivate them. The company had an exclusive elite recreation center only available as a retreat for senior management—no other employee could use the center at any price, and visitors were not accepted. First prize for the competition was chosen to be a two-day retreat for them and their families. Rationally, the operators of the mine harvesters could have rented somewhere just as luxurious for a family holiday, but the exclusivity of the venue made it irresistibly attractive, and the organizers were overwhelmed with entries! The prestige for workers taking a break at the senior management's luxury exclusive resort meant they would do almost anything to try and win. The prize was nothing to the company, as the recreation center was idle much of the time, and the cost of organizing food and accommodation for the winning team and their families was relatively small. A seemingly insignificant non-financial incentive turned out to be very popular with a key group of workers managing one of the critical risks of the enterprise. Another successful example of non-material motivation is awarding a special designated parking space, reserved for the most outstanding employees, right at the entrance of the facility next to the spaces for the senior managers.

One regional manager responsible for the operation and maintenance of turbo-machinery in the power industry gives another example of a non-financial reward: in return for flagging up a significant safety risk, a worker is granted access to senior management. They write a report highlighting the issue in question, and the manager passes the report on to a very senior corporate figure such as the vice president. The employee does not get a quick promotion or a financial reward, but does get recognition for the good work. Next time promotions are being decided, the vice president may well remember the employee's name: a potential longer-term reward to encourage workers to take the initiative.

The head of HSE in a mining company recommends that, when bringing in any non-financial system to reward good safety practice, managers should keep in mind these three criteria: *significance*—the reward should be seen as substantial by employees; *effort and not outcome*[6]—if the employee made a credible effort to stop an incident which still could not be avoided, they should still be rewarded; *timeliness*—the reward must reach the recipient quickly to keep it linked and relevant—rewards made at the end of a year can be less effective.

Respondents also mentioned the following examples of non-material motivation to improve identification and transmission of safety risks:

- workers receive an award at the main annual corporate event at headquarters (all expenses paid);
- awarding medals to outstanding employees with their family members in attendance;

[6] Didier Sornette, Spencer Wheatley, Peter Cauwels, The fair reward problem: the illusion of success and how to solve it, Advances in Complex Systems, 2019, 22 (3) https://doi.org/10.1142/S021952591950005X.

- letters from a company to the families of employees who have excellent safety records in performing hazardous work, and have acted decisively to disclose problems to superiors in their area of responsibility;
- training of outstanding employees at other company enterprises around the country and the world;
- selection of outstanding employees to attend advanced training programs, all at the company's expense;
- specialists who have identified critical problems in an organization are invited to attend international conferences in their specialized field;
- posting photos of the best employees and a brief description of their safety record in the corporate media;
- photos of the best workers posted on the boards of honor at the entrance to site offices (celebrating the achievements of rank-and-file employees and not senior management);
- opportunity for employees to meet with senior managers over an informal lunch to discuss their ideas;
- excursions to other cities and abroad;
- tickets for worker and family to attend major sporting or cultural events;
- extra holiday allowance;
- lifetime health insurance.

The head of an oil production facility recommends asking the views of employees and lower managers before introducing a safety reward system in an industrial company. Each organization will have its own specificities, and a set of policies and recommendations to motivate their employees to disclose risks. It is better to avoid face-to-face surveys where senior management convene the whole workforce to get some idea of the general consensus. These tend to result in employees simply going along with whatever senior managers are proposing, although they may think differently. To get a real answer to a safety question, it is better to conduct anonymous surveys as nobody is afraid of the consequences, so this will invariably produce more honest—and therefore more useful—responses.

The responses of majority of respondents on the topic of non-material motivation vs material stimulation resonate a lot with a theory called "*Motivation crowding theory*",[7] which sets out a very detailed framework to explain why monetary rewards do not work. See also a well-known study of what motivates people by Prof. Frederick Herzberg.[8]

[7] Bruno Frey, Reto Jegen, Motivation Crowding Theory, Journal of Economic Surveys, 2001, 15 (5), pp. 589–611, https://en.wikipedia.org//wiki/Motivation_crowding_theory.

[8] Frederick Herzberg, The Motivation to Work, Wiley, 1959, https://www.slideshare.net/albert2mb/frederick-herzberg-twofactor-hygienemotivator-theory.

Examples of when material incentives are ineffective

The HSE manager of a mining company believes that offering any kind of material reward can cause ill-feeling from an employee's work colleagues—envy, resentment, accusations of careerism—and make their position very difficult. He cites an example in support of this in a company where he once worked. There, the senior management introduced a new reward scheme: employees who disclosed risks would get tablets and mobile phones. In the first month after the launch, about 70 messages were received but, before long, an unanticipated problem emerged and the flow of messages dried up completely. Previously, employees who revealed safety violations and production issues could remain incognito but, with this new scheme of being publicly rewarded for this information, workers quickly realized that it was simply not worth it. These employees may have got a shiny new mobile phone, but this came at a high price—they became virtual outcasts from the rest of the workforce, who despised them for having "*sold their soul*" for the sake of a phone or tablet.

An HSE manager of a manufacturing company gave the following example. One company rewarded any worker who reported any kind of safety hazard in their workplace with US $50 cash. Every month, workers identified thousands of these issues, but this information was worse than useless, as most of the reports were essentially fictitious, or at best distorted. Moreover, the processing of such a huge volume of low-value information noise consumed the resources of labor protection and industrial safety specialists, and made the more significant issues harder to identify.

A consultant in nuclear safety, with long experience in nuclear power plant operations, believes that the reward for disclosure of risk should reflect the amount of likely damage that a critical infrastructure company has avoided due to the timely disclosure of this specific risk. The respondent gave an example of how one organization tried to implement this principle. Many years ago, the respondent was part of a continuous improvement project team. Initially, as a motivation, senior management promised that the team members would receive 10% of the total savings that the improvement project would have brought to the company over the year. However, they had reckoned without anticipating the amazing success of the project's interventions. In the first quarter of the project, the saving was a massive US $3.7 million—leading to a bonus of $370 000, amounting to more than twice the combined annual salaries of the team members. As a result, senior management were forced to abandon their promises—never good for building trust in the integrity of the top brass—and instead each member of the project team was awarded just $1,500.

The last example also demonstrates how careful senior management need to be when determining a fair material reward for the disclosure of critical risks, based on a percentage of potential damage or saved resources. Employees could prevent a serious disaster that would have cost a company billions of dollars—the potential costs of managing and then clearing up the consequences of a major industrial disaster are almost limitless—and senior management would find themselves in the position of having to pay colossal bonuses to a small group of employees who took timely and appropriate action. Others will see these fantastic payouts and some more unscrupulous employees may then spot an opportunity—if they can engineer a similar sequence of events in their area of operations, they may get rewarded in the same way, and the higher the sum the more tempting it will be. An individual or small group of workers could consciously decide not to monitor or report risks at an early stage, but instead engineer a situation whereby critical infrastructure facilities are brought to a pre-critical state—and then, just before a catastrophic failure actually occurs, inform senior management about the presence of a critical risk, rubbing their hands in expectation of receiving a massive reward for supposedly averting a catastrophe. Such material reward systems can thus create the wrong climate within an organization and set the wrong goals for employees who operate critical infrastructure, so that instead of lowering the risk of major safety breaches, the initiative perversely makes serious accidents more likely to occur.

When material incentive for risk disclosure is effective

The HSE head of a fertilizer mining company also warns that managers should be extremely careful about offering financial incentives for accident reduction to employees who operate critical equipment. Better to motivate employees non-financially: public recognition from the top brass, respect and gratitude from colleagues, the admiration of their families, positive PR in corporate media, and so on. Nevertheless, the respondent cites a striking example of the successful use of a financial incentive to encourage employees to report safety violations. At one of their sites, his department analyzed all the accidents that had happened over several decades, trying to identify what mistakes employees had made in the immediate run-up to the accident. 90% of all cases were caused by the same seven mistakes. The most common accident was workers falling from a height when not wearing a safety harness. Armed with this statistical analysis, they began to think how to motivate employees to "*buckle up*" on the job. Numerous instructions, rules and penalties for non-compliance were already in place but, for some reason, none of this was enough to make the workers comply—even though it was their personal safety at stake. The company recognized that they simply did not have enough HSE specialists to have one permanently deployed monitoring every production site. Senior management decided that the only option was to threaten legal prosecution for any further violations. And to spot further infringements, they would use peer monitoring with rewards: any employee would be gifted the equivalent of the average monthly salary at their production site for a picture or video clearly showing another employee or contract worker operating at height without a harness. The evidence had to be sent to the HSE department and their representative would be called in to fully document the breach. Anyone caught violating the safety rules would face immediate dismissal. Two or three months passed and the HSE reps toured the site, but none of the employees had "*handed over*" any of their colleagues, saying that they would never betray a fellow worker. Then one day, the wife of an employee told him they urgently needed a new refrigerator: "*I don t care what you do—we have to have a new fridge*". He started racking his brains for a way to make some money. A few days later, walking through the site, he saw three contract workers up on the gantries with no harnesses in sight. A month's salary for sending evidence of a contractor using unsafe working practices—and here were three of them, practically begging to be filmed! He immediately took some pictures and videos on his smartphone and sent them to HSE. The offending workers were all suspended and banned from returning to the site, while their boss was called in for a dressing-down and issued with a fine. The company employee who had reported the violation duly received his three months' salary—and a grateful wife got her new fridge! He immediately told his colleagues about the episode and his cash reward—did his fellow workers shun him for "*grassing up*" the contractors? Not a bit of it: the next day, there were 80 documented cases of high-altitude safety violations in the HSE inbox! For the next few days, the entire HSE department were working overtime verifying similar cases. Back at the site, supervisors and junior managers began to complain that their employees were finding various pretexts to leave their posts and walk around the site looking for more offenders. The unwritten rule of solidarity that "*you never inform on your mates*" had evaporated overnight, driven out by the smell of hard cash. As promised, rewards were duly paid for each subsequent proven case. There were even cases where an offender tried to bribe his "*informer*" to avoid the sack, promising them the equivalent of the fine if they took the report no further. Most of the identified offenders were contract workers—over the first year, contractors paid the company US$ 270,000 in fines, and a significant part of this was awarded to the employees who had originally recorded the offences. A year or two before the introduction of this system, the enterprise had had an average of 11 unsecured falls from height per year: after operating the system for a year, the incidence dropped to only one case over the following five years. Within a remarkably short space of time, the previously near-universal attitude that it was shameful for workers to report a safety violation to their bosses had radically changed. Wearing a harness while working at height became effectively obligatory in the company, and not doing so became almost taboo—an offense not discussed among employees, rightly punishable by dismissal.

The same approach also worked well for cases of employees trying to fix machinery without shutting it down. However, it did not generally work at all well for reporting electrical risks. It is more difficult for a layperson to understand when electricians are working without turning off the electricity; and most of the work that electricians do on industrial sites is carried out in enclosed spaces where access is prohibited to unauthorized employees. In conclusion, managers are wise to be cautious where considering material financial incentives to influence attitudes to safety reporting. They can work very well when people are looking out for the misdemeanors of other employees whom they do not know, and the system cannot be misapplied to cheat in any way. It is also important that if a serious safety violation offense is proven, the worker concerned is immediately dismissed and does not have the chance to confront the ex-colleague who reported the transgression.

Results of responses to anonymous surveys within the pilot project: *There is a discussion within a company and the participants divide into two groups. The first believes that, to encourage employees to disclose dangerous actions of their colleagues (e.g. working at height without a harness), the management should offer material rewards (e.g. money, valuable gifts,* etc.*)—the most important consideration being to use any means necessary to identify and prevent safety violations. The second group believes that, to encourage employees to disclose dangerous actions by their colleagues, the management should not provide material rewards, because this may threaten the stability of the work teams. Furthermore, reporting safety violations is the professional duty of any responsible employee and should not require any financial incentives. Do you support the beliefs of the first group or the second group?*

	Fully support ideas of 1st group	Rather support ideas of 1st group	Fully support ideas of 2nd group	Rather support ideas of 2nd group	Difficult to answer	Number of respondents
All survey participants	9.6%	21.4%	20.0%	25.4%	23.6%	280
Senior management, heads of departments and directors of sites (middle managers)	8.3%	30.6%	13.9%	36.1%	11.1%	36
Lower managers: deputy directors of sites, chief engineers of sites, heads of workshops, heads and representatives of HSE services at sites	8.7%	21.2%	20.2%	26.0%	24.0%	104
Engineers, foremen, and shop floor employees who operate critical infrastructure at sites	10.7%	19.3%	21.4%	22.1%	26.4%	140

The head of a thermal power plant shares his experience of implementing an employee compensation scheme to reduce lost time injuries (LTIFR or Lost Time Injuries Frequency Rate) as part of a corporate drive to achieve zero injuries —"*target 0*". If there were no accidents or injuries on site in the first two-month period, then all employees would be paid a bonus. If no incidents were registered for a further two months, then the bonus was doubled. Prior to the launch of this system, shop floor employees did not appear to care if someone at the plant was injured, or how many days a worker was absent due to an accident. Afterwards however, any injury began to attract the attention of workers as it affected them all and meant they would lose their bonus. Employees began to ask questions: "*How did this happen? Something went wrong? How badly was the employee injured? How will this injury affect our bonus?*". This remuneration system was recognized as successful, because the number of injuries at the plant dropped significantly, with LTIFR dropping 15-fold in the first five years of the project, and down to zero by the sixth year. To obtain such successful results, the following conditions are necessary: workers need to be honest and trustworthy and fully buy into the scheme; managers need to commit to careful monitoring of worker injuries, despite the fact that along with the injured worker, they risk losing their bonus if the incident is officially recorded. However, it is likely that a system like this may encourage some employees who are injured to cover up what happened in order not to let their colleagues down.

The HSE manager of an oil company shrewdly recommends not rewarding specific employees, but rather entire teams. For example, if a driller at one of the remote fields reports some hazardous practice or safety violation and suggests how to prevent it in future, then the management should reward the whole drilling crew, for example by providing a TV and satellite antenna for their recreational area. Instead of resenting the whistleblower, the rest of the team are more likely to say something along the lines of "*Well, you may have shopped us to headquarters, but it was worth it now we can all watch TV together*". Rewarding the entire work team when one member reports pertinent risk information makes future disclosure a less frowned upon practice, and stops teams from marginalizing a proactive risk-aware employee. It also motivates other more cautious employees to follow the example of their more confident colleagues.

According to the head of HSE in a metallurgy company, direct material rewards should only be given when the risk information leads to improvements in an organization's safety systems. Employees can receive a share of the money a company has saved through introducing new safety solutions—changes that increase labor productivity while maintaining or improving existing safety levels.

Employees need to be clear about what behaviors their company rewards and what it penalizes

The HSE director of an oil company believes that employees need to be exposed to both negative and positive motivations. If an employee deliberately violates safety regulations, they must be strictly penalized in public, up to and including dismissal. However, if they infringe these rules unintentionally, because they were not fully trained or did not understand the risks associated with the work, then a company should invest in their training. When the worker subsequently applies their training and completes their work safely, they should receive acknowledgement and public praise. In this way, employees learn what actions their company will penalize and what it will reward. The same principle should be applied to reporting information about risks: a company should be seen to reward employees for proactive disclosure and penalize them for concealment, distortion and delay.

Consistency is key when dealing with disciplinary issues and sackings. If the transgression is not too serious, senior management should consider giving employees a chance to improve. The "*three strikes and you're out*" approach is common: an employee is given two opportunities to rectify a first instance of unacceptable conduct, but after that will be fired. Employees generally understand the logic and necessity of dismissing employees who repeatedly fail to perform satisfactorily, and will accept it as the ultimate sanction as long as the rules are applied fairly across the board.

Results of responses to anonymous surveys within the framework of the pilot project: *Managers should reward employees in the presence of other workers, but penalize them only in a private conversation and under strict confidentiality.*

	Strongly agree	Rather agree	Rather disagree	Strongly disagree	Difficult to answer	Number of respondents
All survey participants	28.6%	36.8%	23.2%	4.6%	6.8%	280
Senior management, heads of departments and directors of sites (middle managers)	38.9%	36.1%	16.7%	5.6%	2.8%	36
Lower managers: deputy directors of sites, chief engineers of sites, heads of workshops, heads and representatives of HSE services at sites	24.0%	36.5%	25.0%	6.7%	7.7%	104
Engineers, foremen, and shop floor employees who operate critical infrastructure at sites	29.3%	37.1%	23.6%	2.9%	7.1%	140

Interpretation of responses: Although the majority of responders supported public rewarding and private penalization, some others, especially lower managers and shop floor employees, commented that public reprimand of negligent employees is necessary to show the rest of the workforce that their behavior was completely unacceptable. If the violation is left unpunished, then employees are left without clear messaging, and may think that the behavior is acceptable.

Discussion about KPIs and bonus models

The head of HSE at a petrochemical company believes employees start treating safety problems seriously when the reporting of identified risks and faults in equipment performance are included in their KPIs.

The HSE director of an electric power company believes that if senior managers' bonuses are calculated to include markdowns for worker deaths or major accidents, they will be more interested in receiving information about equipment problems or safety violations that could lead to serious accidents at work. He recommends that when a company is looking to improve corporate culture, a comprehensive indicator should be included in the bonus model to motivate action to reduce accidents and injuries. A reasonable starting ratio would be 75% for a decrease in the total recordable injury frequency rate (TRIFR), and 25% for measures to improve work safety conditions—gradually moving towards 25% for the TRIFR and 75% for improving the safety of the workplace. Using this model, workers and managers are progressively penalized less for the fact that accidents have occurred and increasingly for a failure to adequately implement safety improvements that could have prevented future accidents occurring at all.

The HSE director of a metallurgy company does not entirely support this argument. He maintains that it is unwise to include indicators for reducing injuries or accidents in management KPIs, because this will encourage safety problems to be swept under the carpet. Including an indicator for incidents that only happen once and then *do not happen again* is far more effective. If incidents do not recur, this suggests the company has drawn appropriate conclusions from the original incident and introduced risk reduction measures that have proved effective in preventing them. Senior management should always encourage managers and employees to introduce changes to prevent the recurrence of a safety problem. This requires identifying the root cause of the issue—poor work protocols, faulty machinery, lack of training. Specific measures must then be introduced to reduce the risk, which must include the procurement of adequate resources to implement these measures; and they must be maintained in the long term. A "*non-recurrence*" KPI can be effectively extended down to lower-level managers, encouraging them to work closely with specific teams on the shop floor to embed the new safety measures and prevent any recurrence. For example, imagine that in February of a given year, there is a death in the workshop. And suppose that through the company's remuneration scheme, this incident leads to a 20% reduction in the bonus for that shop manager. This will have the effect of removing any motivation for them to change for the better over the remaining 10 months, because at the end of the year their bonus will still be reduced. However, if there is also an additional bonus for *preventing* the recurrence of a similar incident, then the shop manager will be encouraged to immediately look to implement improvements to prevent another death. If these activities are credible and effective, then most of what was deducted —say 10–15% of the manager's total annual bonus—may be returned. A scheme like this was introduced at one oil company as the "*Five Stars*" system. At the beginning of the year, all employees start with five stars (100% of their agreed annual bonus). They can finish the year with up to seven stars if they make relevant suggestions for safety improvements. A leader with 20 subordinates would thus start with 100 stars, and at the end of the year their team tally would be the final total of their combined scores. Stars are deducted for errors and poor safety behavior, but these lost stars can be returned if that employee does not repeat the behavior, thus constantly motivating them to prevent a recurrence. This system also motivates managers to encourage their employees to improve their performance and avoid any mistakes, knowing that their own annual bonus is contingent on how their subordinates get on.

The HSE head of an industrial production company that works with hazardous chemical processes believes that as part of their job descriptions, employees should be required to inform their superiors about critical risks. The KPIs of both employees and managers need to include bonuses for the disclosure of critical risks, but exclude more minor risks in order to minimize information noise. If a one-off risk assessment team finds significant safety issues, they are tasked with running an internal investigation and filing a report to identify causes and set out recommended fully costed measures to mitigate the risk. The risk assessment team's bonus is dependent on the results of the investigation. The report is submitted to the risk management committee, where a decision is made on the allocation of resources and the appointment of executors. When the risk has been eliminated, a summative report is written—usually called the "*lesson*". This is distributed throughout the company, so that employees and managers from all departments can see by example how a particular risk has been successfully addressed. This positive example of one specific department mitigating a given risk motivates others to take the initiative, and reduce critical risks in their own area of operations. With careful planning and successful follow-through, a company can learn how to calmly and cooperatively discuss risk problems, propose and implement solutions, and learn from mistakes.

The choice of remuneration depends on the level of maturity of a company

The HSE vice president of an oilfield services company thinks any channel is good if it transmits accurate and timely critical risk information to the right place, thus saving injuries and lives. He cites several examples where at the beginning of a program to implement better safety communication systems, companies with reactive risk management systems paid employees money for filling out STOP cards—even though a fair proportion of them did not contain any relevant information about serious risks. The main learning point was to encourage employees to always fill out a STOP card when they saw dangerous conditions or actions. Once a company had successfully advanced to more proactive methods, the financial rewards disappeared. The motive for employees to continue improving risk transmission had changed—but the useful habit of quickly filling and sending STOP cards was by then firmly established.

The head of HSE department at an oil company believes that material incentives should be used with caution, and only at the initial stages of implementing a new risk information transmission system. For example, employees should receive monetary rewards—privately of course—for informing about colleagues arriving drunk at the workplace. Material incentives for other safe practice at the workplace, like preemptively stopping production or otherwise preventing accidents, should not be through direct cash payments but rather in the form of gifts: caps, mugs, key fobs, smart-phones, family holidays, and so on. One way he looks at reward incentives is through the application of epidemiological theory. At the start of any initiative, there will only be a small group of "*carriers*" of the new values and, to help create this group, it may be necessary to use material incentives. Slowly, these "*carriers*" will begin to "*infect*" other employees and gradually the new values—in this case, a more positive attitude to open and honest risk communication—develop in the workforce. By this stage, new followers are already being motivated by various non-material incentives: recognition in the group, the desire to receive praise from leaders, and so on. It is important to understand that several approaches may need to be combined for success—it is impossible to overcome all the barriers to introducing a system for open risk communication with just one motivational tool.

The HSE manager of an oil company strongly disagrees with the opinion, held by some respondents, that rewarding employees for disclosing information will always lead to a flood of irrelevant and false messages from careerists and scammers. In his experience, most companies with reactive corporate cultures get little or no information coming up the hierarchy from employees. He maintains that appealing to employees' sense of duty—without any promises of material reward—is fine once the process of reporting risks and problems has been successfully established. Until then, employees should be motivated by both financial and non-material rewards.

According to an HSE director in the petrochemical industry, companies must go through a number of stages when seeking to motivate employees, so that they clearly understand what a company will penalize and what it will reward:

- material demotivation (fear of losing job);
- material motivation (additional income);
- non-material demotivation (blame and condemnation);
- non-material motivation (recognition).

The HSE manager of a metallurgy company believes that it is not just money that is important to people, but also recognition. His father was a military officer. As such, he was paid well above the national average, but the subject of money or income was never raised at home. Rather, they often discussed the importance of duty, of service to the country. When his father received awards for his years of professional service, the son was very pleased and proud of his father, and he is sure his father was pleased that his son was proud of him. The respondent believes that a similar principle of service and duty should be promoted in critical infrastructure companies: they too are serving their country, and many people depend on them doing their job safely and professionally. Having said that, employees must first have a decent income so that they can cover all their basic needs. Only then can they aspire to achieve anything beyond fighting for a loaf of bread to feed their family. With their basic needs met, employees will be ready to come to work not just for the money, but for higher ideals: to look out for others and save lives, to do their professional duty, to be recognized and respected by their colleagues—all the non-material factors that provide enduring motivation. This is what really inspires workers to give their best and gain endorsement by their workmates, their managers, and the wider community.

In general, there is broad agreement among the respondents that what works best in terms of rewards is dependent on the maturity level of a company. Material incentives work better in reactive corporate cultures, and less concrete motivations work best in proactive cultures. In companies where there is a reactive corporate culture and lower wages, it is better to use material incentives first. In companies with a proactive safety culture and higher salaries, it is better to use non-material incentives. In companies in transition, there is a need to employ a mixed reward menu.

It is necessary to consider the cultural characteristics of the country and the norms that exist in the workforce

A risk management consultant in the oil and gas industry insists that before choosing the most effective rewards to encourage employees to report critical risks, senior managers of multinational companies must have a good understanding of the culture of the country where a particular site is located. If the production is located in a collectivist country, then group awards will generally be more effective: the team receives the reward, not just one person, otherwise it may well appear dismissive of the collective work ethic. Senior managers should publicly present a special certificate to all team members in front of the entire workforce. If the production is located in a country where a more individualistic culture is prevalent, then the respondent believes that many workers will not attach much value to team certificates or recognition from their colleagues: they just value more personal recognition.

Accident prevention and risk disclosure is the professional duty of every responsible employee, so additional rewards of any type should not be necessary

The vice president of an electricity company considers that employees should ideally be enticed neither by psychological benefits nor financial gain for actively disclosing risks. The identification of technological risks is the professional duty of every employee working in a critical infrastructure company. The main motivations for employees to discharge this duty are:

- disclosing information about risks, problems or defects in their area of responsibility is the main way that these issues come to the attention of the leadership;
- leaders need this information in order to tackle the issues and reduce accidents;
- this will make their workplace safer, more comfortable and modern, and bring benefits to the whole organization;
- they and their colleagues will have better career prospects, with fewer incidents and shutdowns and a stable flow of work for the company.

People will only start taking the initiative to report safety issues at work if they see that, once these risks and problems are identified, they are promptly and appropriately dealt with. According to the respondent, if there are no other specific incentives for disclosing risks, these problems will only be reported by conscientious workers who need no motivation beyond the fact that it is the right thing to do. It will filter out selfish careerists, or groups who are trying to politicize an organization's problems to use as a weapon against managers or other workers.

Results of answers to anonymous surveys within the framework of the pilot project: *Identifying safety risks and disclosing them to colleagues and management to prevent accidents is the professional duty of every responsible employee, therefore neither material nor non-material rewards for disclosure of risks are necessary.*

	Strongly agree	Rather agree	Rather disagree	Strongly disagree	Difficult to answer	Number of respondents
All survey participants	28.6%	43.6%	18.6%	3.2%	6.1%	280
Senior management, heads of departments and directors of sites (middle managers)	36.1%	27.8%	30.6%	5.6%	0.0%	36
Lower managers: deputy directors of sites, chief engineers of sites, heads of workshops, heads and representatives of HSE services at sites	29.8%	51.9%	13.5%	1.0%	3.8%	104
Engineers, foremen, and shop floor employees who operate critical infrastructure at sites	25.7%	41.4%	19.3%	4.3%	9.3%	140

Other executives interviewed agreed that disclosing risk information to prevent an accident should be viewed as the direct responsibility of every employee of a critical infrastructure company, and the duty of all professionals towards themselves, colleagues, the company, and the wider community.

3.11 Other Recommendations for Improving the Quality of Risk Communication in Critical Infrastructure Companies

In addition to the Top 10 recommendations discussed above, the executives interviewed also suggested the following solutions to improve the quality of risk communication.

The whole company must learn lessons from tragic accidents

Usually, when a tragic industrial accident happens, only a few people at that site will be directly involved, but the lessons learnt from investigation of these incidents must be disseminated throughout the whole company, so that the entire workforce can learn and not make the same error at their facility.

Results of responses to anonymous surveys within the framework of the pilot project: *Do you want to receive information about serious incidents that have occurred across all of your organization's facilities, including a brief analysis of the main reasons behind each one?*

	Strongly agree	Rather agree	Rather disagree	Strongly disagree	Difficult to answer	Number of respondents
All survey participants	39.8%	45.4%	10.4%	1.1%	3.3%	280
Senior management, heads of departments and directors of sites (middle managers)	58.3%	41.7%	0.0%	0.0%	0.0%	36
Lower managers: deputy directors of sites, chief engineers of sites, heads of workshops, heads and representatives of HSE services at sites	42.3%	46.2%	6.7%	1.0%	3.8%	104
Engineers, foremen, and shop floor employees who operate critical infrastructure at sites	33.6%	45.7%	15.0%	1.4%	4.3%	140

Interpretation of responses: senior management demonstrate the greatest interest in obtaining information about serious corporate incidents with an analysis of the causes. The lower the respondents sit in the hierarchy of the company, the less interest they have in this information.

Several respondents from the HSE department of the same oil company reveal that they film "*crash videos*". Their senior managers require contractors to agree that if they are involved in a major incident at a drilling rig, they will make a training film highlighting the mistakes that led to the incident, setting out the lessons that should be learned and how to prevent similar occurrences in the future. Having to make these videos works as a kind of penalty for negligent work. They are shown to workers in production units throughout the company. In making the videos, victims of the accident are always interviewed, so that they can independently reflect on their errors and omissions and give personal and pertinent advice to other drillers on how to avoid making the same mistakes. The interviewers aim to really draw out people s feelings, in the belief that this emotional content will motivate other drillers not to repeat their colleagues' mistakes.

A good example of this was a video made after a fire on a drilling platform. One of the rig operators was trapped in a flaming cabin for 40 s, and survived with bad burns to his hands, but his colleague was burned to death. The survivor talked about his experience, but also said that, if he ever saw a safety violation on the rig, he would never work there again. The interview was professionally filmed and conveyed the surviving driller's emotions very powerfully. The company also wanted to include interviews with the relatives of injured and deceased employees, but in the end decided against asking them to talk about such painful feelings and memories.

One of the most effective ways to reduce injuries is to involve the families of the drillers. They came up with the idea of sending surveillance footage to wives and mothers, showing their husbands and sons violating safety regulations at work. The idea was that close family members could have a greater influence than anyone else on modifying the behavior of workers in such a dangerous industry, urging them to comply with safety rules all the time. Safety violations using equipment as powerful and dangerous as drilling rigs can, in a moment of inattention, permanently disable and kill people, causing tremendous suffering to their families. Another idea was to invite the families of workers to take part in an internal corporate video and broadcast it in the workplace and to workers' homes. Children could tell parents working on a hazardous production site that they are looking forward to them coming home safe and sound.

Sharing the lessons learned from tragic experiences across an industry

According to these last respondents, insufficient knowledge about previous accidents and disasters is a major industry-wide problem, not only in their sector but in critical infrastructure industries at large. At best, a company analyzes and documents its own corporate experience of emergency situations to educate its own workforce and improve safety compliance. However, most do not learn the lessons from mistakes in other organizations, because there is little or no exchange of information between competing companies within a single industry—or between industries—even though they share similar critical risks. Few leaders are willing to disclose their mistakes to competitors, so in many countries no mechanism exists to facilitate the exchange of risk incidence information between companies operating critical infrastructure. Many countries have no single database of major incidents across an industry. The respondents reported that they only learn about incidents in other organizations through friendships and more informal channels, or from contractors who work simultaneously with different oil companies.

An initiative to improve this situation needs to come from regulators or non-profit organizations within an industry, and it should extend to include exchange of information across borders through international industry organizations. For instance, in the nuclear industry the International Atomic Energy Agency (IAEA) shares positive and negative experiences from all nuclear power plants worldwide. There are also other initiatives to gather precursors and near misses through national regulators, international organizations, and academic institutions.[9,10,11] In the European electric power transmission sector, companies

[9] Ali Ayoub, Andrej Stankovski, Wolfgang Kröger, Didier Sornette, The ETH Zurich Curated Nuclear Events Database: Layout, Event Classification, and Analysis of Contributing Factors, Reliability Engineering and System Safety, 2021, Volume 213, https://doi.org/10.1016/j.ress.2021.107781.

[10] Ali Ayoub, Andrej Stankovski, Wolfgang Kröger, and Didier Sornette, Precursors and startling lessons: Statistical analysis of 1250 events with safety significance from the civil nuclear sector, Reliability Engineering and System Safety, 2021, Volume 214, https://doi.org/10.1016/j.ress.2021.107820.

[11] Ali Ayoub, Andrej Stankovski, Spencer Wheatley, Wolfgang Kröger, Didier Sornette, ETHZ Curated Nuclear Events Database, 2020, https://emeritus.er.ethz.ch/nuclear-energy/link.html.

operating the national transmission infrastructure of 35 countries are associated into the European Network of Transmission System Operators for Electricity (ENTSO-E), sharing best practices and accident experience on a large continental scale. In the aviation sector, 193 countries cooperate through the International Civil Aviation Organization (ICAO), sharing information on air safety worldwide and targeting the core areas of global aviation safety planning, oversight, and risk mitigation.

In other critical infrastructure sectors, however, such international exchange of information simply does not exist, or is limited to only the very biggest accidents that receive a lot of media coverage.

The following suggestion was made by the quality director at an international electricity company. National and international industry associations should consider creating industrial safety benchmarks to guide their members operating critical infrastructure. The presence of such benchmarks—which are based on the accumulation and exchange of key information on the very best practice across the industry—will encourage companies to compare themselves against the highest-performing organizations worldwide. It will offer tried and tested ways of improving their safety and reporting processes, and motivate them to raise their standards up to a generally recognized industry benchmark.

Results of responses to anonymous surveys within the framework of the pilot project: *Do you want to receive information about serious incidents (and their causes) that have occurred in your industry in this country and across the world?*

	Strongly agree	Rather agree	Rather disagree	Strongly disagree	Difficult to answer	Number of respondents
All survey participants	33.6%	50.7%	13.6%	1.1%	1.1%	280
Senior management, heads of departments and directors of sites (middle managers)	52.8%	47.2%	0.0%	0.0%	0.0%	36
Lower managers: deputy directors of sites, chief engineers of sites, heads of workshops, heads and representatives of HSE services at sites	41.3%	47.1%	9.6%	1.0%	1.0%	104
Engineers, foremen, and shop floor employees who operate critical infrastructure at sites	22.9%	54.3%	20.0%	1.4%	1.4%	140

Interpretation of responses: senior management show the highest interest in obtaining information about various incidents (and their causes) in their industry. The lower the position of respondents in the hierarchy of the company, the less interested they are in receiving this information.

Intra-corporate opinion polls

Sometimes, leaders genuinely cannot work out what they are doing wrong. One way to combat this ignorance is to conduct an anonymous internal survey every year among employees and managers. Amongst other things, there should be a question about how much they trust managers at various levels, and how willing they are to report safety concerns to their bosses. Once the problematic areas have been identified, managers are in a better position to understand the nature and location of the obstacles, and can then identify which areas to prioritize to improve the situation. Regular anonymous surveys of employees are invariably a useful source of objective feedback within an organization.

Chapter 4
A Pilot Project—Introducing a System for Transmitting Information on Safety and Technological Problems Within a Critical Infrastructure Company

At the end of 2021, the first author of this handbook began a pilot project for an industrial company that is a world leader in its field. By agreement with this company, the following information must remain confidential and not be disclosed to the general public: the company name, the nature of its operations, the countries where it operates, the specific sites included in the project and the names of employees.

The pilot project tested various methods for significantly improving the quality and speed of reporting information about safety and technological problems within the organization. These methods were formulated based on: (I) the recommendations of 100 leaders representing critical infrastructure companies from around the world (Chap. 3) and (II) analysis of the causes of dozens of industrial accidents in different countries of the world, which revealed numerous occasions where information about risks and problems had been concealed or misreported (Sects. 2.1 and 2.2).

The project involved more than 400 employees of the company, from senior management to shop floor employees. During the first few months of the project, shop floor employees and line managers disclosed seven critical risks to senior management that they believed had the potential to lead to accidents resulting in either the death of personnel, long-term decommissioning of production facilities or significant environmental issues. All these risks were quickly addressed by senior management and production site leaders. In several cases, these prompt disclosures and interventions prevented serious incidents from developing. Within the first few months of the introduction of the project, employees disclosed to senior management 104 other safety and technological problems that were compromising the industrial safety of four of the company's production sites. Most of these issues have also now been resolved.

The success of the project indicates that, with suitable information transmission systems in place, shop floor employees and line managers are willing to disclose serious safety and technological problems in their area of responsibility to senior management in order to prevent emergencies.

D. Chernov et al., *Averting Disaster Before It Strikes*,
https://doi.org/10.1007/978-3-031-30772-0_4

This chapter provides a description of the key features of this pilot project, which was carried out over a period of 10 months from December 2021 to October 2022. Rather than taking a chronological approach, the findings of the pilot project are described from a system-based viewpoint, focusing on factors that have shown to improve the quality of risk information transmission. The project findings are, therefore, presented as a catalogue of initiatives and recommendations that can be adopted by any critical infrastructure company looking to fundamentally improve the quality and speed of risk information transmission up the corporate hierarchy.

I. THE MAIN OWNER PRIORITIZES SAFETY OVER PROFIT AND PRODUCTIVITY

Prior to the pilot project, the company experienced several serious accidents that had a negative impact on its finances. In response, the main owner replaced part of the senior management team responsible for production and industrial safety. He set them the following task: *"Further accidents are unacceptable to me, so I am willing to give you the resources and money you need to prevent them. I do not want the company to be focused on achieving production targets at any cost. The most important thing for me is to prevent further injuries or deaths to the workers, and make sure there are no more serious accidents at our production sites".*

The results of the survey of 100 managers from around the world (Sect. 2.3) show that the main reason (cited by 58% of respondents) why managers are reluctant to receive information about safety and technological risks is the high cost of resolving these problems, and the fact that owners and shareholders prioritize financial and operational performance over safety issues.

In this company, it was the main owner who established safety as his priority and undertook to provide senior management with both the finance and the time to control risks. The owner and senior management have repeatedly emphasized to their subordinates that safety is the most important priority across the entire company's operations. Permission was granted to stop production and carry out scheduled, preventive or emergency repairs. The production plan can always be adjusted—the central goal is to control risks, and so prevent accidents and injuries.

Almost certainly, it was the owner's clear statement that he prioritizes safety that encouraged the senior management to adopt a very positive attitude to improving the systems for reporting critical safety and technological problems at the industrial sites entrusted to them.

However, according to senior management, there is a persistent corporate culture among line managers and workers, who continue to believe that the most important goal is to meet the production plan, whatever fine remarks about safety the senior management may say to the contrary. It is possible that some middle managers (heads of production sites and their deputies) are still telling their subordinates that they should not listen to the words of senior management about the priority of safety—*"they are just saying what they know they have to say"*. Instead, workers are encouraged to focus on the implementation of the production plan and hitting their targets. As in many large industrial organizations, the employees and managers of this company grew up in a paradigm where the work focus has always been meeting production targets. Such deeply ingrained beliefs are very difficult to shift in a short time. It will take many years of targeted action by senior management to demonstrate that things have really changed, and that safety truly does take priority over financial and production objectives—or at least that there is a corporate mechanism to ensure an acceptable balance between safety, finance, and production.

Results of responses to anonymous surveys within the framework of the pilot project: *The fulfillment of the production plan is not the main priority of a company —the highest priority is the safety of work and production processes in order to prevent emergency situations and accidents.*

What is the level of awareness of these priorities among managers and employees at various levels in the company?

	Very high awareness of these priorities	Many are aware of these priorities	Some are aware of these priorities	Few are aware of these priorities	No one is aware of these priorities	Difficult to answer	Number of respondents
All survey participants	22.4%	42.0%	16.6%	8.3%	0.3%	10.4%	326
Senior management, heads of departments and directors of sites (middle managers)	43.9%	39.0%	12.2%	4.9%	0.0%	0.0%	41
Lower managers: deputy directors of sites, chief engineers of sites, heads of workshops, heads and representatives of the HSE services at sites	27.8%	48.1%	10.2%	8.3%	0.9%	4.6%	108
Engineers, foremen, and shop floor employees who operate critical infrastructure at sites	14.1%	39.0%	21.5%	9.0%	0.0%	16.4%	177

Interpretation of responses: there is a very significant difference between the answers of senior management and those of shop floor employees, which suggests that the message from the owner and senior management about the priority of the safety of work processes over the execution of the production plan is not yet received/ understood/ accepted by lower-level employees.

Results of responses to anonymous surveys within the framework of the pilot project: *The fulfillment of the production plan is not the main priority of a company —the highest priority is the safety of work and production processes in order to prevent emergency situations and accidents.*

How do things work in reality when choosing priorities in the company?

	Priority is always given to the safety of work and production processes	Priority is usually given to the safety of work and production processes	Priority is usually given to the execution of the production plan	Priority is always given to the execution of the production plan	Difficult to answer	Number of respondents
All survey participants	23.0%	28.8%	31.8%	8.9%	8.0%	326
Senior management, heads of departments and directors of sites (middle managers)	41.5%	24.4%	19.5%	12.2%	2.4%	41
Lower managers: deputy directors of sites, chief engineers of sites, heads of workshops, heads and representatives of the HSE services at sites	27.8%	26.9%	35.2%	5.6%	4.6%	108
Engineers, foremen, and shop floor employees who operate critical infrastructure at sites	15.8%	31.1%	31.6%	10.2%	11.3%	177

Interpretation of responses: the responses from senior management suggest they (unconditionally or usually) give preference to the safety of work and production processes when making decisions, while people at the bottom of the hierarchy are placed in such a position that, regardless of the stated priority of safety, they are often forced to prioritize the production plan.

Lessons learned (from this experience for other critical infrastructure companies that want to fundamentally change the quality and speed of risk reporting within their organization): before implementing any changes, it is necessary to obtain the following agreements from owners and key shareholders: (I) they agree to view their investment in a company as long-term; (II) they are willing to be informed of serious technological and operational problems; (III) they are willing to devote significant resources to solving these problems when they are reported, in order to reduce the likelihood of major accidents and increase the reliability of critical infrastructure in the long term.

If senior management have support from owners and shareholders, then executives will be grateful to receive disclosures from their subordinates about serious risks and problems. For senior management, this information will be viewed as *"good news"*, as it will allow timely identification and mitigation of risks, which in turn will have a positive impact on a company's finances.

If owners and shareholders consider their investments in an organization as short-term and are unwilling to reduce the profit margin of the business—which is clearly likely to occur when accumulated critical risks are addressed—then it is not recommended to launch a similar project there. In the absence of full support from owners and shareholders, implementation is likely to fail and have negative impacts. If senior management are not given enough resources to mitigate any serious risk or problems that their subordinates identify, then employees will quickly lose trust in the new system. They will soon revert to their previous long-standing behavior and keep quiet about safety and technological problems.

II. THE COMPANY HAD PREVIOUS NEGATIVE EXPERIENCE ARISING FROM DELAYED TRANSMISSION OF RISK INFORMATION

The company had experienced a serious accident in the past. In the early hours of the accident, the managers responsible underestimated the scale of the problem. As a result, they timely informed their superiors about an incident, but not about the accident. There was no evidence that key information was deliberately concealed: it was simply that in the beginning, the line managers involved were confident that the incident was local, and that they could manage the situation independently. The real scale of the accident and the significant resources required to manage the escalating consequences only became apparent a day afterwards. It was then that senior managers were informed about real scale of the accident. If the real scale of the accident had been promptly reported to them, then the consequences of the accident would have been much less serious because appropriate decisions could have been made and resources deployed more rapidly.

This company then had a further serious accident where line managers again underestimated the scale of the evolving situation and failed to inform their superiors promptly enough. The result again was a major accident.

Fortunately, neither accident resulted in deaths, but nevertheless they had a significant impact on the company's finances. In critical infrastructure companies, it is almost always much cheaper to deal with risks before they cause an accident, rather than to manage the fall-out from a serious incident.

These previous experiences motivated senior management to find ways to improve communication and action coordination between different levels of management, both before and during an emergency. They openly acknowledged the problem: *"We, as leaders, do not know what is happening at the bottom of the corporate hierarchy. We need to improve critical risk communication".*

Results of responses to anonymous surveys within the framework of the pilot project: *Imagine that employees at an industrial site detect a serious technological risk. Based on your experience, what do you think employees will do when this risk is discovered?*

	Employees will immediately report the risk to their supervisor, who, in turn, will quickly inform the site management. A special group will be promptly established to carry out a risk analysis. If the risk is deemed unacceptable, then production work will be halted, and measures implemented to control the identified situation	Employees will report the risk to their supervisor, but this information will not be transmitted further up the hierarchy. Employees and their manager will try to independently address the risk using their own resources; it is possible that work will be stopped or reduced at the site	Employees will not report the risk to their immediate supervisor, but will try to eliminate the identified risk on their own, and production work will not be stopped	Employees will ignore the risk as they work in a dangerous industrial production facility, where there are already many serious hazards; employees simply do not have time to respond to all potential risks, because they need to keep working and fulfill the production plan	Difficult to answer	Number of responden
All survey participants	38.6%	48.9%	3.9%	2.5%	6.1%	280
Senior management, heads of departments and directors of sites (middle managers)	41.7%	44.4%	8.3%	2.8%	2.8%	36
Lower managers: deputy directors of sites, chief engineers of sites, heads of workshops, heads and representatives of the HSE services at sites	43.3%	44.2%	4.8%	2.9%	4.8%	104
Engineers, foremen, and shop floor employees who operate critical infrastructure at sites	34.3%	53.6%	2.1%	2.1%	7.9%	140

Interpretation of responses: a significant proportion of the respondents across all three groups believe that a serious risk will be promptly transmitted up through the entire management hierarchy of the company. Nevertheless, most lower-level employees believe that information about the risk will not go beyond the level of their immediate supervisors. They, together with their subordinates, will try to address the situation on their own, and not bother their superiors.

Lessons learned: company managers should analyze any previous serious incidents that have occurred within their organization. This will help establish whether there were incidents that occurred after warnings from subordinates had been ignored, and whether there were delays in transmission of critical information up or down the hierarchy after the onset of an emergency. It is also important to analyze a company's successful experiences: times when efficient communication about risks, both before and during emergencies, made it possible to tackle problems before they became too serious. A retrospective analysis of both successful and ineffective management of risk situations within a company will provide invaluable data for all the participants of the new project: it should help them identify specific weaknesses in the current system, and practices that can be expanded where risk transmission worked well, and serious incidents were avoided. It is also recommended to create a special pool of employees and managers who can share positive examples from their own practice that other participants can then develop for their own situation.

III. HOLDING A SPECIAL SEMINAR ON COMMUNICATION AND DECISION MAKING IN THE EVENT OF ACCIDENTS AT LARGE TECHNOLOGICAL FACILITIES

In July 2021, the first author of the handbook received an offer from the company to conduct a seminar on management decisions and communication in the event of major infrastructure accidents. The company's leaders wanted to improve the quality of their team's response to emergency situations and make their communication with external audiences more efficient during emergencies.

As a result, a two-day seminar was held in October 2021, bringing together more than 100 leaders of this company. The research presented demonstrated that the first step to effective emergency management is prompt and reliable reporting about any incident from heads of production sites to senior management at headquarters. This should include detailed, accurate, and transparent information: where, when, and how the incident occurred; an objective analysis of the current and possible extent of the emergency; and their ideas on the best response to the situation, including resources required, production shutdowns and time required before the emergency is under control. The first part of the seminar was devoted to the problems of transmitting risk information up the hierarchy during the early minutes and hours after an accident. Examples of relevant accidents within critical infrastructure organizations were given.[1,2,3] It was emphasized that, when site managers understated the scale of an emergency or concealed information about the reality of the situation at their facility during reporting to a company's headquarters, the situation was made worse. Such concealment of the truth typically leads to: (I) a delayed and inadequate response by senior management and the entire company to the developing crisis; (II) absence of key senior managers and specialists at the scene of the emergency; (III) critical delay of top-level decisions on the allocation of emergency resources for controlling the situation; (IV) an *"information vacuum"* around the accident, which is filled by rumors, misinformation and panic at the site and further afield, for

---

[1] Dmitry Chernov, Didier Sornette, Man-made Catastrophes and Risk Information Concealment: Case Studies of Major Disasters and Human Fallibility, Springer, 2016, https://www.springer.com/gp/book/9783319242996.

[2] Dmitry Chernov, Didier Sornette, Critical Risks of Different Economic Sectors (Based on the Analysis of More Than 500 Incidents, Accidents and Disasters), Springer, 2020, https://link.springer.com/book/10.1007/978-3-030-25034-8.

[3] Dmitry Chernov, Didier Sornette, Giovanni Sansavini, Ali Ayoub, Don't Tell the Boss! How poor communication on risks within organizations causes major catastrophes, Springer, 2022, https://link.springer.com/book/10.1007/978-3-031-05206-4.

example among people living close to the site and the media. It later transpired that this issue was of particular concern to the senior management of this company, after the already mentioned negative experience of two previous serious incidents.

As part of the seminar, an anonymous survey was conducted to hear the opinions of employees regarding the current situation of risk information transmission within the company. The survey showed that in the overwhelming majority of cases, senior management expected that the reports of their subordinates on existing risks would contain distorted or concealed information. This served to further confirm their belief that significant problems existed in the reporting of objective risk information up the corporate hierarchy.

When discussing the results of the seminar and the anonymous survey, the first author of the handbook suggested to senior managers (the Senior Vice President in charge of production at the key industrial site of the company [SVP], and the Vice President in charge of HSE) that they launch a pilot project to radically improve the quality and speed of reporting about safety and technological risks from shop floor employees to senior management. In November 2021, senior management gave the green light for the implementation of this pilot project.

Results of responses to anonymous surveys within the framework of the pilot project: *In your experience, how often are employees afraid of expressing disagreement with their superiors?*

	Constantly	Often	From time to time	Very rarely	Never	Number of respondents
All survey participants	4.9%	42.3%	39.3%	10.7%	2.8%	326
Senior management, heads of departments and directors of sites (middle managers)	2.4%	39.0%	36.6%	22.0%	0.0%	41
Lower managers: deputy directors of sites, chief engineers of sites, heads of workshops, heads and representatives of the HSE services at sites	2.8%	12.0%	44.4%	35.2%	5.6%	108
Engineers, foremen, and shop floor employees who operate critical infrastructure at sites	5.1%	47.5%	36.7%	7.3%	3.4%	177

Interpretation of responses: 86% of all survey participants (all those who responded *"Constantly"*, *"Often"*, *"From time to time"*) considered that employees of the company are afraid to disagree with the opinion of their superiors. This supports the idea that the usual communication model in large industrial organizations is a top-down monologue from managers who make decisions that they expect to be carried out by their subordinates without comment or question. This model discourages a culture of openness when raising any problems and difficulties that employees face in implementing management decisions. Fear of disagreeing with superiors is strongest among shop floor employees, while their immediate superiors (lower management) have a worrying misapprehension that their subordinates are not afraid to raise objections and issues with them. In reality, rank-and-file employees simply prefer not to object to their bosses' decisions and will dutifully implement them to the best of their abilities, although they may disagree with them. This suggests that shop floor employees will even remain silent when they observe risk issues in their own area of responsibility. This is especially true if the problems arise directly because of poor decisions by their immediate supervisors and senior management.

Results of responses to anonymous surveys within the framework of the pilot project: *In your experience, how often do employees hesitate to report problems, risks, and minor incidents to their superiors in their area of responsibility?*

	Constantly	Often	From time to time	Very rarely	Never	Number of respondents
All survey participants	3.1%	30.1%	40.2%	23.6%	3.1%	326
Senior management, heads of departments and directors of sites (middle managers)	2.4%	46.3%	22.0%	26.8%	2.4%	41
Lower managers: deputy directors of sites, chief engineers of sites, heads of workshops, heads and representatives of the HSE services at sites	4.6%	31.5%	34.3%	27.8%	1.9%	108
Engineers, foremen, and shop floor employees who operate critical infrastructure at sites	2.3%	25.4%	48.0%	20.3%	4.0%	177

Interpretation of responses: 73% of all survey participants (all those who responded *"Constantly", "Often", "From time to time"*) admitted that employees hesitate to disclose the problems and risks they observe in their area of responsibility. These results indicate that the company has significant problems in the effective reporting of risk information. Senior management response suggest that they believe their subordinates often distort the real situation in the field, while most lower-level employees believe that this distortion occurs much less frequently. This can be interpreted as follows: most shop floor employees sometimes distort information about observed problems when reporting to superiors. Meanwhile, senior managers are getting information from many departments with hundreds or even thousands of employees—all of whom sometimes distort information. As a result, they often see a discrepancy between reports they receive from their subordinates and what they observe at industrial sites. Therefore, the higher the respondents are in the hierarchy of the company and the more employees they have responsibility for, the more often they will see problems, risks and minor incidents being concealed by their subordinates.

Results of responses to anonymous surveys within the framework of the pilot project: *Who bears most of the responsibility for creating an internal corporate climate where discussion of organizational problems and existing risks is not welcome?*

	Managers	Shop floor employees	Number of respondents
All survey participants	87.5%	12.5%	326
Senior management, heads of departments and directors of sites (middle managers)	95.1%	4.9%	41
Lower managers: deputy directors of sites, chief engineers of sites, heads of workshops, heads and representatives of the HSE services at sites	80.6%	19.4%	108
Engineers, foremen, and shop floor employees who operate critical infrastructure at sites	91.0%	9.0%	177

Interpretation of responses: Answers from the most senior managers (95.1%) correlate well with the responses from the 100 critical infrastructure executives from around the world interviewed between 2018–2021 (Sect. 2.3). In these earlier interviews, 97% of respondents felt that managers bear most of the responsibility for creating a climate in which discussion of organizational problems and existing risks is not welcome.

Results of answers to anonymous surveys within the framework of the pilot project: *It is beneficial within the company not to inform about risks: no one bothers with additional questions, and there are no penalties for concealing risks.*

	Strongly agree	Rather agree	Rather disagree	Strongly disagree	Difficult to answer	Number of respondents
All survey participants	7.5%	25.7%	37.5%	25.0%	4.3%	280
Senior management, heads of departments and directors of sites (middle managers)	8.3%	13.9%	41.7%	36.1%	0.0%	36
Lower managers: deputy directors of sites, chief engineers of sites, heads of workshops, heads and representatives of the HSE services at sites	5.8%	20.2%	39.4%	32.7%	1.9%	104
Engineers, foremen, and shop floor employees who operate critical infrastructure at sites	8.6%	32.9%	35.0%	16.4%	7.1%	140

Interpretation of responses: it is noteworthy that senior management, middle and lower managers insist that the company does not benefit from hiding information about risks, whereas a significant proportion of shop floor employees (more than 40%) believe the opposite. The higher the position of respondents in the hierarchy of the company, the less inclined they are to agree with the statement that it is beneficial for the company not to inform about risks; the lower their position, the more likely they are to support this statement. However, most respondents at all levels of management believe that it is unprofitable to hide information about risks from superiors and colleagues.

Lessons learned: the experience of conducting seminars on how to respond to emergency situations, together with surveys of the participants, shows that the problem of reporting accurate information before and during emergencies is not always obvious to many managers and employees. Prior to the seminars, some did not fully understand the critical importance of communicating objective risk information across the company's hierarchy. Significant number of the seminar participants only began to appreciate the severity of the problem after being made familiar with its consequences in the development of major industrial disasters. Therefore, prior to launching a project to transform a company's culture around communication, it is very beneficial to hold special seminars for key managers and staff to convince them of the critical importance of information transmission, and to illustrate the disastrous consequences of hiding information. It is also recommended that, as part of these seminars, anonymous surveys are run to assess the current situation within the company and across its production sites in regard to the transmission of risk information.

IV. RANKING THE COMPANY'S CRITICAL RISKS TO AID THE SELECTION OF SPECIFIC SITES AND PROJECT PARTICIPANTS

When the decision was made to launch a pilot project, four production sites were selected (from among dozens of possible company plants) where an accident could have catastrophic consequences for the overall production process and the company's finances. The sites were also selected to cover different parts of the production process so that:

- the anonymous surveys of project participants could be analyzed according to the production site they work at, to see if there were significant differences in their corporate cultures. It was also important to identify if there were different reasons for concealing risk information at particular sites.
- it was possible to test different solutions within the pilot project to determine which of them proved effective and which did not. This information was important in selecting the practical solutions that could then be successfully rolled across all the various sites of the company during the subsequent scaling up of the project.

A total of 422 company representatives participated in the project. Approximately 10% of these were senior managers, heads of departments and directors of production sites selected for the pilot project. The other 90% were drawn from different levels of management across the four production sites and included lower-level managers and shop floor employees who regularly managed critical facilities, where failure could inflict serious losses for the company.

Participants were selected from the four production sites as follows:

- general director of the site (head of the plant);
- all deputies of the general director of the site supervising production, industrial safety, logistics, procurement and warehouse;
- head of the HSE services at the site, as well as their subordinates;
- heads of key production workshops and their deputies, and heads of sections of these workshops;
- engineers, foremen and shop floor employees operating critical infrastructure in these workshops.

Approximately 5% of the combined workforce of the four production sites became participants in the project.

Lessons learned: there is a danger of excessive information noise when shop floor employees and line managers are reporting data about safety and technological risks to senior management. When setting up similar pilot projects, in order to reduce information noise to a minimum, it is recommended to choose for reporting those production sites in which accidents can have the most serious consequences for a company. The number of managers and employees who operate and manage the most critical infrastructure is limited. By selecting around 5–10% of the whole workforce of the selected production sites, senior management will not be overwhelmed by having to engage with thousands of employees. Executives will be able to meet the participating employees face to face, and quickly establish the process of getting a faster and more accurate flow of information about critical and serious risks up through the company's hierarchy.

When scaling up solutions across a company, it is also recommended to start with only 5–10% of the employees who manage the most dangerous production processes across an organization. It is not recommended to scale up too rapidly by trying to immediately include every employee. There is a serious threat that this would overwhelm a company's capacity by triggering a huge wave of messages about various technological risks and problems—the criticality of which will, for the most part, be low. The priority is to get the most critical and serious risks under full control. Only then will it work to gradually expand the circle of employees and managers who have the authority to report information about risks in their area of responsibility to senior management through specially established direct channels.

V. IN-DEPTH INTERVIEWS WITH SENIOR MANAGERS AND MANAGERS OF PRODUCTION SITES SELECTED FOR THE PILOT PROJECT

In December 2021, the project leader began conducting in-depth interviews with the company's senior managers to understand why the company had problems with internal risk communication, and how they imagine a successful intra-organizational risk communication system would look as a result of the project. In-depth interviews allowed the project leader to immerse himself in the company's culture and activities, and understand the hopes and expectations that the managers had for the project.

Lessons learned: from the outset, it is important to determine the main senior managers who will be involved in the project for many months. At this initial pilot stage, only a proportion of the senior managers were involved, and had a key influence on how the project unfolded. Retrospectively, it became clear that it would have been better to draw in all the most senior managers who, in one way or another, could influence how the project evolved in the future. They should all be invited to the initial launch meeting, attend the introductory educational seminars, participate in in-depth interviews, and so on. If some are excluded from the discussion and decision-making during the pilot project, they are more likely to oppose the ideology and direction of the project in the longer run, when it is rolled out company-wide. The experience of the pilot project showed that any managers not involved from the outset might well continue to behave as they always have done—reinforcing the old fear among employees about bringing "*bad news*" to their superiors, so that risk information continues to be concealed. This can seriously undermine the entire project, and the problems of risk communication within the organization will simply resurface.

Therefore, a list of all senior managers who could potentially be included in the project should be discussed with the head of a company. If the project later requires the involvement of some new senior managers, then they must undergo special training and be brought up to speed so that they share the ideology of the project before being authorized to make any decision affecting it.

VI. HOLD A LAUNCH MEETING WITH SENIOR MANAGEMENT AND MANAGERS OF PRODUCTION SITES SELECTED FOR THE PILOT PROJECT

When in-depth interviews with senior managers have been completed and the main sticking points within the company have been identified, the next step is to hold a launch meeting of all senior managers included in the project, as well as the managers of the production sites selected for the pilot.

At this launch meeting in December 2021, the company's SVP[4] told the audience why the company had decided to support the pilot project. The SVP also outlined the key project goals and expectations, and asked his subordinates to assist the project leader in making the project a success. The pilot project leader then presented the work plan for the coming months. Questions were taken from the audience, which were answered by the SVP of the company and the project leader. From that moment on, the pilot project was officially launched.

[4] SVP: Senior Vice President in charge of production at the key industrial site of the company.

Lessons learned: prepare the key points of the project and have the head of a company present them. This is important, since the entire management team will see that the implementation of this project is a priority for the company's top brass. This will have a positive impact on the successful implementation of the project, since lower-level managers will try to follow the direction that has been set by the company's head.

VII. HOLD A SERIES OF SEMINARS FOR SENIOR MANAGEMENT, MANAGERS AND EMPLOYEES OF THE SITES SELECTED FOR THE PILOT PROJECT

In the opinion of the authors of the handbook, holding a full day interactive educational seminar is the most effective way to convey the main principles of the project to the participants. It also provides an ideal opportunity to hear first-hand why the participants think risk information is concealed in the company, and what they think can be done to improve the quality and speed of risk communication and thus reduce accidents. During an in-depth interview, one senior manager of the company said: *"People really appreciate that they are being listened to. Listening is key. It's great that you [the pilot project leader] will go to the workforce and ask them what they think is the best thing to do to change the situation regarding the reporting of risk-related information. I hope that at the very least people will open up a bit—this in itself would be a great achievement—but if we can change how they think—well, that would be a real breakthrough. This project is a unique experiment. Our company has never talked to its employees like this before. We have always just told them what to do, and never been interested in hearing what they had to say"*.

As part of the pilot project, the project leader conducted 15 seminars for 422 participants:

- 3 seminars for senior management, heads of departments of the company and heads of production sites selected for the pilot project (10% of participants);
- 5 seminars for lower-level managers: deputy directors, heads of workshops, and heads of HSE services at the selected sites (25% of participants);
- 7 seminars for engineers, foremen and shop floor employees who operate critical infrastructure at the selected sites (65% of participants).

Below is the structure of this interactive full day seminar. It ran from 9 am—6 pm with an hour's lunch break and two 20 min coffee breaks. For the seminars, a 300-page presentation was prepared, which included the following sections:

1. **Demonstration of the relevance of the problem of concealing information about risks in critical infrastructure companies**

 - *Major management errors recorded in recent large-scale industrial accidents*. Drawing on 15 years of research by the first author of the handbook on management decisions and communication in emergencies, ten major management errors have been identified that are seen over and over again in many disasters, regardless of the country. One of these is the suppression of objective information about emergencies reported by subordinates through the corporate hierarchy of a company—including communication with external audiences, such as regulatory authorities, local populations, media, and so on. Examples were given of large and well-known accidents where it has been established that information about the true scale of the incident and the real state of affairs at the scene of the emergency were concealed.[5]
 - *Examples of information about risks being concealed before a disaster*. Some of these accidents were then discussed in more detail. The reasons risk information was suppressed were analyzed, as well as how this failure in communication had led to the accidents.[6] To highlight the relevance of the problem, the accidents discussed were selected from the same industry as the pilot project company. Naturally, this helped the seminar participants engage with the discussion: they were studying the negative experiences of comparable organizations, facing similar kinds of production pressures and problems as their own company.
 - *Negative experience of this company in reporting information about critical risks before and during emergencies*. The presentation then turned to the two incidents that had recently occurred in this company, where problems had been identified with the transmission of risk information (the details of these were mentioned earlier).

2. **Parting words of the company's SVP**

The company's SVP recorded a special video message shown to all the pilot project participants at the seminars:

"Dear colleagues!

We operate a critical industrial infrastructure.

It is important for all of us to work together to prevent emergencies by avoiding a critical build up of negative events. We need to be proactive and control serious risks as effectively as we can so as to prevent accidents occurring.

---

[5] Dmitry Chernov, Didier Sornette, Giovanni Sansavini, Ali Ayoub, Don't Tell the Boss! How poor communication on risks within organizations causes major catastrophes, Springer, 2022, https://link.springer.com/book/10.1007/978-3-031-05206-4.

[6] Dmitry Chernov, Didier Sornette, Giovanni Sansavini, Ali Ayoub, Don't Tell the Boss! How poor communication on risks within organizations causes major catastrophes, Springer, 2022, https://link.springer.com/book/10.1007/978-3-031-05206-4.

The safety of employees and the reliability of our critical infrastructure is the most important priority for our company.

The implementation of the production plan should only proceed when the safety of employees and production processes has first been secured.

The company owners and shareholders have provided me and the heads of the production sites with the necessary authority and resources to adjust production plans in order to prevent critical problems from developing. We are also ready to take prompt action to stop production if the safety of workers and production processes cannot be guaranteed. All this is aimed at preventing emergencies and accidents occurring at our production facilities.

I also want to say, personally as well in my role as the director of your facility, that we actively want to hear any information you have about safety, technological, and production issues that you observe in your area of responsibility, however difficult or serious the problem might appear.

This project is aimed at improving the quality and speed of the transmission of critical risk information from employees at production sites to the company's executives.

We need to ensure that we, as executives, are informed early on about problems on the ground so that we can respond to them promptly. We do not want delays in receiving this information as the longer a problem is left, the harder it becomes to solve. We are all determined to work together to stop critical events developing and prevent emergencies.

I would like to make it clear that nobody will be penalized for voluntary disclosure of information about critical risks. On the contrary, we will be grateful to you for these communications, and guarantee to assist in solving the problems you have helped identify. All the senior managers who participate in this project will also praise rather than punish any subordinates reporting problems on the ground, and will help employees tackle the issues. All senior and site managers are committed to providing local managers with the resources they need to tackle any safety problems and production issues that their subordinates report to them.

I ask you to work proactively, learn to identify risks at an early stage, then report them promptly to me personally or to your line managers, so that critical situations do not have a chance to develop. Never be afraid to disclose risks in your area of work to your superiors—you can be confident that your reports will be welcomed and carefully analyzed, and you will then be given the necessary authority and resources to solve the problem you have helped to identify.

I ask that you take an active part in this seminar. Provide an honest and objective assessment of the state of affairs in your own area of responsibility. Do your best to share any ideas that you think could improve the quality and speed of transmitting information about risks so that together we can prevent serious incidents and emergencies from occurring.

Thanks a lot! I wish you fruitful work!".

This appeal contained several key messages from senior management to subordinates:

- owners and senior management agree that safety is the most important corporate priority;
- owners and senior management are ready to stop production in order to prevent accidents if a critical risk is detected;
- senior management and site managers (middle management) actively want to receive accurate, objective information about existing risks and problems that could lead to a serious problem;
- senior management have the resources available to address critical and serious issues reported by project participants;
- senior management promise not to penalize any employee for disclosing information about risks but on the contrary to praise them.

3. Anonymous survey No. 1—the reasons why risk-related information is concealed within the organization

As part of this seminar, an anonymous online survey was conducted to understand: (I) why employees and managers at various levels in the company have difficulty transmitting risk information to their line managers, and (II) why managers at various levels have difficulty receiving information from their subordinates about risks and problems.

Participants were asked to scan a QR code into their smartphones or follow a link to an online survey page, where they were asked to answer 40 questions about why risk-related information was concealed within the organization:

1. **In your experience, how often are employees afraid of expressing disagreement with their superiors?** (*Constantly* | *Often* | *From time to time* | *Very rarely* | *Never*)
2. **In your experience, how often do employees hesitate to report problems, risks, and minor incidents to their superiors in their area of responsibility?** (*Constantly* | *Often* | *From time to time* | *Very rarely* | *Never*)
3. **Some employees appear reluctant to inform managers about problems such as equipment failure, mistakes in their work or inability to achieve their targets. Why is this happening?**

   3.1. The income of some employees is linked to reaching production targets. Reporting risks to management can jeopardize things here, as production will probably be halted, delaying the payment they usually receive when the targets are reached. *(Strongly agree | Rather agree | Rather disagree | Strongly disagree | Difficult to answer)*

   3.2. Fear of blame and punishment from supervisors. Employees believe they will be held accountable for any problem they report to their supervisors. *(Strongly agree | Rather agree | Rather disagree | Strongly disagree | Difficult to answer)*

   3.3. Employees are afraid of losing income and spoiling their career prospects (including being fired) because they could look incompetent if they report a problem in their area of responsibility to their superiors. *(Strongly agree | Rather agree | Rather disagree | Strongly disagree | Difficult to answer)*

   3.4. In many organizations it is simply not customary to discuss risks and problems with managers. *(Strongly agree | Rather agree | Rather disagree | Strongly disagree | Difficult to answer)*

3.5. Fear of destroying relationships with colleagues or line managers. *(Strongly agree | Rather agree | Rather disagree | Strongly disagree | Difficult to answer)*

3.6. Additional burden and responsibility. Some employees fear that managers will require them to take responsibility for any problem they report, in addition to their current workload. *(Strongly agree | Rather agree | Rather disagree | Strongly disagree | Difficult to answer)*

3.7. Employees do not fully understand risks in their area of responsibility and lack the training or experience to accurately assess the criticality of the situation. *(Strongly agree | Rather agree | Rather disagree | Strongly disagree | Difficult to answer)*

3.8. Employees believe it is pointless to transmit information about risks to managers, because all similar previous warnings have failed to produce any kind of response. *(Strongly agree | Rather agree | Rather disagree | Strongly disagree | Difficult to answer)*

3.9. Employees are afraid to appear disloyal to a company, be labeled as troublemakers who are"*rocking the boat"*, or give the impression that they think they are better than everybody else. *(Strongly agree | Rather agree | Rather disagree | Strongly disagree | Difficult to answer)*

3.10. Some employees are over-confident in their own abilities and believe they can solve the problem on their own, without requesting the support of their superiors. *(Strongly agree | Rather agree | Rather disagree | Strongly disagree | Difficult to answer)*

3.11. Industrial safety indicators and the bonus system within a company work to keep risks and problems concealed when reporting to superiors. *(Strongly agree | Rather agree | Rather disagree | Strongly disagree | Difficult to answer)*

3.12. Employees are afraid to take the initiative as it can produce unpredictable and risky results: *"why stick your face into a hornet's nest", "don't put your head above the parapet unless you want it shot off", "slow and steady wins the race". (Strongly agree | Rather agree | Rather disagree | Strongly disagree | Difficult to answer)*

3.13. Some employees only want to show themselves in the best possible light to their superiors. *(Strongly agree | Rather agree | Rather disagree | Strongly disagree | Difficult to answer)*

3.14. Unwillingness to upset superiors by reporting negative news about risks within an organization: "*superiors can take information like this very personally... it is like the employee is blaming them for the problem because of earlier bad decisions they made*". *(Strongly agree | Rather agree | Rather disagree | Strongly disagree | Difficult to answer)*

3.15. Some employees are indifferent to any risks they might notice, being simply too lazy to take positive action, and like to believe that with a bit of luck nothing serious will come of it anyway. *(Strongly agree | Rather agree | Rather disagree | Strongly disagree | Difficult to answer)*

4. **Why do you think managers often do not want to hear bad news from employees about matters like observed risks and problems in an organization and the need for additional investments, like equipment upgrades, to create safer production processes?**

4.1. Shareholders set tough and ambitious financial and operational goals for the management, which do not make any allowance for the additional—often high—costs that may be necessary to deal with problems identified by employees. *(Strongly agree | Rather agree | Rather disagree | Strongly disagree | Difficult to answer)*

4.2. The income of some managers is tied to the implementation of production plans. An adequate response to most risks requires a halt to production. This threatens the successful implementation of the production plan, as targets are difficult to adjust downwards, however necessary the stoppage. *(Strongly agree | Rather agree | Rather disagree | Strongly disagree | Difficult to answer)*

4.3. When a manager receives information about a problem that requires additional resources to rectify, he will have to report this bad news to his superiors. If he cannot sort the problem out at his level, he may find this has a negative impact on his career. *(Strongly agree | Rather agree | Rather disagree | Strongly disagree | Difficult to answer)*

4.4. Excessive bureaucracy interferes with the practical solution of any identified problems. Managers face complex corporate procedures when going through budgeting and investment committees, so issues are resolved very slowly. *(Strongly agree | Rather agree | Rather disagree | Strongly disagree | Difficult to answer)*

4.5. There are limited resources available for managers to solve problems. *(Strongly agree | Rather agree | Rather disagree | Strongly disagree | Difficult to answer)*

4.6. Leaders are afraid of being seen as incompetent and being held accountable for their previous bad decisions that created the current problems. *(Strongly agree | Rather agree | Rather disagree | Strongly disagree | Difficult to answer)*

4.7. Managers believe that once a problem has been reported to them, they are automatically responsible for solving it. *(Strongly agree | Rather agree | Rather disagree | Strongly disagree | Difficult to answer)*

4.8. Managers expect employees to solve problems on their own when they occur in their area of responsibility. *(Strongly agree | Rather agree | Rather disagree | Strongly disagree | Difficult to answer)*

4.9. Managers prefer not to be made aware of identified risks so as not to become legally liable if these eventually lead to an incident. *(Strongly agree | Rather agree | Rather disagree | Strongly disagree | Difficult to answer)*

4.10. Leaders do not want to deal with difficult issues that require them to step out of their comfort zone. *(Strongly agree | Rather agree | Rather disagree | Strongly disagree | Difficult to answer)*

4.11. Managers are people too – they prefer to receive good news, not bad. *(Strongly agree | Rather agree | Rather disagree | Strongly disagree | Difficult to answer)*

4.12. Managers consider problems reported by employees to be unimportant. *(Strongly agree | Rather agree | Rather disagree | Strongly disagree | Difficult to answer)*

4.13. Managers have short-term contracts and are focused on achieving short-term results, so do not want to get involved in solving serious problems that require years of effort, because the results of these efforts will not be visible until several years after they have left their position. *(Strongly agree | Rather agree | Rather disagree | Strongly disagree | Difficult to answer)*

4.14. The generally accepted culture of behavior for leaders permeates entire industries: leaders insist that employees bring them mostly good news about successes and achievements, and problems are solved by employees without bringing them to the attention of managers. *(Strongly agree | Rather agree | Rather disagree | Strongly disagree | Difficult to answer)*

4.15. When employees bring information about problems to their managers that they should then disclose to regulators, this is likely to mean an immediate increase in the number of site inspections, which in turn means additional time, cost and stress to manage this extra burden. Managers can therefore be reluctant to pass the information on to appropriate regulators. *(Strongly agree | Rather agree | Rather disagree | Strongly disagree | Difficult to answer)*

5. **Who bears most of the responsibility for creating an internal corporate climate where discussion of organizational problems and existing risks is not welcome?** *(Managers | Shop floor employees)*
6. **How do you rate the level of trust of employees towards their managers in your enterprise?** *(Very high | Medium high | Medium | Low | None | Difficult to answer)*
7. **How do you rate the level of trust of managers towards their employees in your enterprise?** *(Very high | Medium high | Medium | Low | None | Difficult to answer)*
8. **Do you think that the managers and employees in your company are in a hurry?** *(Rarely | Sometimes | From time to time | Often | Constantly | Difficult to answer)*

9. **In your organization is there a lot of pressure on managers and employees to implement the agreed production plan?** *(No | Some pressure | Yes | Difficult to answer)*
10. **Middle and lower managers have the power to stop the work of a workshop, and even the entire operation of a plant, if critical risks are identified. Employees are given the right to refuse to perform unsafe work. In your experience, how often do managers and employees exercise these rights?** *(Very often | Often | Occasionally | Rarely | Never | Difficult to answer)*
11. **The fulfillment of the production plan is not the main priority of a company—the highest priority is the safety of work and production processes in order to prevent emergency situations and accidents.**

    11.1. What is the level of awareness of these priorities among managers and employees at various levels in the company? *(Very high awareness of these priorities | Many are aware of these priorities | Some are aware of these priorities | Few are aware of these priorities | No one is aware of these priorities | Difficult to answer)*

    11.2. How do things work in reality when choosing priorities in the company? *(Priority is always given to the safety of work and production processes | Priority is usually given to the safety of work and production processes | Priority is usually given to the execution of the production plan | Priority is always given to the execution of the production plan | Difficult to answer)*

    11.3. Have you been made aware of the main critical risks of your enterprise? *(High awareness | Medium awareness | Low awareness | No awareness | Difficult to answer)*

It is worth noting that not all seminar participants took part in the anonymous online survey. Out of 422 people who attended the 15 seminars, 326 (77% of the seminar participants) answered these questions. This suggests that a significant proportion of managers and employees (a quarter of all respondents) are afraid or unwilling to express their opinion on difficult issues around relations between managers and employees in a company even within the anonymous online survey.

4. Discussion on the reasons behind risk information concealment

The participants of the seminar were shown the analysis of dozens of industrial disasters conducted by the authors of this handbook between 2013–2022 (summarized in Sects. 2.1 and 2.2). In addition, they were shown a study of the reasons behind concealment of risk-related information based on in-depth interviews with 100 critical infrastructure executives conducted in 2018–2021 (Sect. 2.3).

After that, participants were presented with the results of anonymous survey No. 1 conducted earlier within this same group. This gave them an opportunity to examine the reasons for risk information concealment within various large industrial companies, and also to analyze the main reasons why proper risk information transmission was sometimes blocked or impeded in their own company.

5. Group work No. 1—how the problem of risk information concealment can be solved in this company

The participants were divided into several small groups and moved to different areas of the seminar hall. They were asked the following questions, to answer as a group:

- Paint an ideal picture of an effective process for communicating technological risks from shop floor to senior management within your organization.
- What are the current barriers/challenges to rapidly reporting risk-related information from the bottom up in your organization? Where is information lost? Why is information lost?

- What needs to be done to improve the quality of risk-related information that is reported from the bottom up in your organization? Offer practical solutions.
- What barriers/problems currently exist to effectively mitigating technological risks identified in your organization?
- What needs to be done to improve the quality of ongoing mitigation measures for technological risks identified in your organization?
- Suggest how to build a high level of employee trust in managers—in other words, what should managers do to gain the trust of their subordinates?
- Do you agree with the statement that preventing accidents by identifying technological risks and disclosing these risks to colleagues and management is the professional duty of responsible employees, so neither tangible nor intangible rewards are needed for disclosing risks?
- If the company does decide on rewards for disclosure, what are the pros and cons of financial incentives for employees disclosing information about technological risks? Identify the pros and cons of non-material rewards for employees disclosing this information.
- Describe what you think is the best way of rewarding employees for disclosing information about technological risks in your organization.
- Instead of reporting a safety issue to the head of a production unit, employees shoot a video and upload it onto social networks. Who is to blame for this situation? How can management motivate employees to report risks to their managers, rather than exposing this sensitive information to the public?

The seminar participants were given approximately 40 min to answer these questions. The responses of the group members were recorded in writing on large sheets (flip-chart format). Then each group delegated a representative to present the group's results to the seminar's participants, and an audio recording was made of these presentations. The project leader later analyzed and collated all this data and created a list of possible solutions that should be explored within the project.

6. How the *"cover-up"* problem can be solved: 10 key recommendations from in-depth interviews with 100 executives around the world

The seminar participants were shown the results of the 2018–2021 study described in detail in Chap. 3 of this handbook.

7. How the problem of concealment can be solved: a thermodynamic model for the transmission of information about risks

The seminar participants were shown the thermodynamic model presented in Recommendation 5 of Chap. 3, which makes a direct link between the growth of employee trust in managers and an increase in the quality and speed of communication about risks. This model builds on the following recommendation: *"Senior management should build an atmosphere of trust and security, so that employees feel safe to disclose risk-related information"*.

After that, a video was shown to the project participants, presenting an appeal from the company's SVP: *"We all need to strive to build trusting relationships with our subordinates. This is the only way we can start the process of transmitting really important information about risks from the bottom up, from departments to senior management, and horizontally, between different departments. I say again: do not be afraid to bring me and your superiors information about serious problems—we will deal with these and calmly solve them together! I am ready to play my part in the analysis of the critical information that you will be reporting, and then making decisions to solve them. I will also personally thank everyone who voluntarily discloses serious safety and technological risks and thus helps prevent serious incidents".*

This message helped convince the project participants that senior management were focused on building trust in relationships with their subordinates, and ready to fully commit to finding effective solutions to safety and technological problems.

8. **Anonymous survey No. 2—how the problem of risk information concealment can be solved**

The purpose of this survey was to understand what needs to be done to fundamentally change the situation in respect to communication of risk-related information at the pilot production sites, by creating simple, practical, and easily implemented solutions.

Seminar participants had to scan a QR code on their smartphones or follow a link to the online survey, where they had to answer next 40 questions on what practical steps could be taken to improve the quality and speed of risk communication in the company:

12. **How do you increase the level of trust of employees in managers, and motivate them to report risks upwards?**

    12.1. A leader must be authoritarian (e.g. be the sole decision-maker and carry overall responsibility for a company's performance) in order to meet the difficult challenges of managing a critical infrastructure company. *(Strongly agree | Rather agree | Rather disagree | Strongly disagree | Difficult to answer)*

    12.2. A leader should be more democratic (e.g. share decision-making, delegate some power to active employees, take joint responsibility for the results of the unit's work with them). *(Strongly agree | Rather agree | Rather disagree | Strongly disagree | Difficult to answer)*

    12.3. A trusting relationship between a manager and their subordinates is necessary to create an environment where if feels safe to share information about existing problems. *(Strongly agree | Rather agree | Rather disagree | Strongly disagree | Difficult to answer)*

    12.4. It is beneficial within the company not to inform about risks: no one bothers with additional questions, there are no penalties for concealing risks. *(Strongly agree | Rather agree | Rather disagree | Strongly disagree | Difficult to answer)*

    12.5. A high level of employee trust in managers leads to improved transmission of information about risks throughout a company; and conversely, low trust leads to a lack of willingness to transmit risk information to superiors. *(Strongly agree | Rather agree | Rather disagree | Strongly disagree | Difficult to answer)*

12.6. In order for employees to trust managers, their actions should match their words. *(Strongly agree | Rather agree | Rather disagree | Strongly disagree | Difficult to answer)*

You were shown the "*Thermodynamic model of risk information transmission*". This explains how a company's conscious policy of building trust between managers and employees leads to an increase in the quality and speed of communication about risks in the organization, while a low level of employee trust in managers negatively affects the quality of risk information transmitted through the corporate hierarchy.

12.7. Do you understand the thermodynamic model of risk information transmission that has been presented to you? *(Fully understand | Mostly understand | Don't understand very well | Don't understand at all | Difficult to answer)*

12.8. Do you agree with the principles of this model? *(Strongly agree | Rather agree | Rather disagree | Strongly disagree | Difficult to answer)*

12.9. What stage do you feel your organization is at in the framework of the "*Thermodynamic Model of Risk Information Transmission*"? *(Steam (warm relationship) | Water (transition state) | Ice (cold relationship) | Difficult to answer)*

Below are pairs of statements that have opposite meanings. Please rate which statement from each pair you most agree with *(0—fully agree with the statement on the left, 10—fully agree with the statement on the right).*

12.10. Excessive caution when dealing with managers and colleagues does not hurt / Most managers and colleagues can be trusted

12.11. Most managers and colleagues are willing to cheat if the opportunity presents itself / Most managers and colleagues try to be honest

12.12. Most managers and colleagues care only about themselves / Most managers and colleagues are happy to help others

12.13. Most managers and colleagues are very closed-minded / Most managers and colleagues are very open

12.14. For the most part, work is stress, conflict and punishment / For the most part, work is good relationships with colleagues, cooperation and freedom from punishment

12.15. Imagine that employees at an industrial site detect a serious technological risk. Based on your experience, what do you think employees will do when this risk is discovered? *(1. Employees will immediately report the risk to their supervisor, who, in turn, will quickly inform the site management. A special group will be promptly established to carry out a risk analysis. If the risk is deemed unacceptable, then production work will be halted, and measures implemented to control the identified situation | 2. Employees will report the risk to their supervisor, but this information will not be transmitted further up the hierarchy. Employees and their manager will try to independently address the risk using their own resources; it is possible that work will be stopped or reduced at the site | 3. Employees will not report the risk to their immediate supervisor, but will try to eliminate the identified risk on their own, and production work will not be stopped | 4. Employees will ignore the risk as they work in a dangerous industrial production facility, where there are already many serious hazards; employees simply do not have time to respond to all potential risks, because they need to keep working and fulfill the production plan | 5. Difficult to answer)*

12.16. Based on your experience, can employees in the company come to their superiors and talk about problems in their area of responsibility, confident that they will not be punished for bringing this information, and that suitable resources—money, time, personnel—will be provided to address the issue? *(1. Yes, employees can go to their superiors and calmly discuss problems without fear of punishment and get the resources to solve them | 2. Yes, employees can go to their superiors and disclose problems. They will not be punished, but they will not be given the required resources to solve them | 3. Yes, employees can come to their superiors and talk about problems—but they will be punished and will not be given resources to solve them | 4. No, nothing will be changed, no one will provide any resources, and employees will be forced to solve problems in their area of competence independently | 5. Difficult to answer)*

13. **Information transmission channels**
    13.1. What channels should be developed at your production site for transmitting information about risks from the bottom of the corporate hierarchy? *(1. It is only necessary to develop a risk communication channel based on the traditional chain of command corporate hierarchy (employees report information to their supervisor, who reports to his/her line manager, and so on up the chain from shop floor to senior management) | 2. It is only necessary to develop an alternative channel (any employee can send information about risks directly to senior management, bypassing the traditional hierarchy, through smartphone apps, hotlines, mailboxes, meeting with executives, etc.) | 3. Both channels need to be developed (transmitting information about risks through the traditional hierarchy and through an alternative channel) | Difficult to answer)*
    13.2. If a company decides that it is necessary to develop both channels—communicating information about risks through the traditional hierarchy and through an alternative channel—then which channel should be the main priority for development? *(1. Greater priority should be given to the channel using the traditional chain of command | 2. Greater priority should be given to the alternative channel | 3. Develop both channels with equal priority | Difficult to answer)*
    13.3. Alternative communication channels used to transmit risk information between shop floor employees and senior managers should keep the identity of the employee raising the issue secret—i.e. be anonymous. *(Strongly agree | Rather agree | Rather disagree | Strongly disagree | Difficult to answer)*
    13.4. Alternative communication channels used to transmit risk information between shop floor employees and senior managers should identify the employee raising the issue—i.e. not be anonymous but include personal identification details. *(Strongly agree | Rather agree | Rather disagree | Strongly disagree | Difficult to answer)*
    13.5. How do you feel about this system for cascading the transmission of risk information up the hierarchy? A shop floor employee observes a risk or problem and informs their immediate supervisor. If within a certain time there has been no response, then that employee has the right to contact their line manager's superior. If this still fails to produce any feedback, then the employee has the right to inform the next level of management up—all the way up to the site director and even beyond to the CEO of a company. This system is intended to encourage the traditional management hierarchy to act more quickly to address problems raised by shop floor employees, without going straight to alternative (emergency) channels of communication between shop floor employees and senior management, where the traditional management hierarchy is immediately bypassed. *(Strongly agree | Rather agree | Rather disagree | Strongly disagree | Difficult to answer)*
    13.6. Do you want to receive information about serious incidents that have occurred across all of your organization's facilities, including a brief analysis of the main reasons behind each one? *(Strongly agree | Rather agree | Rather disagree | Strongly disagree | Difficult to answer)*
    13.7. Do you want to receive information about serious incidents (and their causes) that have occurred in your industry, in this country and across the world? *(Strongly agree | Rather agree | Rather disagree | Strongly disagree | Difficult to answer)*
    13.8. As part of the pilot project, it is planned to create a smartphone app so that employees can quickly transmit information about observed risks up the corporate hierarchy to senior management. This raises a technical question: what operating system does your phone run on? (If you have several phones with different operating systems, then mark all the systems that are used on your phones.) *(Google (Android) | Apple (iOS) | Huawei (HarmonyOS) | Microsoft (Windows Phone) | different operating system | I don't have a smartphone)*
    13.9. As part of the pilot project, it is planned to present the solutions developed by around 500 employees and managers who attended the seminars. What is the most convenient format for you to participate in this presentation? *(1. Personal participation in a special event where senior management will talk about decisions within the project; opportunity for attendees to*

ask questions directly | 2. Receive a link to a video on your smartphone, which explains what solutions were developed within the project; online chat to answer any questions | 3. View specially prepared video with colleagues at the facility; ask questions to a representative of HSE service | 4. Difficult to answer)

14. **Rewards and punishments**
    - 14.1. How do you view the adequacy of penalties at your enterprise for various misconduct offenses? *(Excessively hard | Hard | Adequate/Fair | Soft | Excessively soft)*
    - 14.2. Managers should not penalize employees for their mistakes, but look for systemic flaws in the work of an organization that may have created an unsafe situation for employees. *(Strongly agree | Rather agree | Rather disagree | Strongly disagree | Difficult to answer)*
    - 14.3. It is justified to penalize employees only if they have deliberately violated safety rules or neglected their duties. *(Strongly agree | Rather agree | Rather disagree | Strongly disagree | Difficult to answer)*
    - 14.4. When employees report a problem to a manager, they then share the responsibility for solving it. If employees keep a problem to themselves, they are taking full responsibility for whatever happens. *(Strongly agree | Rather agree | Rather disagree | Strongly disagree | Difficult to answer)*
    - 14.5. Managers should create a workplace climate where the voluntary admission of an error by an employee does not then result in them being penalized. *(Strongly agree | Rather agree | Rather disagree | Strongly disagree | Difficult to answer)*
    - 14.6. Any employee—worker or manager—found to have concealed important risk information should face serious consequences, including possible dismissal. *(Strongly agree | Rather agree | Rather disagree | Strongly disagree | Difficult to answer)*
    - 14.7. In a company where a voluntary risk disclosure to superiors is never penalized, employees or managers who deliberately hide information about risks in their area of responsibility should face dismissal. *(Strongly agree | Rather agree | Rather disagree | Strongly disagree | Difficult to answer)*
    - 14.8. Managers should reward employees publicly in the presence of their co-workers, but penalize them only in private conversation and with total confidentiality. *(Strongly agree | Rather agree | Rather disagree | Strongly disagree | Difficult to answer)*
    - 14.9. Managers should not offend or humiliate employees, as this can have a negative impact on productivity and damage communications. *(Strongly agree | Rather agree | Rather disagree | Strongly disagree | Difficult to answer)*
    - 14.10. Identifying safety risks and disclosing them to colleagues and management to prevent accidents is the professional duty of every responsible employee. Therefore, neither material nor non-material rewards for disclosure of risks are necessary. *(Strongly agree | Rather agree | Rather disagree | Strongly disagree | Difficult to answer)*
    - 14.11. There is a discussion within a company and the participants divide into two groups. The first believes that, to encourage employees to disclose dangerous actions by their colleagues (e.g. working at height without a harness), the management should offer material rewards (e.g. money, valuable gifts, and so on). The priority is to identify and prevent safety violations, by any means necessary. The second group believes that to encourage employees to disclose dangerous actions by their colleagues, the management should not provide material rewards. This may threaten the stability of the work teams—and in any case, reporting safety violations is the professional duty of any responsible employee, and should not require any financial incentives. Do you support the beliefs of the first group or the second group? *(Fully support ideas of 1st group | Rather support ideas of 1st group | Rather support ideas of 2nd group | Fully support ideas of 2nd group | Difficult to answer)*

14.12. Choose what you believe to be the most appropriate methods of reward for disclosing information about safety and technological problems (choose only one answer). *(Material rewards only | Non-material rewards only | Combination of material and non-material rewards | No rewards needed—disclosure is the professional duty of employees | Difficult to answer)*

14.13. If the company decides that it needs to combine material and non-material rewards for disclosing information about safety and technological problems, what should be prioritized? *(Combination of mostly material and some non-material rewards | Combination of mostly non-material and some material rewards | Equal priority to material and non-material rewards | Difficult to answer)*

14.14. Choose what you consider to be the most appropriate and effective non-material rewards for employees disclosing information about technological risks (multiple answers can be selected).

- Public commendation from senior management to an employee or work team
- Letter of gratitude sent from senior management to an employee's family acknowledging his/her contribution to the safety of the site
- Award of diploma personally signed by senior management
- Award of special honorary order (corporate medal)
- Joint photo with senior management
- Article in the corporate media featuring the employee/team who prevented a serious emergency
- Inclusion in the list of best employees at the site
- Lavish dinner for distinguished employee/team with senior management
- Weekends at the corporate recreation center for all family members
- Allocation of annual personal parking spot alongside top company executives.
- Professional internships at other company enterprises or additional professional training (including abroad)
- Videos featuring the outstanding contribution to company safety by the employee/team. Broadcasting these videos to other employees across all the company sites.
- Tickets for all family members to attend major sporting events or concerts
- Additional days off
- Opportunity to become a mentor, passing on their successful experience to other employees within the company

After conducting these online surveys, results were immediately shared with all seminar participants.

9. Group work No. 2—how the problem of risk information concealment can be solved in this company

The second group work session followed the same structure as the first. The seminar participants were asked the following questions:

- How is it possible to "*unfreeze*" the corporate hierarchy and *"melt the ice"* in the interactions between managers and employees? The "*Thermodynamic model of risk information transmission*" would describe this as a successful transition from *"cold"* to *"warm"* relations. The goal of this thawing of relations would be to motivate employees to voluntarily disclose critical risks to their superiors.
- The absence of penalties for the voluntary disclosure of technological risks is one of the fundamental principles for improving communication about risks

within an organization. However, this can create problems. How can the organization avoid creating an atmosphere where employees feel they can act with impunity and where kindness (i.e. lack of punishment) is perceived as weakness? Offer practical solutions.
- How can lower managers be motivated to actively address the problems that subordinates bring to them?
- What is a simple step-by-step protocol (three to five steps) that all employees could follow when they discover a critical risk that may require an urgent coordinated response?
- Alternative channels for transmitting information about risks are auxiliary channels. The main channel through which information about technological risks should be reported is the traditional hierarchy of management control. List the existing risk communication channels in the company that follow this traditional pattern. How can the quality of reporting using these channels be improved?
- Part of this project will involve creating a special smartphone app to function as an alternative channel for transmitting information about technological risks. Is it also worth considering an app to support the existing channels for transmitting risk information along the traditional hierarchy? What are the pros and cons of this approach?
- What do you think should be the criteria for the success or failure of this pilot project to improve the quality and speed of risk information transmission? How can these criteria be measured in practice?

The experience of conducting seminars for various levels of management showed that the most active participants in group work No. 2 were senior managers, and the least active were shop floor workers. After holding two out of the seven planned seminars for shop floor employees, group work No. 2 was excluded from the presentation, due to the obvious fatigue of shop floor employees towards the end of the session.

10. Group work No. 3—reveal to senior management the most critical problems of your enterprise

Some senior managers of the company, who were interviewed as part of in-depth interviews at the very beginning of the pilot project, were skeptical that their subordinates would reveal anything serious or problematic during these special seminars. They were convinced that the workforce held a firm belief that it was better not to risk disclosing bad news to their superiors.

However, once employees during the seminars had seen the extreme emergencies that poor risk communication can cause in an industrial company, they seemed to have overcome this reluctance. As a result, participants across the board showed a willingness to share problems they had encountered within the company.

Group work No. 3 generated the most active and productive responses from shop floor workers and lower managers. They were asked to disclose to the senior management and the site director any serious problems and risks that: (I) they had observed within their specific area of responsibility, or in the wider production process; (II) had the potential to create a critical problem in the near/medium term; and (III) they believed should receive urgent attention from senior management.

Senior management promised the participants of the seminars that no one would be penalized for honestly disclosing any of this information, that a detailed analysis would be made of all the risks or issues raised, and that rapid corrective measures would be taken wherever they were needed. These statements were included in the following video message from the SVP to the seminar participants:

"Dear colleagues!

I am sure that the seminar has helped us all recognize the extreme importance of prompt and accurate reporting of risk-related information throughout the management hierarchy. This can prevent a critical development of events and avoid an emergency.

We will need time to evaluate the results of the work of the groups involved in these introductory seminars—and based on this analysis, to create effective practical solutions to improve the quality and speed of the transmission of risk information over the coming months.

However, I believe that we should not waste precious time! I would like to invite you all, right now in this seminar, to inform me and the site managers about any serious problems and risks that you are already aware of, anywhere across the production site or in your own workshops, that you think could have the potential to create a serious problem in the near future.

Tell us what the senior management should urgently pay attention to. Share this information in your group or personally with Dmitry [leader of this pilot project], who I have complete trust in. All this information will then be handed on to me personally.

Once again, I personally guarantee that no one, including your superiors, will be punished or blamed for sharing this information! A detailed analysis of all the disclosed risks will be carried out and, where necessary, corrective actions will be taken.

All this will help us save lives, increase the reliability of our production facilities, and also save our jobs and ensure the stability of your incomes for many years to come.

Thank you very much for your help, concern, and contribution to the safety of our company!".

After listening to this message, the seminar participants divided into groups, went to different areas of the hall and began to discuss the risks that they wanted to disclose to senior management, writing them down on flip-chart papers.

There was no limit to the number of relevant issues that could be raised—all were welcome. However, it was crucial that the participants ranked the risks and problems raised in their group. They were ranked on scales of 5 down to 1, by *criticality*—how serious was the possible danger, 5 being highly critical and 1 least critical—and by *urgency*—how soon the problem could escalate, 5 being the most urgent and 1 the least. This helped senior management to immediately identify the most pressing issues, and where they should first direct their attention. An issue scoring 5 on criticality and 5 on urgency obviously required their immediate intervention!

Running this exercise over 15 seminars revealed seven critical and urgent risks, and a further 104 significant technological, production and organizational problems —all of which had an existing or potential negative impact on risk management and industrial safety across all four industrial sites.

Interestingly, several seminar participants stated the importance of having a seminar facilitator (the pilot project leader) who is an independent scientist, reporting only to the company's SVP. After each seminar, the facilitator sent all the information about the disclosed risks to the SVP alone.

It is also important that separate seminars were held for various levels of management (i.e. shop floor employees did not attend a seminar with their line managers). This allowed seminar participants to disclose risks and problems they were aware of without fear of being identified by their superiors.

Six critical risks were revealed during this group work No. 3, while the seventh was revealed by one of the participants after the seminar had ended and everyone else had left. A lower-level manager approached the pilot project leader and said that he would like to disclose a critical risk, but did not want his colleagues to know. He explained that he is a new manager and does not want to endanger his relationships with his colleagues. The pilot project leader agreed, and the manager shared what he knew about the risk. Senior management arrived at the site the next day. The risk was indeed recognized as critical, and work began immediately to address it. As part of this corrective action, the peers of the lower-level manager who had disclosed the risk were faced with an increased workload because the situation was so urgent. Workers began to guess who had revealed the risk, but they had no evidence. The situation was successfully contained, to the satisfaction of the lower-level manager, his colleagues, the head of the facility, and senior management. No one was penalized for having allowed this critical risk to have developed. Instead, all the workers were told that situation like this must not happen again—risks need to be identified at an early stage and promptly reported to the head of the facility, and control measures need to be taken and/or referred up the hierarchy if additional support is required. Later, the manager involved was invited to receive an award at a special company ceremony (discussed below)—but he chose not to attend, preferring to maintain his anonymity.

It is worth comparing the results obtained during the seminars with the results of other channels for reporting risks and problems in the company. There were three such channels at the pilot plants: a problem-solving board, boxes for anonymous reporting of problems and risks, and a helpline. The last two channels were not popular with employees—the number of messages sent was minimal. Many more messages were sent through the problem-solving board, but these were also related to minor production problems. According to the manager who oversees these problem-solving boards, employees are mostly reporting problems with the cleanliness of industrial premises. No critical and serious risks have ever been reported through these three channels. By comparison, holding special seminars for a narrow circle of selected employees and managers who manage the critical infrastructure at the sites is a much more effective way to get information about critical and serious technological or production issues.

11. Anonymous survey No. 3: feedback from seminar participants

At the end of each seminar, a final online survey (five answers) was conducted, aiming to assess how helpful and relevant the session had been for the participants.

Results of responses to anonymous surveys within the framework of the pilot project: *Prior to this seminar, did you have a clear understanding that the problem of employees concealing risk information from their superiors is one of the main causes of serious industrial accidents?*

	Clear understanding	Some understanding	No understanding	Difficult to answer	Number of respondents
All survey participants	51.0%	40.0%	8.0%	1.0%	252
Senior management, heads of departments and directors of sites (middle managers)	67.6%	29.4%	2.9%	0.0%	34
Lower managers: deputy directors of sites, chief engineers of sites, heads of workshops, heads and representatives of HSE services at sites	51.1%	40.0%	7.8%	1.1%	90
Engineers, foremen, and shop floor employees who operate critical infrastructure at sites	46.9%	42.2%	10.2%	0.8%	128

Interpretation of responses: the lower the respondents are in the company's hierarchy, the less they realize that the problem of employees concealing risk information from their superiors is one of the main barriers to preventing serious industrial accidents. This shows that senior management need to constantly emphasize to their subordinates the importance of communicating objective information about any safety and technological problems they encounter in their area of responsibility. Leaders should also reassure the workforce that they very much welcome this information, and that nobody will be blamed or penalized for disclosing problems, however serious. On the contrary, they will have the management's gratitude for taking prompt action that may well prevent an emergency, and resources will be made available to address the issues they have helped identify.

Results of responses to anonymous surveys within the framework of the pilot project: *How relevant was the content of the seminar for you? (1 = totally irrelevant; 10 = 100% relevant)*

	1 = totally irrelevant 10 = 100% relevant											
	1	2	3	4	5	6	7	8	9	10	Average score	Number of respondents
All survey participants	0.0%	0.0%	0.4%	0.8%	2.4%	3.2%	10.7%	15.5%	12.7%	54.4%	8.9	252
Senior management, heads of departments and directors of sites (middle managers)	0.0%	0.0%	0.0%	0.0%	0.0%	0.0%	8.8%	8.8%	11.8%	70.6%	9.4	34
Lower managers: deputy directors of sites, chief engineers of sites, heads of workshops, heads and representatives of HSE services at sites	0.0%	0.0%	0.0%	0.0%	3.3%	3.3%	7.8%	21.1%	12.2%	52.2%	8.9	90
Engineers, foremen, and shop floor employees who operate critical infrastructure at sites	0.0%	0.0%	0.8%	1.6%	2.3%	3.9%	13.3%	13.3%	13.3%	51.6%	8.7	128

Results of responses to anonymous surveys within the framework of the pilot project: *How ready are you to use the knowledge and skills that you have gained from this seminar in your day-to-day work? (1 = absolutely not ready; 10 = 100% ready)*

	1 = absolutely not ready 10 = 100% ready											
	1	2	3	4	5	6	7	8	9	10	Average score	Number of respondents
All survey participants	0.0%	0.0%	0.0%	0.4%	2.8%	4.0%	11.1%	13.5%	13.5%	54.8%	8.9	252
Senior management, heads of departments and directors of sites (middle managers)	0.0%	0.0%	0.0%	0.0%	0.0%	0.0%	8.8%	2.9%	14.7%	73.5%	9.5	34
Lower managers: deputy directors of sites, chief engineers of sites, heads of workshops, heads and representatives of HSE services at sites	0.0%	0.0%	0.0%	0.0%	3.3%	6.7%	8.9%	14.4%	10.0%	56.7%	8.9	90
Engineers, foremen, and shop floor employees who operate critical infrastructure at sites	0.0%	0.0%	0.0%	0.8%	3.1%	3.1%	13.3%	15.6%	15.6%	48.4%	8.8	128

Results of responses to anonymous surveys in the framework of the pilot project: *It is necessary to run explanatory trainings and seminars, to explore the consequences of distorting risk information transmitted through the corporate hierarchy. Many employees simply do not realize that this is one of the key problems within industrial safety, and do not understand what catastrophic consequences it can have. Special seminars can advance the understanding of this urgent problem and offer managers and employees opportunities to "break the ice" around improving communication on difficult topics.*

	Strongly agree	Rather agree	Rather disagree	Strongly disagree	Difficult to answer	Number of respondents
All survey participants	57.0%	38.0%	2.0%	0.0%	2.0%	252
Senior management, heads of departments and directors of sites (middle managers)	88.2%	11.8%	0.0%	0.0%	0.0%	34
Lower managers: deputy directors of sites, chief engineers of sites, heads of workshops, heads and representatives of HSE services at sites	54.4%	41.1%	2.2%	0.0%	2.2%	90
Engineers, foremen, and shop floor employees who operate critical infrastructure at sites	50.8%	43.8%	1.6%	0.8%	3.1%	128

Results of responses to anonymous surveys within the framework of the pilot project: *I am ready to recommend this seminar to work colleagues. (1 = not ready at all; 10 = 100% ready to recommend)*

	1 = absolutely not ready 10 = 100% ready											
	1	2	3	4	5	6	7	8	9	10	Average Score	Number of respondents
All survey participants	0.4%	0.4%	0.0%	1.2%	2.8%	1.2%	5.6%	11.1%	14.3%	63.1%	9.1	252
Senior management, heads of departments and directors of sites (middle managers)	0.0%	0.0%	0.0%	0.0%	0.0%	0.0%	0.0%	8.8%	17.6%	73.5%	9.6	34
Lower managers: deputy directors of sites, chief engineers of sites, heads of workshops, heads and representatives of HSE services at sites	0.0%	0.0%	0.0%	1.1%	4.4%	1.1%	7.8%	12.2%	13.3%	60.0%	9.0	90
Engineers, foremen, and shop floor employees who operate critical infrastructure at sites	0.8%	0.8%	0.0%	1.6%	2.3%	1.6%	5.5%	10.9%	14.1%	62.5%	9.0	128

The results of anonymous survey No. 3 show that conducting interactive seminars is a very effective method of (I) demonstrating the significance of the widespread problem of employees suppressing and distorting information about risks; (II) getting feedback about the reasons why this problem occurs in a given company; (III) getting first-hand feedback about possible solutions to the problem; (IV) obtaining information on a wide range of existing safety and production risks within the organization, which can then be rated for criticality and urgency.

Lessons learned: Based on a survey of 100 critical infrastructure executives, a recommendation was made that senior management should regularly visit industrial sites to identify and address serious safety and technological problems (see Chap. 3, Recommendation 7: *"Use multiple channels for obtaining risk information"*). Conducting seminars like this, prior to a site visit by executives from headquarters, allows senior management to have an up-to-date list of the most serious current problems of the facility. This will help them to be properly prepared for visiting the facility: executives can demonstrate to their subordinates that they are already aware of the most critical issues, and immediately announce suitable remedial measures to address them. Senior managers visiting the facility are unlikely to meet with all employees at the site but should focus their attention on a selected 5–10% of the workforce who manage the critical facilities. If these employees have already undergone special training by participating in these seminars, they will already understand the vital importance of communicating honest and accurate risk information to their superiors. Therefore, they should be primed and ready to share this with the top brass. This will make the visits much more productive and efficient and improve safety management across the production process. Trained employees should also be granted access to a special corporate smartphone app that will enable them to promptly inform their superiors about any operational risks that come to their attention during their daily work (the development of this smartphone app will be discussed later).

It is important to mention that these seminars are essential at the start of the project. At this stage, hundreds of senior managers and their subordinates need to be shown the importance of prompt and accurate upward risk reporting, and trained in a new way to respond if they detect critical or serious risks. During the subsequent implementation of the project over several years, the emphasis will shift. The main alternative channels for reporting risk-related information will now be the special smartphone app (distributed among staff trained by means of such seminars), and direct communication between senior management and production site workers during their visits. In these later years of the project, new seminars should be organized only at new production sites, or for the initial training of new personnel at the existing pilot sites.

VIII. TAKE ACTION ON ALL IDENTIFIED CRITICAL AND URGENT RISKS

When the pilot project began, its main objective was formulated as follows: to encourage shop floor employees to promptly transmit information about safety and production risks to senior management, so that remedial action can be taken before a serious situation can develop. The focus was on improving the speed and quality of risk reporting up the corporate hierarchy.

Therefore, from the very beginning of the project, there was a separation of tasks. The project leader concentrated on ways of improving the quality and speed of communication about risks and problems (in other words, how to motivate employees and line managers at the pilot sites to disclose this information to senior management), while senior management took responsibility for providing solutions to the identified risks and problems.

From the outset, it was agreed that when participants disclosed problems at the seminars, it would be the responsibility of the project leader to document the issues raised and send this information directly to the SVP of the company, who would then decide with other executives on the best course of action for each problem. Accordingly, immediately after each seminar, the project leader emailed the company's SVP a list of the identified critical risk issues. The following day, he would send a second list detailing risks and issues of a lower level of criticality and urgency.

Through this process, senior management were able to take swift control of seven critical and/or urgent situations. In some cases, their prompt intervention solved problems that could very quickly have resulted in a major incident. In two of the seven cases, the relevant site directors had not previously been aware that such a problem even existed at their facility.

Summing up, the pilot project immediately proved its effectiveness and value by bringing these serious critical risks to the attention of senior management so that remedial action could be quickly taken, preventing the problems escalating into an emergency.

Lessons learned: from the retrospective analysis of how these critical problems were solved, it appears very important that the project leader visits production sites together with senior managers, and that they assess critical risks revealed during the last seminar. By personally taking part in this inspection, the project leader can observe firsthand the communication between senior management and the employees who had reported critical risks during the seminar.

This realization came a few months after the seminars began, when the project leader accidentally became aware that one of the senior managers—a highly qualified expert in the production process—was asked by the company's SVP to analyze one identified critical risk at the pilot plant. For unknown reasons, this senior manager had not been included in the initial list of the pilot project participants. As a result, he had not undergone the specific project training, and was unaware of the project's ideology and the newly introduced methods of risk communication. Arriving at the plant and assessing the situation (which was indeed very critical), he proceeded to condemn the lower-level manager who had uncovered this critical risk and requested urgent assistance from senior management to address the situation.

The truth was that an external contractor had aggravated the problem, and the lower-level manager judged that he lacked the resources to tackle this unexpected issue in his area of responsibility. To make matters worse, the head of the plant was on vacation when the problem came to light. Therefore, following what he had learnt at the seminar, the lower-level manager decided to approach the company's SVP directly and without delay to request assistance in tackling the problem. As a result of his actions, the critical risk was promptly mitigated, and a potential accident averted. As it turned out, this risk was the most serious of the seven identified in the pilot project seminars. But instead of receiving positive feedback and recognition, the manager was publicly criticized by the senior manager for failing to do his job properly.

By simply blaming the lower-level manager, rather than making a detailed analysis of the true causes of the problem, the senior manager endangered the successful implementation of the pilot project across the entire site. For a start, the lower-level manager who disclosed the risk would be sure to keep quiet about any safety issues in his area of responsibility in the future, and try and tackle them on his own. But the damage did not stop there: the dressing-down given by the senior manager was also witnessed by the lower-level manager's colleagues. It is likely that they drew similar conclusions about the dangers of informing their superiors about production problems, despite the SVP's pledge not to blame employees for communicating difficult issues. Once the head of the plant had returned from vacation and conducted his own investigation, he personally thanked the lower-level manager for his courage and promptness in revealing a serious risk and preventing an accident. He also urged him not to be fearful of disclosing further problems, but to continue to be proactive in caring about safety. Nevertheless, this incident appears to have had a negative impact on the willingness of the head of the plant to disclose new risks and problems to headquarters. One indication of this is the fact that, under various pretexts, the head of the facility did not allow the pilot project leader to organize further seminars at his plant. Finally, he explained his reluctance by saying that further seminars would likely lead to the disclosure of new problems: this would just attract more attention from senior management with no guarantee that the information would be well received and might well leave his plant workers being blamed and criticized again.

This incident offered valuable lessons:

- At the beginning of the project, the pilot project leader underestimated how important it is to oversee the process of addressing any serious problems that come to light. If the project is to succeed, it is essential that, as well as improving the process of risk information transmission, the process of solving the problems should be well prepared and organized. It was impossible to immediately delegate the latter process exclusively to the company's management. Due to corporate inertia and habit, managers may well simply repeat the negative behaviors that had led to risk communication problems in the first place. In other words, the project must be perceived as a concrete opportunity for new ways of addressing problems across the entire company. At its outset, this new approach should be agreed upon and formalized with senior executives, making it clear to them how they are now required to respond to subordinates who report risks and problems.

- All managers who have a role in mitigating identified risk problems must undergo special training and accept the ideology of the project—especially the stipulation that any employee voluntarily disclosing risks in their area of responsibility must always receive positive feedback from their superiors. It is absolutely essential that employees feel safe to disclose risks to their superiors and are motivated to continue doing so.
- The project leader must travel with the team of senior managers to production sites where critical risks have been identified. This should be continued until the project leader has confidence in the senior management to properly analyze the reasons why any risks have arisen, give positive feedback to those who disclosed the information, and conduct their interactions in a way that reduces employees' fears about risk reporting and encourages them to inform their superiors if they become aware of further problems.
- To counteract the misguided negative feedback, it would have been enormously helpful to organize an urgent visit by the company's SVP to the production site. This would have allowed him to thank the lower-level manager personally and publicly for his correct and courageous actions, and encourage all workers to follow this example and willingly disclose risks to their managers. A visit could also have reassured the head of the site that promptly identifying and reporting critical risks would help and not hinder career prospects—while on the other hand, suppression of risk information is going to be dealt with very severely. Unfortunately, the project leader did not become aware of this incident until a few months afterwards, so no such visit was possible. However, the lower-level manager who identified the concerned critical risk was one of the employees that the SVP selected to publicly commend during the pilot project's inaugural event (discussed below).
- After this incident, the project leader found the opportunity to briefly speak to the senior manager involved, to share the ideology and objectives of the project, and explain the damage that negative feedback could inflict on its overall success. The senior manager agreed to attend a special full day senior management seminar led by the project leader. But when it came to it, citing a heavy workload, he was unfortunately only able to attend about 40% of the seminar. This experience suggests that only managers who have completed the special training, and fully support the project's ideology and principles, should be involved in solving the problems that are identified through the seminars. The training of lower-level employees should only begin when all involved senior managers have completed theirs.

IX. PROGRESS IN ADDRESSING LESS CRITICAL PROBLEMS

In addition to critical and urgent risks, more than a hundred less serious issues became known during the seminars.

When employees show trust by reporting risks or problems in response to requests from their senior management, they act in the belief that prompt action will be taken to solve, or at least reduce, these problems. If senior management cannot do so, they must explain this to their subordinates, and reassure them that—for the time being at least—shop floor workers can continue to work safely even though such risks are present. Employees need to see progress in dealing with the problems they have helped identify. If not, eventually they will stop believing that things are changing and will revert to their old habits of concealing risks and problems to avoid trouble landing on their own heads.

As part of the pilot project, it was important to understand the limits of the existing corporate management system to successfully address these various issues.

The management system may be able to show at the pilot stage that it can quickly and effectively solve all the problems that come up, regardless of their level of criticality. In that case, when it comes to scaling the project up to include additional production sites, it will be safe to encourage employees to disclose all production risks and problems they are aware of, and not restrict this to just the most serious.

Conversely, if the company cannot address all the problems reported during the pilot project, then, it is advisable not to promise future project participants that any issue they identify at their sites will be quickly dealt with. Indeed, this has the potential for raising unrealistic expectations. The SVP's message to the participants of the seminars contained the following clause: *"A detailed analysis of all the disclosed risks will be carried out and, where necessary, corrective actions will be taken"*. The project initiators were careful not to guarantee seminar participants that every problem would be dealt with. Instead, a more realistic approach was taken. They were assured that any issue raised in the seminars would be brought to the attention of senior management and carefully analyzed, but that only the most critical and urgent would be resolved straight away.

During the pilot project, the following problem-solving issues were encountered at the four sites.

After each seminar, information about the risks and issues disclosed was passed on to the company's SVP. He acquainted himself with this list of identified problems, and then delegated actions to his HSE deputy. The HSE department studied the problems in more detail, collecting any additional information they required, and then began to plan and execute appropriate remedial action plans.

As it turned out, many of the problems identified were related to the production department of the company. Gradually, departments responsible for production, logistics, and procurement became involved in addressing the various problems. The number of seminars held at the production sites grew as the number of problems identified inevitably increased. The project leader held a special educational seminar for 20 heads of departments. Here he outlined in detail the principles and objectives of the project and called on them to get closely involved in addressing the problems that had now been identified at four production sites. Eventually the number of problems identified by employees grew so large (in total, 104 were raised during the 15 seminars) that it was decided to create a special working team of company executives and production unit managers. This included various senior managers (some of whom did not receive the special project training), along with the relevant department heads, and all the site managers at the selected sites (all of whom did attend the training).

Functions of the special working team:

- comprehensive assistance to production sites to tackle identified problems;
- prompt and expert assessment of critical and urgent problems: members of the team visited the location; all equipment involved was examined to determine the potential risks from its continued operation; equipment was decommissioned when necessary; the company's senior officials and any relevant departments were immediately informed about the presence of critical and urgent issues;
- creation of temporary joint working groups to manage specific identified problems (e.g. equipment repair programs; scaling back of production targets and plans; resolving employment of workers operating the faulty equipment in cases of lengthy shut downs; repairs and replacements etc.);
- securing additional finance and other resources when required;
- conducting a qualitative analysis to determine the reasons a problem had occurred;
- maintaining a database of identified problems and actions and coordinating this information with the company's database of safety and technological risks.

To conclude, the company management system was able to quickly and effectively respond to seven extremely critical risks (where serious accidents were likely to be imminent) and address a further 25 major risks across the four pilot facilities. The heads of the sites were able to resolve most of the moderate risks (approximately 50) using each facility's own resources. However, more than six months after the start of the project, about 25 serious problems remained unresolved. It became apparent that a considerable proportion of these hold-ups were the result of accumulated malfunctions within the company's organizational and technological structures over many years. The *"treatment"* of these deep-seated problems would require painful decisions at the highest level, restructuring of the entire organization, many years of efforts by senior management, and the allocation of significant extra resources.

In the opinion of the project leader, there were also other reasons why these problems were still not addressed more than six months after the launch of the pilot project:

A. The company's SVP participated in all key meetings of the project, but with the exception of the special working team meetings where solutions to the identified problems were discussed. Retrospective analysis suggests that, if the SVP had participated regularly in these meetings and immersed himself in delivering solutions, then this would have increased the involvement of the entire senior management team in resolving these issues—which, while not being urgent or critical, were still important in the medium or longer term.

B. Not all production-related issues can be resolved quickly. The head of one production site (middle management) offered the following explanation: *"Many heads of the facilities are aware of existing risks, but we cannot immediately address all the problems and minimize the risk of accidents, because our resources are limited. Therefore, we can only make repairs sequentially. We are manually redirecting resources to rectify the most critical risks, but other important risks have to wait their turn. My colleagues and I really do want to reduce all the risks at once, but this is simply impossible, because all production cannot be stopped at one go—we can only withdraw a part of the production capacity at a time. I don't want to continue working while risks are there: in fact, I don't want risks to exist at all. I want everything to be working perfectly, everything in harmony, nothing worn or defective. However, everything depends on resources: time, money, the availability of contractors to carry out the repairs, logistics of new equipment and the existence of a realistic production plan allowing us to temporarily shut down some production processes for the repairs. In the end, we have no option but to constantly prioritize what needs to be repaired first, and this is a never-ending process".*
C. Senior managers are extremely busy with regular responsibilities. As result, it is difficult for them to find additional time to address significant organizational and production problems that have accumulated over many years and, to some extent, have been tolerated. Their rectification requires time and effort, and often only produces safety improvements in the longer term. Most of the unresolved major problems revealed during the project, however, required long-term solutions.

Results of responses to anonymous surveys within the framework of the pilot project: *Do you think that the managers and employees in your company are in a hurry?*

	Constantly	Often	Sometimes	Rarely	Difficult to answer	Number of respondents
All survey participants	19.9%	23.0%	32.5%	18.1%	6.4%	326
Senior management, heads of departments and directors of sites (middle managers)	26.8%	24.4%	36.6%	9.8%	2.4%	41
Lower managers: deputy directors of sites, chief engineers of sites, heads of workshops, heads and representatives of HSE services at sites	20.4%	28.7%	27.8%	12.0%	11.1%	108
Engineers, foremen, and shop floor employees who operate critical infrastructure at sites	18.1%	19.2%	34.5%	23.7%	4.5%	177

Interpretation of responses: the higher the respondents are in the company hierarchy, the more they feel overloaded and always in a hurry. They feel they just do not have the time to analyze problems and risks in detail, so they and their subordinates often ignore them. The authors of this handbook examined 20 major accidents[7] where intra-organizational concealment of information about risks was established as a major factor in causing disaster. The analysis showed that, in 12 out of 20 disasters (60%), managers and employees were reported to be constantly in a hurry. In the organizations where these accidents occurred, employees were often urgently required to implement a range of tasks, which forced them to ignore many of the risks that were shown later to have caused serious incidents.

D. Some senior managers and department heads were overwhelmed by current production tasks. Instead of giving the production sites the comprehensive assistance they had promised, they were forced to delegate the job of solving some risk issues to their subordinates. As a result, responsibility for a sizable proportion of these problems shifted to the heads of the pilot production sites, who did not have the additional resources required to correct these issues. Such downward delegation went against a central principle of the project: the guarantee that senior management would assist the production sites to address their problems, so that employees and lower managers would not be afraid to report problems in their area of responsibility. As a result, these problems became a real headache for the heads of the sites. Some of them did not understand why they should risk disclosing problems at their sites to senior management as part of a pilot project, if in the end they were left to tackle these problems alone, without practical or financial support from executives. Meanwhile, some dutifully tried to follow their bosses' instructions and tackle the problems the project had brought to light. But with senior management now preoccupied with *"more important"* matters, the site managers were very reluctant to admit to the leadership that (I) a significant number of their production site problems could only be solved with the direct involvement of senior managers; (II) they had neither the resources nor the authority to deliver effective solutions; (III) it was at best misguided, and at worst a dereliction of duty, for senior managers to be delegating responsibility back down to site level for tackling such serious production problems.

[7] Dmitry Chernov, Didier Sornette, Giovanni Sansavini, Ali Ayoub, Don't Tell the Boss! How poor communication on risks within organizations causes major catastrophes, Springer, 2022, https://link.springer.com/book/10.1007/978-3-031-05206-4.

Results of responses to anonymous surveys within the framework of the pilot project: *Why do you think managers often do not want to hear bad news from employees about matters like observed risks and problems in an organization and the need for additional investments, like equipment upgrades, to create safer production processes?*

Managers expect employees to solve problems on their own when they occur in their area of responsibility.

	Strongly agree	Rather agree	Rather disagree	Strongly disagree	Difficult to answer	Number of respondents
All survey participants	16.3%	46.6%	25.2%	7.4%	4.6%	326
Senior management, heads of departments and directors of sites (middle managers)	14.6%	48.8%	19.5%	14.6%	2.4%	41
Lower managers: deputy directors of sites, chief engineers of sites, heads of workshops, heads and representatives of HSE services at sites	19.4%	40.7%	27.8%	8.3%	3.7%	108
Engineers, foremen, and shop floor employees who operate critical infrastructure at sites	14.7%	49.7%	24.9%	5.1%	5.6%	177

E. The main reason the leaders of the company are reluctant to receive information about risks and problems from their subordinates is the excessive bureaucracy involved in getting anything done about them. The very serious issues identified during the seminars required extraordinary efforts from senior management to get the remedial plan through the bureaucracy of this very large industrial company.

Results of responses to anonymous surveys within the framework of the pilot project: *Why do you think managers often do not want to hear bad news from employees about matters like observed risks and problems in an organization and the need for additional investments, like equipment upgrades, to create safer production processes?*

Excessive bureaucracy interferes with the practical solution of any identified problems. Managers face complex corporate procedures when going through budgeting and investment committees, so issues are resolved very slowly.

	Strongly agree	Rather agree	Rather disagree	Strongly disagree	Difficult to answer	Number of respondents
All survey participants	42.6%	35.9%	12.9%	1.5%	7.1%	326
Senior management, heads of departments and directors of sites (middle managers)	48.8%	29.3%	19.5%	2.4%	0.0%	41

(continued)

(continued)

	Strongly agree	Rather agree	Rather disagree	Strongly disagree	Difficult to answer	Number of respondents
Lower managers: deputy directors of sites, chief engineers of sites, heads of workshops, heads and representatives of HSE services at sites	46.3%	36.1%	10.2%	1.9%	5.6%	108
Engineers, foremen, and shop floor employees who operate critical infrastructure at sites	39.0%	37.3%	13.0%	1.1%	9.6%	177

Interpretation of responses: a very large rate of agreement to this question across all groups indicates that this company has an excessively bureaucratic system of budgeting, procurement, repairs, ordering internal corporate services, and so on. This has a negative impact on both the speed of response to identified risks and the quality of decision-making. Another factor that has a negative impact here is the geographical remoteness of the company's production sites, so delivering new equipment for example can be very slow. Curiously, this reason was not mentioned even once during the interviews of the 100 critical infrastructure executives conducted in 2018–2021. This suggests that each critical infrastructure company may have some unique combination of factors that have a negative impact on the willingness of managers to hear about and respond to risks.

F. The geopolitical turbulence in 2022 caused economic difficulties in many countries around the world. This company also faced serious new challenges, which required immediate decisions from the top brass. Senior managers may have decided that they had to prioritize these challenges because the company's immediate survival depended on a quick response. They could not simultaneously devote time and resources to addressing the chronic safety and technological problems the seminars had revealed—and the benefits from solving these chronic problems may not be apparent for years.
G. This geopolitical turbulence also put pressure on the company to initiate massive cost reductions, so some risk issues were left unresolved due to the company's financial situation.
H. The reluctance of some managers to acknowledge that safety and technological problems really do exist, may indicate that earlier managerial mistakes were made. This may threaten their authority in the eyes of their superiors, subordinates and colleagues, casting doubt on their competency and other leadership decisions.

   Real-life examples of refusing to acknowledge serious problems can be revealing. In one of the pilot facilities, all three seminar groups highlighted the same major problem caused by recent innovations in production management. At the company-wide level, these changes were generally recognized as a bad decision. At some production sites, the return to the previous management model—which had worked successfully for decades—was already under way. However, one of the senior managers, responsible for a particular aspect of the production innovations, refused to admit that they had caused any problems—even when all three seminar groups independently raised the same issue. This manager had invested much time and effort into introducing the change, so it was hard for him to face the safety and efficiency problems that the change had also introduced. Rather than admitting the problem, he simply deleted it from the list of urgent issues identified during the seminars. Nevertheless, during a meeting of the special working team, he promised to run a promotion campaign among his subordinates that would show the benefits of the new production system, and convince them to stop demanding a return to the previous model. Clearly, he was finding all the public discussion of the problems caused by the new model, and the complaints it generated among lower-level employees, very threatening. When the project will be scaled up across other company sites where he has significant control, the company's SVP and many of his fellow managers would be there—and, of course, they would see the dissent among employees against his wish to push the new model through. He decided he would deal with this by writing to the pilot project leader: in future, when scaling the project to other enterprises under his control, he himself would organize the seminars. This really showed just how far he was prepared to go to control and limit the risk information that employees could report to the company's SVP, bypassing the traditional hierarchy in general and himself in particular. It might also indicate his wish to be the first to receive risk information, so that he could control which problems he would allow to be sent higher up the corporate hierarchy. It would seem that the fear of having their own previous mistakes made public can make senior managers completely refuse to admit the existence of serious problems—even when they are raised by the majority of their subordinates. This, of course, will make any attempt to mitigate those problems impossible.

Another group of middle managers from one of the production sites selected for the pilot project were also eager to show their senior managers that there were no serious problems at their site. They underestimated the criticality and urgency of some of the problems that their employees had highlighted. They dismissed some problems as irrelevant and crossed them off the list of issues to be addressed immediately, despite the objections of various departments of the company arguing that they were important. This experience suggests that the special working team of the project, although it will consist mostly of senior managers, must also include highly qualified employees in the field who are independent of site management. These employees will keep top managers informed about the real situation in a production site, through constant communication with those who have disclosed risks. This is necessary to provide senior management with a second opinion on the criticality and urgency of reported problems and feedback on how the management of a production site is progressing in solving them.

Results of responses to anonymous surveys within the framework of the pilot project: *Why do you think managers often do not want to hear bad news from employees about matters like observed risks and problems in an organization and the need for additional investments, like equipment upgrades, to create safer production processes?*

Leaders are afraid of being seen as incompetent and being held accountable for their previous bad decisions that created the current problems.

	Strongly agree	Rather agree	Rather disagree	Strongly disagree	Difficult to answer	Number of respondents
All survey participants	16.0%	36.2%	31.6%	7.4%	8.9%	326
Senior management, heads of departments and directors of sites (middle managers)	17.1%	43.9%	29.3%	9.8%	0.0%	41
Lower managers: deputy directors of sites, chief engineers of sites, heads of workshops, heads and representatives of HSE services at sites	13.9%	27.8%	38.9%	13.0%	6.5%	108
Engineers, foremen, and shop floor employees who operate critical infrastructure at sites	16.9%	39.5%	27.7%	3.4%	12.4%	177

Interpretation of responses: Senior and middle managers are most sensitive to the fact that problems reported to them by their subordinates may have been caused by their previous poor management decisions.

I. Tackling serious production problems may also create conflict with other senior managers, because public discussion of such problems within the company is likely to call into question the wisdom of previous management decisions. For this reason, some managers who were delegated the responsibility for tackling serious problems identified in the pilot project were unwilling to really delve into what had caused these problems and how they might be solved.

J. Some leaders appeared unwilling to take personal responsibility for tacking serious problems, due to the threat to their own careers if they made mistakes. Responsibility for coordinating solutions to issues identified within the project was constantly being handed on from one senior manager to another during the first six to eight months after the launch of the pilot. This continued until the SVP chose one senior manager as the main person responsible for coordinating the solution of all the problems that had arisen.

Results of responses to anonymous surveys within the framework of the pilot project: *Why do you think managers often do not want to hear bad news from employees about matters like observed risks and problems in an organization and the need for additional investments, like equipment upgrades, to create safer production processes?*

Managers believe that once a problem has been reported to them, they are automatically responsible for solving it.

	Strongly agree	Rather agree	Rather disagree	Strongly disagree	Difficult to answer	Number of respondents
All survey participants	16.3%	47.9%	23.0%	5.5%	7.4%	326
Senior management, heads of departments and directors of sites (middle managers)	17.1%	46.3%	24.4%	12.2%	0.0%	41
Lower managers: deputy directors of sites, chief engineers of sites, heads of workshops, heads and representatives of HSE services at sites	17.6%	43.5%	25.0%	11.1%	2.8%	108
Engineers, foremen, and shop floor employees who operate critical infrastructure at sites	15.3%	50.8%	21.5%	0.6%	11.9%	177

Interpretation of responses: A high percentage of agreement among all levels of respondents suggests that managers feel overwhelmed by their workload and would rather not receive information about problems from the field, so they do not have to take responsibly for solving them. Leaders would much prefer that employees in the field somehow find a way to solve these issues on their own, without attracting the attention of their superiors.

K. Unwillingness to take responsibility for addressing specific risk problems—some senior managers were clearly reluctant to take on the responsibility for investigating and implementing solutions to the problems identified during the seminars.

Summing up, within the framework of the pilot project, the company's existing management system coped effectively with finding ways of reducing or removing the seven critical production risks that came to light. However, the company was not fully prepared for the very large number of less critical but still serious production and organizational problems that the pilot project seminars revealed. Some of these had been building up for years, and the overall corporate response to these was much less successful, with some left unresolved.

The authors believe that the best way to scale up a similar project across new company sites is to begin by asking all participants for information they have about existing risks and problems but include a warning that not all these problems can be tackled at once. Major problems that do not actually pose a direct threat to operational safety cannot always be resolved quickly. An honest dialogue between executives and employees about corporate priorities, and the unavoidable limitations on addressing every problem, is essential to avoid unrealistic expectations.

There is a more radical solution—restructuring the company's entire management system to enable it to tackle all the identified problems. However, such a big decision would require huge investment by the leadership. This would only be justified if the company found itself unable to respond adequately even to critical risks identified during the project.

Lessons learned: During the pilot project, the majority of the risks identified by the project participants were addressed. Some further conclusions can be drawn:

- Expand the project's objectives. The central goal remains: "*to ensure prompt transmission of information about critical safety and technological problems from shop floor employees to senior management, in order to prevent emergencies from arising*". However, a further crucial goal needs to be added: "*to create a corporate mechanism for promptly dealing with all significant safety and technological problems disclosed to senior management*".
- From the onset of the project, the company involved should have in place a mechanism for dealing with all the risks and problems that are reported. This should include:
  - a list of the leaders who will take responsibility for dealing with safety and technological problems into the future;
  - specific training for senior managers;
  - creation of a pool of technical experts who can rapidly diagnose the relative criticality of any safety and production risks;
  - access to additional resources to enable solutions to be swiftly and properly implemented;
  - agreeing on specific regulations to govern how managers must act when addressing production risks and communicating with the employees who revealed the issues.
- When conducting seminars at production sites, it is advisable to be very specific about how disclosed risk information will be managed and recorded:
  - seminar participants are asked to disclose any information they have about critical and urgent safety and production risks that could escalate out of control and cause accidents and emergencies. It should be explicitly understood that, by requesting this information, senior managers are assuming full responsibility for taking action to bring these risks under control;
  - seminar participants are also asked to disclose any information they have about major (but less critical) production and organizational problems that could have a negative impact on the labor protection and industrial safety within their facility and across the whole company. It is also recommended to discuss both the positive and negative experiences of the company in solving similar problems in the past. It should be openly acknowledged that not all the issues raised can be resolved, but that senior managers will use this information to improve their understanding of more critical safety issues;
  - Identify and list the less critical problems at the facility that the production site managers can solve without assistance from headquarters.

- A gradation of risks and problems by severity and urgency allows the company to prioritize and immediately begin the process of addressing the issues that have been disclosed:
  - critical safety and production risks will be taken under the immediate control of senior management due to their potentially catastrophic consequences;
  - major problems will be analyzed, and priority given to tackling the most dangerous ones as quickly as possible;
  - moderately serious problems can be made the responsibility of site directors and addressed promptly.
- Resolving as many issues as quickly as possible will:
  - demonstrate to staff that risk issues they raise can be swiftly dealt with;
  - convince employees that senior managers are serious about solving problems and improving everyone's safety and security;
  - encourage employees to continue to disclose new risks to their superiors.
- It is important that a company CEO takes part in the initial discussion around safety and technological problems at a given site, in order to show the entire management hierarchy that this issue is now a priority for the leadership. The message is made clear: *"If the head of a company is personally involved in tackling the problems identified by the employees, then managers at every corporate level should also take an active part and do all they can to reduce these risks"*. Immediately after each project seminar, it is advisable to call a meeting with senior managers, the special working team of the project, and the site heads. The project leader uses this opportunity to present the collective feedback of the seminar participants. The meeting then moves on to discuss the identified problems in detail, looking at priorities, practical solutions, and responsibilities. After that, the CEO delegates roles and allocates additional resources to relevant executives and production sites. Regular monthly follow-up meetings under the leadership of the CEO may also be needed to review progress in solving the identified issues and making necessary adjustments.

X. PROCESSING THE RESULTS OF RESPONSES TO THE ANONYMOUS SURVEYS

Responses to the 85 questions included in the three anonymous surveys were collected during all 15 pilot project seminars. The responses were then grouped and processed.

The first analysis was based on where respondents were located—whether at one of the four selected production sites or at headquarters. This made it possible to analyze responses by site to see if there were any differences in their attitudes and corporate cultures.

The second analysis was based on the respondent's position within the company's hierarchy: (I) senior management, heads of departments and directors of sites (middle managers); (II) deputy directors of sites, chief engineers of sites, heads of workshops, heads, and representatives of HSE services at sites (lower managers); (III) engineers, foremen, and shop floor employees who operate critical infrastructure at sites. In the opinion of the authors of this handbook, this was the most informative breakdown, as it gave a good indication of the predominant opinions at each level of the hierarchy, including preferred solutions for managing the issues (some results of this second analysis are presented in the handbook).

XI. CREATION OF A SMARTPHONE APP AS AN ALTERNATIVE CHANNEL FOR TRANSMITTING INFORMATION ABOUT RISKS

The participants' survey responses helped give a sense of what kinds of methods and systems should be developed as part of the pilot project, to improve the transmission of risk-related information from shop floor employees to senior management.

Results of answers to anonymous surveys within the framework of the pilot project: *What channels should be developed at your production site for transmitting information about risks from the bottom of the corporate hierarchy?*

	It is only necessary to develop a risk communication channel based on the traditional chain of command (employees report information to their supervisor, who reports to his/her line manager, and so on up the chain from shop floor to senior management)	It is only necessary to develop an alternative channel (any employee can send information about risks directly to senior management, bypassing the traditional hierarchy, through smartphone app, hotlines, mailboxes, meeting with executives, etc.)	Both channels need to be developed (transmitting information about risks through the traditional hierarchy and through an alternative channel)	Difficult to answer	Number of respondents
All survey participants	20.4%	7.1%	67.5%	5.0%	280
Senior management, heads of departments and directors of sites (middle managers)	13.9%	0.0%	86.1%	0.0%	36
Lower managers: deputy directors of sites, chief engineers of sites, heads of workshops, heads and representatives of HSE services at sites	25.0%	5.8%	60.6%	8.7%	104
Engineers, foremen, and shop floor employees who operate critical infrastructure at sites	18.6%	10.0%	67.9%	3.6%	140

Results of responses to anonymous surveys within the pilot project: *If the company decides that it is necessary to develop both channels—communicating information about risks through the traditional hierarchy and through an alternative channel—then which channel should be the main focus of development?*

	Greater priority should be given to the channel using the traditional chain of command	Greater priority should be given to the alternative channel	Develop both channels with equal priority	Difficult to answer	Number of respondents
All survey participants	34.3%	15.0%	44.6%	6.1%	280
Senior management, heads of departments and directors of sites (middle managers)	50.0%	5.6%	41.7%	2.8%	36
Lower managers: deputy directors of sites, chief engineers of sites, heads of workshops, heads and representatives of HSE services at sites	32.7%	20.2%	39.4%	7.7%	104
Engineers, foremen, and shop floor employees who operate critical infrastructure at sites	31.4%	13.6%	49.3%	5.7%	140

Most of the project participants supported the development of a smartphone app that would allow shop floor employees to immediately inform higher-level managers about any safety or production risks they encountered, bypassing slower communication channels via the traditional management hierarchy. The SVP acknowledged the following: *"To be honest, these days the traditional hierarchy of management does not work well when it comes to transmitting objective information about safety and technological problems to senior management... Therefore, an alternative channel for communicating this information using a dedicated smartphone app should be created. Until a culture of prompt and honest transmission of information about risks along the management hierarchy is established, a company is much better off having both traditional and alternative communication channels"*.

Progress on developing the app began after the seminars and included several key features. The app can be installed on employees' personal smartphones, but use of these is forbidden where there is a risk of explosion. Employees working in these areas will be issued with a special explosion-proof smartphone. Anyone using the app needs to be registered and sign in using their own corporate login and password. All information reported through the app will be stored on the company's servers.

Initially, it was decided that only 5–10% of the company employees at each pilot site would be included in the scheme. It would be limited to those who (a) operated production facilities defined on site risk maps as critical, and (b) had taken part in the seminar described earlier on the problems of transmitting risk information.

Limiting the number of users was deliberate, to minimize information noise which could otherwise have obscured more important risk information. If thousands of company employees swamped the system with minor issues that senior management could not possibly address, this could raise doubts about the company's ability to address critical issues. Struggling with all these minor issues could also negatively impact the motivation and attention of senior management to tackle critical problems at the production sites.

The employees given access to the app were first asked to assess any risk they were already aware of in terms of criticality and urgency. Critical and very serious problems were to be sent directly to senior management at headquarters. Significant and moderate problems should be reported to the appropriate site managers, and simultaneously to the special working team at headquarters. This initial grading of observed problems, and two-channel system for reporting them, helped to transmit the information as quickly as possible to the specific managers who would be responsible for addressing a given problem. Critical and very serious problems more likely to require significant additional resources were rapidly brought to the attention of senior managers.

The following scale was adopted to assist employees using the app in grading the criticality of safety and technological problems they had observed:

	Criticality of the problem
	Moderate problem (level 1): • risk of minor injuries (reversible health impacts) • threat of failure of non-primary, auxiliary equipment, which can be promptly replaced or repaired • threat of a minor environmental incident that does not extend beyond production site
	Significant problem (level 2): • threat of serious injury to employees (irreversible health impacts) • threat of equipment failure without stopping production • threat of a minor environmental incident that extends beyond production site
	Very serious problem (level 3): • threat of a fatal and/or group incident • threat of a temporary shutdown in production • threat of a serious environmental incident with a wider geographical impact
	Critical issue (level 4): • threat of mass fatalities of employees and the general population • threat of long-term or permanent closure of production site due to serious damage caused by accident • threat of a serious environmental incident with impacts at national or international level

To assist users in searching and identifying critical and serious problems, the home page of the app contains information about the most likely dangerous situations and issues that might arise in the workshop and production site where a particular employee works, arranged according to their position in the hierarchy.

These lists were created by the head and chief engineer of each production site with help from the company's production department.

Using the app, any employee can promptly inform managers about three types of observed irregularities:

I. An ongoing safety or production problem (moderate, significant, very serious, critical);
II. Pre-emergency situation (Level 5)
III. An incident/accident already in progress.

Severity of observed irregularities		Message recipients	Type of message possible
I. Problem	Moderate problem (Level 1)	• Head of site, chief engineer of site • Head of workshop where the sender works • Special working team (HQ)	• Can include personal data • Can report problem anonymously
	Significant problem (Level 2)	• Head of site, chief engineer of site • Head of workshop where the sender works • Special working team (HQ)	• Can include personal data • Can report problem anonymously
	Very serious problem (Level 3)	• Senior management (production department) • Heads of company functional departments • Head of site, chief engineer of site • Head of workshop where the sender works • Special working team (HQ)	• Can include personal data • Can report problem anonymously
	Critical issue (Level 4)	• SVP of the company • Senior management (production department) • Heads of company functional departments • Head of site, chief engineer of site • Head of workshop where the sender works • Special working team (HQ)	• Report must include personal data • Anonymous report is not permitted
II. Pre-emergency situation (Level 5)		• SVP of the company • Senior management (production department) • Heads of company functional departments • Head of site, chief engineer of site • Head of workshop where the sender works • Special working team (HQ)	• Report must include personal data • Anonymous report is not permitted
III. Incident/accident already in progress		• SVP of the company • Senior management (production department) • Heads of company functional departments • Head of site, chief engineer of the site • Head of workshop where the sender works • Special working team (HQ)	• Report must include personal data • Anonymous report is not permitted

Employees granted permission to use the app have the right to take photos and video or audio clips of safety and technological problems and send them straight through to their superiors. Using the app, employees can also report accidents and incidents outside the site they work at—for example, on their way to or from work in a single-industry town where several company production sites are located. This speeds up the identification and elimination of emergencies across all company facilities.

Anonymity can be maintained for senders who may be anxious about the possible personal consequences of disclosing information about safety and technological problems. The company still retains a culture of fear among employees and managers at various levels around the disclosure of risk-related information (see tables below). This probably stems from previous experience of penalties, conflicts with colleagues, discrimination against whistleblowers, the failure of managers to address the problems identified, etc.

Results of responses to anonymous surveys within the framework of the pilot project: *Some employees appear reluctant to inform managers about problems such as equipment failure, mistakes in their work or inability to achieve their targets. Why is this happening?*

Fear of blame and punishment from supervisors. Employees believe they will be held accountable for any problem they report to their supervisors.

	Strongly agree	Rather agree	Rather disagree	Strongly disagree	Difficult to answer	Number of respondents
All survey participants	17.2%	48.8%	23.0%	9.2%	1.8%	326
Senior management, heads of departments and directors of sites (middle managers)	14.6%	43.9%	29.3%	12.2%	0.0%	41
Lower managers: deputy directors of sites, chief engineers of sites, heads of workshops, heads and representatives of HSE services at sites	19.4%	49.1%	19.4%	10.2%	1.9%	108
Engineers, foremen, and shop floor employees who operate critical infrastructure at sites	16.4%	49.7%	23.7%	7.9%	2.3%	177

Interpretation of responses: the company appears to have a strong tendency to blame specific managers and employees when they disclose information about problems to their superiors, instead of looking for the root causes of these problems (corporate goals, higher management decisions, inadequate provision of shop floor resources, etc.).

Results of responses to anonymous surveys in the framework of the pilot project: *Alternative communication channels used to transmit risk information between shop floor employees and senior managers should keep the identity of the employee raising the issue secret—i.e. be anonymous.*

	Strongly agree	Rather agree	Rather disagree	Strongly disagree	Difficult to answer	Number of respondents
All survey participants	25.7%	40.4%	21.4%	8.6%	3.9%	280
Senior management, heads of departments and directors of sites (middle managers)	19.4%	33.3%	36.1%	11.1%	0.0%	36
Lower managers: deputy directors of sites, chief engineers of sites, heads of workshops, heads and representatives of HSE services at sites	16.3%	41.3%	24.0%	14.4%	3.8%	104
Engineers, foremen, and shop floor employees who operate critical infrastructure at sites	34.3%	41.4%	15.7%	3.6%	5.0%	140

Interpretation of responses: a high percentage of agreement to this question among shop floor employees suggests that they are reluctant to give their identities when reporting risk information, perhaps because of previous negative experience.

Employees must feel confident that they will not suffer negative personal consequences if they disclose information about problems. Attitudes will change—and then only slowly—when employees see repeated evidence that the company's priorities have changed. They need to see that there will be no penalties for reporting problems; that on the contrary, employees are praised and rewarded for being proactive, and encouraged to maintain their vigilance; that the problems they report will be addressed. The project team was focused on shifting these attitudes at the production sites and reducing fear among the workforce of the consequences of transmitting *"bad news"* to their superiors. Gradually, as employee confidence in senior management grows, there should be a reduction in the proportion of anonymous communications, and more employees willing to identify that they are the source of the report.

Results of responses to anonymous surveys within the framework of the pilot project: *Alternative communication channels used to transmit risk information between shop floor employees and senior managers should identify the employee raising the issue—i.e. not be anonymous but include personal identification details.*

	Strongly agree	Rather agree	Rather disagree	Strongly disagree	Difficult to answer	Number of respondents
All survey participants	23.6%	42.5%	22.9%	5.4%	5.7%	280
Senior management, heads of departments and directors of sites (middle managers)	33.3%	47.2%	16.7%	2.8%	0.0%	36
Lower managers: deputy directors of sites, chief engineers of sites, heads of workshops, heads and representatives of HSE services at sites	25.0%	34.6%	25.0%	7.7%	7.7%	104
Engineers, foremen, and shop floor employees who operate critical infrastructure at sites	20.0%	47.1%	22.9%	4.3%	5.7%	140

Interpretation of responses: respondents lower in the hierarchy of the company seem slightly less inclined than their bosses to support the creation of non-anonymous channels for transmitting information about the problems of the organization.

As shop floor employees mostly prefer to send anonymous risk information, it is important to make this possible when communicating problems of lower criticality, i.e. Levels 1–3 (moderate, significant, and very serious problems). The personal data of employees choosing to transmit anonymously must be guaranteed to remain permanently concealed from everyone else, including senior management; the IT and security departments of the company can credibly guarantee this. When reporting a Level 4 problem (a critical issue) and pre-emergency situation (Level 5) then it is impossible for the sender to remain anonymous, because the company may well need to get back to them directly and immediately to prevent the situation escalating into a full-scale emergency.

Without going into the fine details of how the smartphone app works, it is useful to highlight some key features regarding the processing of incoming messages.

(I) Information about Level 3 (very serious) and Level 4 (critical) problems is not immediately sent to the SVP and senior management of the company but does go automatically to headquarters via representatives of the special working team. At the request of the heads of the pilot production sites, details of these more critical issues are double checked by the chief engineer or facility director within 4–12 h.[8] There had been previous instances in the

[8] This message validation model is a trial version within the pilot project. It has both supporters and critics among the senior management of the company. To decide on the effectiveness of the model, it should be tested within 3–6 months of the launch of the smartphone application.

company when shop floor employees informed senior management about what they believed were Level 3 and 4 problems, but these were later recognized to be only Level 2 (significant) or even Level 1 (moderate) problems. To prevent further overestimates of criticality, these reports should be first assessed by the facility managers, who must provide senior management with their assessment of the situation within 4–12 h of receiving a high criticality message. This must include both how the shop floor employee or lower-level manager who sent the initial message view the situation, and the site management's analysis. If senior management have not received this second report from the chief engineer or facility director within 4–12 h, then the sender's original message will automatically be forwarded to the company's senior management, as per the agreed protocol for Level 3 and Level 4 problems.

(II) If the chief engineer or facility director does not agree with the criticality assessment of the original message and downgrades it from a Level 4 or 3 to Level to 2 or 1, then the initial sender has an opportunity to voice their disagreement with the revised assessment by filing an appeal. The appeal is automatically transferred to the special working team at headquarters. They must arrange a prompt visit to the site by a group of company specialists, and if necessary external experts, for a final decision on the actual level of criticality of the identified problem. Neither the sender nor the site managers can dispute the final assessment of the experts. The sender's original message will be forwarded according to the protocol to be taken into consideration for the final criticality assessment.

(III) The production site management may decide, after assessing a message of Level 1 or 2 criticality, that the problem reported should be increased to Level 3 or 4. If so, it is taken out of the hands of the facility and automatically sent up to the company's senior management to be dealt with in accordance to the agreed protocol. The sender of the initial message does not have the opportunity to challenge this raised criticality assessment.

(IV) Representatives of the special working team will reply to the sender of the report on the app's chat, to inform them about the progress in addressing the revealed problem. If need be, they can ask the sender further questions to improve their understanding of the situation. Based on their analysis, they will then message the sender with a brief outline of their planned solution. If the sender has reservations about the proposed solution, sees a delay in its application, or considers that the problem has only been partly addressed, then the sender can use the app to resend the problem message marking it "*Problem not resolved*". If this happens, then a senior member of the special working team takes over the responsibility for coordinating the response. This might involve further assessment, with or without a site visit, and initiating additional remedial action. Once the problem has been resolved to the satisfaction of the senior member of the special working team, the executive can decide to close the case. The sender cannot then challenge this decision.

(V) All messages about pre-emergency situations (Level 5) or an already developing incident/accident are immediately sent to the company's senior managers as per the protocol above. The heads of the sites affected will of course be involved in tackling the situation, but senior management will make the preliminary assessment of these messages, because of the potential severity of the consequences and the need for urgent intervention.

Every time the system helps to prompt identification of a very serious or critical issue, allowing swift and effective intervention to prevent a serious incident, senior management should use the app to spread the success story. They can message the pilot project participants or a targeted subgroup—for instance, the workforce at that site, or all senior and middle managers in the company—by push notifications.

One of the objectives in creating the smartphone app was to facilitate the prompt transmission of photo and video material from the scene of any incident to senior management. It is very important that employees only send this sensitive—and probably volatile—information using the app and are not tempted to post it on social networks or pass it on to the media. The company makes it clear that unauthorized sharing of such audio, video, and photo information on social networks or to the media is completely unacceptable and will result in an investigation, and possibly the dismissal of the employees concerned. At the same time, prompt transmission of serious safety and production issues up the hierarchy will allow the company's emergency services to urgently address the situation, minimizing serious incidents and reducing the scale of emergencies. Employees who send these messages can potentially save the lives of their co-workers. The aim is to encourage employees to see themselves as active, responsible, and caring employees rather than bloggers.

XII. CREATION OF ADDITIONAL OPERATING FUNDS AT THE PRODUCTION SITES

To address severe Level 3 and 4 problems identified through the app, all the operational resources available to the company's senior management may be required. The precise solution and resources will depend on the nature and extent of each individual problem, and this will be determined primarily by the company's SVP and the special working team leaders on a case-by-case basis.

In order to effectively resolve moderate (Level 1) and significant (Level 2) problems identified through the app, the company allocated significant additional funds to each pilot project production site. The site directors were given full authority to make purchases as they saw necessary to speed up the response to Level 1 and 2 issues. This allowed them to bypass the bureaucracy of the standard company procurement system. The size of each fund was determined by the SVP according to individual site requirements and the size of the facility. This process has the additional benefit of demonstrating that the SVP trusts the site managers in their ability to solve most of the moderate and significant problems disclosed through the app at their own facility. These budgets are revised every year and increased where necessary. Any unused budget can be carried over from one year to the next without reduction.

Establishing these additional operating funds:

- encourages site directors to identify and address moderate and significant problems at their facility;
- speeds up the resolution of many minor site problems to the benefit of all shop floor employees;
- saves both managers' and employees' time by simplifying financial processes and avoiding the bureaucracy that was involved in working through the standard company procurement system;
- reduces the number of requests from production sites to senior managers for permission/assistance to solve their onsite issues, as they can now tackle most of these without input from headquarters.

XIII. NO PENALTY FOR DISCLOSURE OF INFORMATION ABOUT RISKS. PENALTY FOR CONCEALMENT OF RISK INFORMATION

As part of an in-depth interview, a senior manager said: *"The first time a leader yells at his subordinates is the last time he will hear the truth from them"*. The entire philosophy of the project is focused not on penalizing employees and managers who make mistakes, but on helping them so that they do not repeat the same mistake in the future, and will openly admit any errors they do make and report them to management.

Results of responses to anonymous surveys within the framework of the pilot project: *How do you view the adequacy of penalties at your enterprise for various misconduct offenses?*

	Excessively hard	Hard	Adequate/ Fair	Soft	Excessively soft	Number of respondents
All survey participants	2.5%	22.5%	70.4%	3.9%	0.7%	280
Senior management, heads of departments and directors of sites (middle managers)	0.0%	13.9%	72.2%	13.9%	0.0%	36
Lower managers: deputy directors of sites, chief engineers of sites, heads of workshops, heads and representatives of HSE services at sites	3.8%	21.2%	72.1%	1.9%	1.0%	104
Engineers, foremen, and shop floor employees who operate critical infrastructure at sites	2.1%	25.7%	68.6%	2.9%	0.7%	140

Interpretation of responses: the lower the respondents are in the company hierarchy, the more inclined they are to rate the existing penalty system as harsh.

Results of responses to anonymous surveys within the framework of the pilot project: *Managers should not penalize employees for their mistakes, but look for systemic flaws in the work of an organization that may have created an unsafe situation for employees.*

	Strongly agree	Rather agree	Rather disagree	Strongly disagree	Difficult to answer	Number of respondents
All survey participants	31.4%	44.6%	19.3%	2.5%	2.1%	280
Senior management, heads of departments and directors of sites (middle managers)	36.1%	47.2%	16.7%	0.0%	0.0%	36
Lower managers: deputy directors of sites, chief engineers of sites, heads of workshops, heads and representatives of HSE services at sites	26.9%	51.9%	15.4%	4.8%	1.0%	104
Engineers, foremen, and shop floor employees who operate critical infrastructure at sites	35.7%	42.1%	16.4%	0.7%	5.0%	140

Results of responses to anonymous surveys within the framework of the pilot project: *It is only justified to penalize employees when they have deliberately violated safety rules or neglected their duties.*

	Strongly agree	Rather agree	Rather disagree	Strongly disagree	Difficult to answer	Number of respondents
All survey participants	62.9%	31.8%	3.6%	0.4%	1.4%	280
Senior management, heads of departments and directors of sites (middle managers)	75.0%	25.0%	0.0%	0.0%	0.0%	36
Lower managers: deputy directors of sites, chief engineers of sites, heads of workshops, heads and representatives of HSE services at sites	64.4%	29.8%	2.9%	1.0%	1.9%	104
Engineers, foremen, and shop floor employees who operate critical infrastructure at sites	58.6%	35.0%	5.0%	0.0%	1.4%	140

Results of responses to anonymous surveys in the framework of the pilot project: *Managers should create a workplace climate where the voluntary admission of an error by an employee does not then result in them being penalized.*

	Strongly agree	Rather agree	Rather disagree	Strongly disagree	Difficult to answer	Number of respondents
All survey participants	42.1%	49.6%	5.7%	0.7%	1.8%	280
Senior management, heads of departments and directors of sites (middle managers)	50.0%	47.2%	2.8%	0.0%	0.0%	36
Lower managers: deputy directors of sites, chief engineers of sites, heads of workshops, heads and representatives of HSE services at sites	45.2%	51.0%	1.9%	1.0%	1.0%	104
Engineers, foremen, and shop floor employees who operate critical infrastructure at sites	37.9%	49.3%	9.3%	0.7%	2.9%	140

During the seminar training for management, and in all the supporting documents for the pilot project, it is essential to emphasize that employees who voluntary disclose safety and technological problems must always get positive feedback from higher managers. This can be in the form of thanks, rewards and public praise. No employee who voluntarily reports an issue should in any way be held personally responsible, reprimanded, or have their bonuses docked—however serious the problem that has arisen or the outcome of the disclosure. It must also be considered unacceptable for senior management to express dissatisfaction with the director of a production site where problems have been reported voluntarily. Most unresolved risks at industrial facilities are the result of previous long-term underinvestment in the modernization of infrastructure, previous over-ambitions corporate goals, and so on. In most cases, production site directors have for many years had limited resources to address safety and production risks.

Senior management should immediately and repeatedly reassure facility heads that:

- no one is looking to find employees or managers to blame for reported problems;
- senior management and site managers are allies and must work together to manage the risks at their facilities and prevent emergencies;
- senior management are ready to devote their time and attention to gathering details of all critical risks, and securing the resources required to address the underlying issues.

This reassurance will stop site managers from seeing the pilot project as a threat to their career and position, with information about risks and problems reaching senior management independent of the traditional management hierarchy. They will then be less likely to try and prevent their subordinates from reporting critical problems to senior management via the smartphone app. Rather than showing displeasure with their most proactive and conscientious workers, site managers should praise them for giving honest feedback about problems and thank them for information that could prevent disasters.

Results of responses to anonymous surveys in the framework of the pilot project: *In a company where a voluntary risk disclosure to superiors is never penalized, employees or managers who deliberately hide information about risks in their area of responsibility should face dismissal.*

	Strongly agree	Rather agree	Rather disagree	Strongly disagree	Difficult to answer	Number of respondents
All survey participants	28.6%	47.9%	16.1%	2.5%	5.0%	280
Senior management, heads of departments and directors of sites (middle managers)	50.0%	38.9%	8.3%	2.8%	0.0%	36
Lower managers: deputy directors of sites, chief engineers of sites, heads of workshops, heads and representatives of HSE services at sites	24.0%	50.0%	19.2%	1.9%	4.8%	104
Engineers, foremen, and shop floor employees who operate critical infrastructure at sites	26.4%	48.6%	15.7%	2.9%	6.4%	140

Interpretation of responses: the lower the respondents are in the company hierarchy, the less inclined they are to fully agree that it is necessary to dismiss employees or managers who deliberately conceal information about risks in their area of responsibility.

The company's official supporting documents for the pilot project stipulate that any employee—regardless of corporate level—found concealing information about safety and technological problems or incidents will face disciplinary measures, which might include dismissal. It must be crystal clear to every employee that deliberately withholding information about risks can have very serious personal consequences.

Some managers were worried that if they did not discipline their subordinates for safety violations reported voluntarily, they could be perceived as weak, leading some to think they could act with impunity. The company's senior management explained: *"We do not agree that, if an employee comes to us and admits a problem or mistake in his area of responsibility, then he must always be completely forgiven. This could provoke the following problem. A second employee might think: 'I also made a mistake, but nothing bad happened. So I'll go confess now, and I won't be punished'. And then a third one will think: 'I'll go and say, sorry, dear boss, I made a mistake. I won't do it again'. For the first time, the manager may be happy to let things ride, but this becomes difficult if similar mistakes continue to be repeated by the same employees. Surely it is necessary to understand why the employee made that mistake. Look at what really happened—was it an organizational problem, faulty machinery, poor instructions, or maybe simple human error,* etc. *Only from an objective analysis will it become clear where the responsibility and fault really lies. Only then can the truth be uncovered and employees fairly reprimanded. The word 'justice' is key here. Management should only penalize subordinates according to their degree of true responsibility, so that the whole workforce can see that the system is treating everyone fairly".*

Based on the responses of senior leaders, the following provisions were adopted:

- Every critical and very serious problem disclosed should be investigated in detail to understand the systemic root causes that led to the problem, and to accurately determine the relative responsibilities of each manager and employee involved.
- When systemic causes and specific failures in responsibility are identified, employees are informed of their degree of responsibility for the situation, but still no one is penalized if the risk was disclosed voluntarily. Everyone is encouraged to learn from their mistakes, so as to not repeat them.
- Fair and proportionate punishment is applied only when employees and managers repeat mistakes they have already been warned about and were not previously punished for.

XIV. ANALYSIS OF ROOT CAUSES OF CRITICAL AND VERY SERIOUS PROBLEMS

The special working team of executives meet regularly to analyze the causes behind criticality Level 3 and 4 problems. Analysis of level 1 and 2 problems is carried out at the discretion of the site directors.

When conducting an analysis of a particular issue, managers should try to answer the following questions:

- How did the corporate management system allow the problem to occur?
- What top-level decisions contributed to the problem?
- Why could managers and employees in the field not promptly and effectively manage the problem?
- What responsibility do the individual managers and employees involved bear for the issue becoming critical?
- How can company procedures be improved to prevent a similar problem from occurring again?

The results of this analysis are documented as a free-form report and sent to the managers and employees involved, both at headquarters and at the site(s) where the problem was identified.

If the analysis reveals shortcomings in the work of specific managers or employees at the site, no disciplinary measures are taken. Instead, the information is shared directly with them, and changes that both the company and the employee can make to avoid a recurrence are discussed.

If the analysis establishes that either the managers or the employees have repeated previously identified errors, then they are issued with a warning. If the same errors are revealed to have caused the same problem for a third time, then disciplinary measures are taken, up to and including dismissal.

XV. EMPLOYEE REWARDS FOR DISCLOSURE OF INFORMATION ABOUT RISKS

Results of responses to anonymous surveys within the framework of the pilot project: *Choose what you believe to be the most appropriate methods of reward for disclosing information about safety and technological problems (choose only one answer).*

	Material rewards only	Non-material rewards only	Combination of material and non-material rewards	No rewards needed — disclosure is the professional duty of employees	Difficult to answer	Number of respondents
All survey participants	7.9%	4.3%	57.5%	25.7%	4.6%	280
Senior management, heads of departments and directors of sites (middle managers)	0.0%	2.8%	80.6%	16.7%	0.0%	36
Lower managers: deputy directors of sites, chief engineers of sites, heads of workshops, heads and representatives of HSE services at sites	4.8%	5.8%	58.7%	28.8%	1.9%	104
Engineers, foremen, and shop floor employees who operate critical infrastructure at sites	12.1%	3.6%	50.7%	25.7%	7.9%	140

Results of responses to anonymous surveys within the framework of the pilot project: *If the company decides that it needs to combine material and non-material rewards for disclosing information about safety and technological problems, what should be prioritized?*

	Combination of mostly material and some non-material rewards	Combination of mostly non-material and some material rewards	Equal priority to material and non-material rewards	Difficult to answer	Number of respondents
All survey participants	31.4%	27.9%	30.4%	10.4%	280
Senior management, heads of departments and directors of sites (middle managers)	13.9%	50.0%	33.3%	2.8%	36
Lower managers: deputy directors of sites, chief engineers of sites, heads of workshops, heads and representatives of HSE services at sites	27.9%	32.7%	29.8%	9.6%	104
Engineers, foremen, and shop floor employees who operate critical infrastructure at sites	38.6%	18.6%	30.0%	12.9%	140

Interpretation of responses: the lower the respondents are in the company hierarchy, the more highly they regard material rewards. The higher the position of the respondents, the more important they regard non-material rewards. The most likely explanation of this variation is that managers have built a career in a critical infrastructure organization, so they value non-material rewards such as praise from a senior manager as they believe this will assist their career progression. On the other hand, shop floor employees are more concerned with securing an adequate income, so they set greater value on material rewards, especially financial.

One of the most controversial issues in the pilot project was how to reward employees and lower managers for their proactive and timely action in voluntarily disclosing information about safety and technological problems, about pre-emergency situations to avoid escalation, or promptly reporting an ongoing emergency.

It was decided that the company would never offer employees personal financial and material rewards (i.e. hard cash or personal presents like smartphones, TVs, cars, flats, etc.) for disclosing information about safety and technological problems. Such payments have several negative aspects as discussed in Chap. 3, Recommendation No. 10: *"Reward employees for disclosure of safety and technological risks"*. These were confirmed by the group work during the site seminars.

Advantages of personal material rewards for specific employees (from results of seminar group work):

- High motivation to become involved—many employees will immediately join the search for risks and problems at their sites.
- Fast transmission of risk information and a high volume of risks reported.
- Training of employees to communicate risk information and remove psychological barriers to voluntary, objective disclosure.

Disadvantages of personal material rewards for specific employees (from results of seminar group work):

- Information noise: employees will try to find as many problems as possible. There is a threat of information overload with many insignificant problems that still have to addressed.
- Senders will demand rewards for trivial messages. A high rejection rate will lead to loss of confidence in the project.
- Deliberately bringing equipment to a critical state: some employees will be tempted to allow/ encourage machinery to reach a pre-critical state in order to receive a reward for reporting it. Smaller monetary rewards may not be enough to incentivize employees to disclose risks, whereas a large reward may motivate some employees to go as far as deliberately damaging equipment.
- Envy from colleagues.
- Employees will spend too much time *"risk hunting"* and this could reduce productivity.

After discussion with senior management and with all seminar participants, a mixed reward model including material and non-material rewards was agreed. This was designed to maximize incentives, and convince employees that reporting safety and technological problems was worth their while and would be generously rewarded by the company.

A key factor with financial rewards paid to employees for disclosing information is that they can only be spent within the company's operations (i.e. tokens or *"virtual money"*). This means employees who have uncovered issues spend their rewards on professional equipment, tools, and extras to make their immediate workplace and that of their colleagues safer, more comfortable, and more productive.

Type of message		Employee motivation for disclosure of risk	Reward model
I. Problem	Moderate problem (Level 1)	Senders can help mitigate risks and problems in their area of responsibility by obtaining resources from the additional operating fund This helps to secure the incomes and job security of themselves and their colleagues	Various non-material rewards Sender of the best message received in a month (judged in competition) is given a virtual check for US $1000. This can be spent (with site director's approval) on the purchase of equipment and extras for the sender's workplace to make it safer, more comfortable, and more productive
	Significant problem (Level 2)	Senders can help mitigate risks and problems in their area of responsibility by obtaining resources from the additional operating fund This helps to secure the incomes and job security of themselves and their colleagues	Various non-material rewards Sender of the best message received in a month (judged in competition) is given a virtual check for US $2000. This can be spent (with site director's approval) on the purchase of equipment and extras for the sender's workplace to make it safer, more comfortable, and more productive
	Very serious problem (Level 3)	Senders can save themselves and their colleagues from serious injury and even death, as well as protect their jobs and income	Public expression of gratitude to the sender from the site director Various non-material rewards Sender is given a virtual check for US $10,000 for the purchase (with site director's approval and support of three colleagues) of equipment and extras for the sender's workplace to make it safer, more comfortable, and more productive
	Critical problem (Level 4)	Senders can save themselves and their colleagues from serious injury and death, as well as protecting all employees' jobs and income by maintaining the long-term resilience of the company's critical sites They are also protecting the safety and livelihoods of more than 100,000 people living near the company's industrial facilities	Public praise and gratitude is given by the SVP — the highest award the project can bestow Various non-material rewards Sender is given a virtual check for US $20,000 (up to a max US $60,000 in exceptional cases) for the purchase of equipment, vehicles and extras for the sender's workplace to make it safer, more comfortable, and more productive. Sender needs to obtain spending approval from five colleagues as well as the site director
II. Pre-emergency situation (Level 5)		By informing senior management about pre-emergency situation, senders could save themselves and their colleagues from serious injury and death, as well as protecting all employees' jobs and income by maintaining the long-term resilience of the company's critical sites They are also protecting the safety and livelihoods of more than 100,000 people living near the company's industrial facilities	Public praise and gratitude is given by the senior manager of the company's production department Various non-material rewards
III. Incident/accident already in progress		By promptly informing the entire management hierarchy about an incident/accident, senders have initiated an urgent emergency response, saving the lives of employees and public and reducing the overall severity of the incident and the amount of environmental damage caused	Personal public gratitude from one of the heads of the company's internal emergency services Sender is presented with the corporate *"Rescuer"* award and a formal letter of thanks from the company's leaders

The amount of remuneration is determined according to the severity of the identified problem assessed and confirmed by the special working team.

It is important that the sender cannot independently decide what equipment to purchase. This must be done in agreement with other members of the workforce and the site director. This ensures that the sender does not become an outcast. Colleagues of the sender can feel they are involved in the process and can see that there is no benefit in deliberately ignoring risks. It also discourages employees from bringing production facilities to a pre-critical state in the hope of receiving a significant reward that they can spend on themselves: the sender's colleagues could prevent this at the outset, well before the equipment has been driven to a pre-critical state.

Payments are made from the site director's additional operating fund to make the process as simple as possible and minimize bureaucracy and delays. The site director has access to significant financial resources and rewarding employees for reporting major safety concerns is a legitimate use of these.

If messages about problems are sent anonymously, then the sender cannot receive any reward or recognition for their actions. However, the sender of the message can choose to reveal their name at any time, and then receive their deserved reward.

The company's risk information transmission system is focused on motivating employees and managers to preemptively disclose information about problems in their area of responsibility to prevent serious incidents and emergencies. The level of remuneration for reporting the problem decreases the closer the problem comes to causing an emergency situation. This encourages employees and managers operating critical infrastructure to focus on preventative control of existing problems, and to report any risks they encounter to their superiors before the situation reaches a dangerous point.

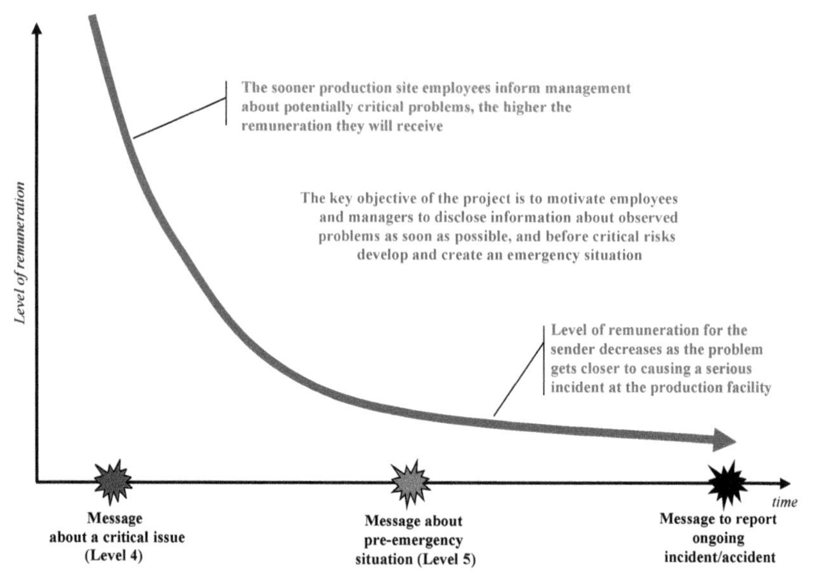

Three out of four of the pilot production site directors agreed with the following scheme of virtual money, which could be spent on equipment and extras for the sender's workplace to make it safer, more comfortable, and more productive. Nevertheless, the fourth plant director stated firmly that he is worried about giving senders any kind of financial and material reward: he thinks that even with virtual money, there is a real threat of equipment being deliberately pushed towards, or negligently allowed to reach, a pre-critical state in order to report a more serious situation and get the reward. He insisted that he will reward his staff only by non-material means. Finally, the senior management came to a compromise that all the reward schemes mentioned should be tested within the pilot production sites over the next 6–12 months, according to the respective preferences of the plant directors. This will give the project team more information about the most effective ways to motivate staff at the plants to disclose risks. Based on this experience, the best reward menu will be selected and implemented as the project is extended to the other production sites.

Along with these material rewards, various non-material incentives and acknowledgments were developed to recognize the valuable contribution of employees or work teams who report safety and technological problems. Below is a list of the ones selected for use by the seminar participants. The three most popular responses at each level of management are highlighted in green.

Results of responses to anonymous surveys within the framework of the pilot project: *Choose what you consider to be the most appropriate and effective non-material rewards for employees disclosing information about technological risks (multiple answers can be selected).*

	All survey participants	Senior management, heads of departments and directors of sites (middle managers)	Lower managers: deputy directors of sites, chief engineers of sites, heads of workshops, heads and representatives of HSE services at sites	Engineers, foremen, and shop floor employees who operate critical infrastructure at sites
Public commendation from senior management to an employee or work team	58.6%	81.0%	87.0%	53.0%
Professional internships at other company enterprises or additional professional training (including abroad)	37.9%	50.0%	58.0%	33.0%
Diploma personally signed by senior management	36.1%	47.0%	61.0%	27.0%
Special honorary order (corporate medal)	35.0%	53.0%	61.0%	25.0%
Inclusion in the list of best employees at the site	34.3%	50.0%	57.0%	26.0%
Tickets for all family members to attend major sporting events or concerts	32.9%	56.0%	56.0%	24.0%
Letter of gratitude sent from senior management to a employee's family acknowledging his/her contribution to the safety of the site	32.5%	33.0%	47.0%	30.0%
Additional days off	30.7%	28.0%	31.0%	39.0%
Weekends at the corporate recreation center for all family members	29.6%	44.0%	42.0%	28.0%
Article in the corporate the employee/team who prevented a serious emergency	24.3%	44.0%	42.0%	17.0%
Opportunity to become a mentor, passing on their successful experience to other employees within the company	20.0%	42.0%	34.0%	15.0%
Joint photo with senior management	16.8%	25.0%	26.0%	14.0%

(continued)

(continued)

	All survey participants	Senior management, heads of departments and directors of sites (middle managers)	Lower managers: deputy directors of sites, chief engineers of sites, heads of workshops, heads and representatives of HSE services at sites	Engineers, foremen, and shop floor employees who operate critical infrastructure at sites
Videos featuring outstanding contribution to company safety by employee/team. Broadcasting these videos to other employees across all the company sites	9.6%	31.0%	16.0%	7.0%
Lavish dinner for distinguished employee/team with senior management	8.6%	11.0%	15.0%	6.0%
One-year allocation of personal parking spot alongside top company executives	8.6%	11.0%	13.0%	7.0%
Total respondents	280	36	104	140

XVI. HOLDING AN INAUGURAL EVENT TO PRESENT THE SOLUTIONS CREATED DURING IMPLEMENTATION OF THE PILOT PROJECT

When the pilot project begins, it is agreed that once all the above elements have been established (the app, the reward schemes, the additional operating funds, etc.) it would be beneficial to mark the occasion with an inaugural event for the project participants. The event is guided by the project leader, and attended by the SVP and key senior managers of the company's production and HSE departments. It is held in the largest venue in the city, which has the capacity to accommodate 250–400 project participants.

The central task for the SVP and the senior managers at this event is to convince the project participants that the top officials of the company really do want to receive prompt and accurate information from their subordinates about critical and serious safety and technological problems. This is because getting such information in time can prevent these problems from escalating into emergencies, thus saving people's lives and the integrity of production facilities.

It is also important that senior managers take the opportunity to personally re-emphasize the key aspects of the pilot project to all the assembled participants.

- Senior management and site directors will quickly address any critical or very serious problem brought to their attention. To demonstrate how this has already been successfully achieved, videos are shown of how the seven critical problems revealed by participants in the pilot project were successfully dealt with.
- The company has had successes (and difficulties) in addressing 104 other less critical problems disclosed by the pilot project seminar participants.

- Senior management have sufficient resources allocated by the owners and main shareholders to address the most critical and serious issues that are identified.
- The site directors have additional resources available to them for addressing significant and moderate issues at their facility.
- The causes of critical and very serious problems will always be fully investigated to identify the underlying organizational flaws that created the issues and to prevent the repetition of the same errors in the future. Senior management will explain how this analysis is carried out, as well as the penalties employees can expect to receive if they repeat the same mistakes and fail to learn from past negative experiences.
- The company guarantees that it will never penalize employees for mistakes if these were voluntarily and promptly disclosed to their superiors. However, severe penalties will be applied to anyone deliberately concealing information about risks.
- A scheme has been agreed to reward employees for the disclosure of information about risks. This will be explained in detail.

In the presence of hundreds of project participants, the SVP and senior management of the company then personally reward those employees who have already uncovered critical and very serious problems, thus preventing several accidents and incidents. All project participants are urged to follow the example of these valued and proactive employees.

As part of preparing for this event, the IT department pre-installs the application on the smartphones of all project participants. At the end of the event, the senior management urge all project participants that when they return to their everyday job, they should start using the app to quickly transmit information about any problems and risks that they observe.

For those employees who cannot attend this inaugural event, the company's PR department will provide a short video covering all the main points discussed. This is posted on a dedicated web page on the company's intranet. This web page also includes details of all the principles and solutions developed within the framework of the project so far. All project participants receive a link to this page in three ways —an SMS, a push notification in the project smartphone app, and on corporate email. (Not all project participants, especially shop floor workers, have a personal corporate email.)

Results of responses to anonymous surveys within the framework of the pilot project: *As part of the pilot project, it is planned to present the solutions developed by around 500 employees and managers who attended the seminars. What is the most convenient format for you to participate in this presentation?*

	Personally participate in a special event where senior management will talk about decisions within the project; opportunity for attendees to ask questions directly	Receive a link to a video on your smartphone, which explains the solutions that were developed within the project; online chat with answers to questions	Watch a specially prepared video with colleagues at the facility; ask questions to a representative of the HSE service	Difficult to answer	Number of respondents
All survey participants	45.0%	36.2%	11.8%	7.0%	280
Senior management, heads of departments and directors of sites (middle managers)	60.0%	35.0%	5.0%	0.0%	36
Lower managers: deputy directors of sites, chief engineers of sites, heads of workshops, heads and representatives of HSE services at sites	52.9%	25.9%	9.4%	11.8%	104
Engineers, foremen, and shop floor employees who operate critical infrastructure at sites	37.1%	43.5%	14.5%	4.8%	140

Interpretation of responses: managers are often trying to build a career in the company, therefore more likely to be interested in building good relationships with higher managers by personally attending a special event. Shop floor employees are more focused on earning a regular income than on a career in the company, therefore show less interest in attending events where senior management are in attendance.

XVII. TRAINING OF INTERNAL TRAINERS

Within the 15 seminars, a special seminar was held for 33 managers working at various levels within the company's HSE department. In addition to familiarizing these managers with the ideology and details of the pilot project, the task was to find potential internal trainers among the seminar participants. As a result, eight people expressed a desire to undergo further special training under the guidance of the project leader, so that they could conduct similar seminars in the future, when scaling up this project at other production sites.

XVIII. HOLDING NEW SEMINARS AT THE COMPANY'S OTHER INDUSTRIAL SITES

The project team agreed that the scaling up of the project will begin when the procedures and solutions mentioned above have been successfully tested at the four pilot production sites. When scaling up the project, 5–10% of employees will be selected from the workforce of each new site. These employees should be responsible for the operation of critical site infrastructure and have extensive experience in their role. The project leader will conduct training seminars for site managers, workshop supervisors, and other lower-level managers, as well as for representatives of the HSE department at each new site. Specially trained internal instructors will conduct training seminars for the selected engineers, foremen, and shop floor employees who operate critical infrastructure.

Four to six months after these new seminars, the company's SVP, together with other senior managers, will hold an inaugural event (as described above) to reward employees who have uncovered new critical risks at their sites and helped prevent disasters. The company's management will also use this opportunity to encourage all employees to always be alert in identifying risks and problems in their area of responsibility, and immediately informing their superiors about these issues using the dedicated smartphone app.

According to the authors of the handbook, it will take this large industrial company approximately three years to roll this project out across all sites, and significantly improve the quality and speed of reporting accurate risk information at each site.

PRELIMINARY RESULTS OF THE PILOT PROJECT

The authors would like to cite a letter of recommendation provided by the company's senior vice president (referenced earlier as SVP), who is responsible for the production at the main industrial complex of the company. This was based upon the results achieved over the first 10 months of the pilot project.

October 27, 2022

Letter of Recommendation

(hereinafter – the company) is the world's largest producer the biggest producers in the world.

In the period from December, 2021 – October, 2022, Dr. Dmitry Chernov successfully implemented a research & practice project (hereinafter – the project) at four production units of the company aimed at increasing the speed and quality of information transmission about critical technical & operational issues from frontline employees to executives, in order to prevent emergencies at the production sites.

Results of the project's implementation:

- 422 employees of participated in 15 seminars: company executives and the managers of the production units selected for the project, as well as supervisors, technical specialists and frontline employees. **The participants of the seminars, conducted by Dr. Dmitry Chernov, revealed to executives 7 critical (high) risks with the potential to develop into an emergency with personnel fatalities, long-term shut-down of production facilities or serious environmental damage.** The executives and managers of promptly took control over the risks. During the seminars, 104 technical & production issues which were having an adverse effect on the risk management and industrial safety of the production units were also revealed to company executives. The majority of these issues were also taken under control by a special team of company executives and production unit managers;
- **a sociological study based on 300+ anonymous responses from the seminar participants** revealed the factors which were complicating upward risk communication, and showed the changes needed to improve the speed and quality of risk information transmission within the company hierarchy;
- based on the information gained at the seminars **recommendations were made to company executives on a strategy to comprehensively improve the speed and quality of risk communication at the production sites of** ;
- **a mobile app for prompt risk reporting from frontline employees to company executives** was created and launched into programming;
- **a remuneration scheme for employees voluntarily reporting risks and issues to superiors** was created (including both tangible and intangible rewards);
- **rules for the communication of critical technical & operational issues** were developed, **aimed at emergency prevention within the hierarchy of the company**. This document provides the procedures for frontline employees and managers to follow at every stage of risk communication, risk evaluation, investigation, risk control, rewarding and the sharing of lessons learned within the wider company;
- some other actions were also taken to improve the speed and quality of risk communication at production units.

The project results show that the frontline employees and low-level managers of production units selected for the project are willing to communicate critical risks at the sites to the executives of the company in order to prevent disasters. The project has paid for itself many times over as risks revealed during the seminars were swiftly brought under control, allowing the company to avoid a number of emergencies.

While implementing future projects to improve the transmission of risk information in other industrial companies, I would recommend Dr. Dmitry Chernov to pay special attention to the value of developing an effective corporate mechanism to support a prompt response to any technical & production issues revealed by seminar participants and mobile app users.

I believe that the solutions proposed by Dr. Dmitry Chernov are unique. I recommend his studies and services to all of my colleagues from critical infrastructure companies around the world who are seeking to fundamentally improve the speed and quality of technical & operational risk communication at their production operations.

Senior Vice-President –

TESTING IDEAS AND SOLUTIONS IN OTHER CRITICAL INFRASTRUCTURE COMPANIES

The authors are open to cooperation with other critical infrastructure companies worldwide to test ideas and solutions for improving the quality and speed of critical risk information transmission and for preventing accidents. They are ready to provide critical infrastructure companies with the results of their many years of research and their extensive experience in managing similar projects in other companies. They are also willing to spend lengthy periods at the location of a company's industrial sites to lead and supervise the implementation of new scientific and practical projects.

The authors are confident that a radical improvement in the quality and speed of communicating information about safety and technological problems is possible in companies operating critical infrastructure. This can be achieved by the open exchange of experience between industrial companies around the world, by the systematization of project design and implementation, and by the publication and distribution of best practices. This will help to establish tested mechanisms for proper and timely intra-organizational risk communication for a global community of risk management specialists. The goal of all of this is clear: to prevent industrial accidents and disasters from occurring in the first place, to save people's lives, reduce environmental damage, and increase the resilience of critical infrastructure facilities.

The authors believe that an international corporate standard for building a risk information transmission system within critical infrastructure companies could be established within approximately 10–15 years. To this end, the first author plans to devote his future scientific and consulting career in the field of risk management to developing such an international corporate standard, through the experience of implementing further scientific and practical projects in various critical infrastructure companies around the world.

Additional information: dmitrychernov@riskcommunication.info and dmitrychernov@mail.ch or dmitrychernov@ethz.ch.

Conclusion

Having analyzed dozens of accidents and seen so many cases of information about risks being concealed, it seemed to the authors that this problem would have required enormous organizational effort to solve. However, the interviewed industry leaders made numerous and very promising suggestions on how to radically improve the transmission of information about risks in simple ways. It turns out that senior managers just need to show their subordinates that they want to listen and are willing to work with them to solve complex safety, technological and organizational problems. The key factor influencing the willingness of management to hear about problems and solve them is the position of the owners and shareholders. They must be prepared to reduce their appetite for short-term profits from the operation of critical infrastructure and view their assets as long-term investments. It is the owners and shareholders who must agree on the allocation of resources to management—so that the top officials of a company, having received information about the presence of a critical risk, can immediately make a decision to control and reduce it. Without the resources, senior managers can do nothing when their subordinates come to them with information about a company's problems.

Leaders must always remember that it is impossible to fully manage all the risks of a company: they will never have enough resources to do this. Therefore, they need to prioritize. Risk prioritization means focusing primarily on the control of the critical risks that can destroy a company. Once the main critical risks are under control, a company can look at less serious challenges.

For subordinates at all levels to feel safe in disclosing risks and problems to their superiors, they must be sure that no one will punish them for an objective report. It takes time and lots of positive examples of risk disclosure for most subordinates to believe that they will not be hurt if they report risks to their superiors.

Employees must also be sure that the problems they are highlighting will be solved sooner or later, otherwise many will feel there is little point in reporting them. Therefore, to gain the trust of their subordinates, senior management must be willing to address and solve the problems that are reported. If there is trust in senior management, even the most cautious employees will have the confidence to give

D. Chernov et al., *Averting Disaster Before It Strikes*,
https://doi.org/10.1007/978-3-031-30772-0

feedback to their bosses on how they see the situation in a company. The experience of implementing the pilot project has shown that ordinary employees and lower managers willingly disclose to senior management the critical and serious risks that are in their area of responsibility. The main thing is the sincere desire of the leaders of critical infrastructure companies to receive information about such risks, and secure the resources to promptly resolve the identified problems.

The goal in writing this handbook was to provide executives operating critical infrastructure with practical tools and solutions, so that they can improve the quality of risk communication in their companies. Better information makes for better decisions, and these in turn have an impact on reducing the likelihood of severe accidents at industrial facilities. The authors hope that this handbook will help prevent major emergencies and save many lives.

Discussion: Automating the Collection of Information About Equipment Operation, and the Prospects for Artificial Intelligence in the Operation of Critical Infrastructure

Automating the collection of information about equipment operation

Most of the interviewees are positive about developing automated systems to collect complex information using sensors attached to critical equipment, which continuously transmit feedback on their condition and operation to headquarters. The obvious advantage of such automation is its ability to reduce the influence of the subjective human factor: once it is set up and running reliably, there is no further need for the manual collection, processing, and transmission of information about critical risks through the traditional management hierarchy. It is very important that in such automated systems there is no manual data entry, to exclude the possibility of any manipulation of data at different levels. Most of the respondents stressed that the degree of automation of information collection depends primarily on economic feasibility. The main criterion for assessing the feasibility of introducing an automated system should be the level of risk that is removed as a result. For critical risks, several different mechanisms of automatic monitoring and control should be implemented. The most important condition for the effective operation of any feedback system is the completeness and reliability of the information that is collected. Based on the information received, managers will make decisions on controlling the risks identified. Assessing feedback from the system to inform a management decision is often impossible without human involvement, so at this stage, it is possible only talk about automation up to a certain level.

An oil production facility manager considers that the underdevelopment of automatic monitoring systems makes it easier to hide minor incidents and injuries at the grassroots level.

D. Chernov et al., *Averting Disaster Before It Strikes*,
https://doi.org/10.1007/978-3-031-30772-0

The head of a thermal power plant believes that the main obstacle to automating processes in a critical infrastructure company is the negative attitude to change of the employees and managers on the ground. Senior managers need to understand from the outset that the process of replacing current employees with qualified automation specialists will be very difficult, and may have undesirable consequences for a company.

The HSE manager of a manufacturing company believes that one of the main problems facing the implementation of automated solutions in his country is that the unions do not like technology. They are vehemently opposed to anything that is technologically advanced, which they suspect is being introduced primarily with the agenda of reducing the influence of the workforce—and the unions—on the operation of critical infrastructure companies.

The HSE manager of a production company which uses hazardous chemical processes considers that currently most monitoring systems are still human-oriented, and the imperfection of the data collection and entry process allows field employees to enter information that suits their interests. It should be borne in mind that many shop floor employees will oppose a system that automatically sends information about their errors, and about faults in the equipment they are responsible for. After all, if in the company there is a culture of searching for someone to punish for any problem that arises, then automation will just allow people to be punished even more effectively. Hence, if they can, people will resist automation by distorting the information that is entered into the system. This means that unless the corporate culture is changed, automation will not help management get a better understanding of the situation on the ground. It is necessary to make sure that employees are not afraid of their errors, or defects in the equipment they operate, coming to the attention of the boss.

The ideal model is when employees want to communicate issues arising in their area of expertise to management, in order to get the additional resources they need to address them. Only then will employees not try to distort what is transmitted through the automated system. It is very important that employees and managers are thinking in the same way, that they are working towards shared production and safety goals, and that there are no penalties for informing management about problems—whether through the traditional hierarchy or the continuous feedback from the automated system.

According to this respondent, senior managers can also resist setting up automated systems: if information about problems on the ground is sent directly to the board of directors, those managers know they will have to make immediate decisions to resolve the problems identified by the system. They would rather be able to continue withholding information about problems in the field, to avoid being held responsible for solving them or being blamed in the event of an incident. On the other hand, some executives who are pursuing aggressive production and financial plans have noticed an unintended consequence of these systems which suits their agenda. An alternative channel of direct automatic feedback is clearly designed to prevent managers from going beyond the physical limits of the equipment, beyond which there could be a major accident. Nevertheless, having that feedback also allows managers who wish to push the equipment as close as possible to those limits to do so.

The executive at an electric power company offered to clearly define some terms. In determining the life cycle of operation of a critically important technological object, it is customary to distinguish two main technological processes and, accordingly, the management of these processes:

- *production exploitation*—is the production of a product; in the electric power industry, this is the production of electricity and heat;
- *technological exploitation*—is the safe maintenance in working order of tangible assets including equipment, tools, lifting equipment, machinery, rigging and so on.

Production exploitation is now almost completely automated at all levels in critical infrastructure companies. Thus the main ongoing task is to improve systems and technologies for the automation of technological exploitation, including the system for collecting primary information.

For various reasons, the management of technological exploitation is very heterogeneous, so the task of choosing the most acceptable technologies and deciding what level of automation will work is a complex and urgent one. The main reason for this is that there is a wide variety of equipment designs that perform the same function, so the amount and nature of information required to assess their current and future technological state varies greatly. The task is complicated by the fact that almost all the design calculations that underlie this assessment are probabilistic in nature. For these and other reasons, it may not yet be possible to definitively assess what will work, and thus to plan the management of technological equipment solely based on automatic collection and processing of information.

Automated systems can record and inform decision-makers about the occurrence of equipment breakdown; but a simple monitoring of indicators cannot generally give a clear answer to the question of why the breakdown occurred and what needs to be done to eliminate it. This can only be done by highly qualified employees, based on their diagnosis of the condition of the failed equipment. In many cases, it is necessary to disassemble the failed equipment to determine the true causes of the failure. In this regard, it is very important that these highly qualified experts, after inspecting and diagnosing the condition of the equipment, enter objective information about its technical condition into the information system. Once this data is on the system, it should immediately become available to anyone in a company involved in decision-making on how best to restore or replace failed equipment.

According to this respondent, the main problem with ensuring the reliability of data in the system is the possibility that managers or employees may attempt to influence these experts in order to distort or embellish the information they enter about the technical condition of equipment.

As a rule, attempts to pressure the experts to distort the collected information have the following reasons:

- equipment repairs may require unplanned costs outside the agreed budget;
- equipment repairs may lead to a shortfall in profits below the agreed production target;

- some managers responsible for the operation of out-of-order equipment will want to hide their miscalculations;
- in some companies there is a system of punishment of both employees and managers for admitting breakdowns or accidents.

Thus, some managers and employees may be tempted to prevent accurate information about the condition of failed equipment getting on to the system. One way to avoid this is to set the system up so that information about a given piece of equipment can be added but not replaced, and no manual entries can be made anonymously. An additional tool to minimize the possibility of information being distorted due to pressure on those making the entries is to ensure that they are independent of those managers who are responsible for operating the equipment. Ideally, a group of technical experts will report to the board of directors and be independent in their assessments from senior, middle and lower management. Experts should be held accountable only for the accuracy and correctness of the equipment condition assessments for a company, and not for why the equipment performance has become worse or better. The reliability of information that requires human processing cannot be improved only by technological methods—it requires the use of organizational methods. In the same way, the risk of information being concealed cannot be dealt with by even tighter control and complication of the information entry process, but only by a systemic approach: changing corporate goals, providing the resources for quick repair, abolishing punishments for unintentional mistakes or omissions, and so on.

For all aspects of technological exploitation, including monitoring, the ratio between what can be handled by automatic systems and what requires expert human intervention depends very significantly on the type of equipment a company is operating.

If the equipment consists of thousands of pieces of the same type, then to analyze the causes and consequences of its failure, one can use the accumulated data on hundreds of similar faults that have occurred at other facilities. With this huge array of data on regular operating modes and previous deviations of the same type of equipment, the expert-analytical system can be very highly automated. It also enables the use of artificial intelligence (AI) developments[1] to analyze trends in the operation of the equipment and predict its failure, and hence to minimize the impact of subjective factors. AI has long been used in industrial process control. The energy production processes at most steam turbine and gas turbine power plants are now controlled and monitored automatically, often using AI technology. The control system itself chooses the optimal operating modes with continuous

[1]The use of the term "*artificial intelligence*" taken as a generic concept may be exaggerated here. Formally, AI refers to the general ability of computers to imitate and reproduce human thought and find "*intelligent*" solutions for real-world problems. In above discussion, machine learning is likely to be more relevant, as it corresponds to the technologies and algorithms by which patterns can be identified and decisions can be operationalized. Through experience (either unsupervised or supervised), machine learning algorithms can improve themselves. Machine learning is how a computer system may develop its AI.

feedback from the machinery, and the human operator only formulates tasks and oversees how the equipment is performing the program set by the system. An expert analytical system, enhanced by AI, can predict the likelihood of failure of any component piece of equipment.

According to the executive at an electric power company, the opposite is true for much of the equipment installed at critical energy facilities—hydroelectric power plants, drilling platforms, and so on—which are unique, or one of a very small number of similar installations.[2] Due to a very limited sample of comparable equipment, it is impossible to use statistical methods of data processing and AI technology.[3] The incoming flow of data is insufficient to inform reliable predictions, or give an accurate assessment of the probability of different outcomes. Thus, most numerical methods for working with random events will give a less accurate result than expert judgment. In this kind of situation, automated and AI-assisted systems can only act as a support to inform the assessment of a skilled and experienced human analyst. The professional assessments of people who have been designing, analyzing or operating this equipment on a daily basis for many years come first. These human assessments cannot be replaced with automated diagnostic systems. Nevertheless, the automation of the process of monitoring and diagnosing

[2]While the present authors respect the opinion of this interviewee, the present authors suggest that this view is incomplete and outdated. Taking the example of nuclear energy, there has been a long-held view in the industry that each civil nuclear plant is unique, so that it requires its own individual probabilistic safety assessment (PSA) accounting for its specific characteristics and configuration. In this view, the plant specific PSA cannot be used for another plant. To illustrate a problem with this view, consider human health. No one will contradict the statement that each of us is unique in his/her genome, transcriptome, proteome, microbiome and so on. One would then derive the logical consequence that our uniqueness requires individual specialized diagnostics and specific medical treatments that cannot and should not be used for others. In reality, while there is a trend towards more personalized medicine, there are sufficiently many common processes and factors to allow for a generic medicine. Similarly, actuarial life statistics works very well to provide efficient insurance pricing. Returning to civil nuclear energy, over the last decade two of the present authors have shown that it is possible to develop a generic PSA for multiple plants, and transfer knowledge between them. Thus the claim that the uniqueness of each plant rules out such transfer of learning is old dogma. The authors are optimistic that such a generic cross-plant approach could be very useful in the future development of efficient systems for risk management and communication, transforming the way risks are monitored, understood and managed in critical infrastructure. See Ali Ayoub, Wolfgang Kröger, and Didier Sornette, Generic and adaptive probabilistic safety assessment models: Precursor analysis and multi-purpose utilization, Nuclear Engineering and Technology 54 (8), 2924–2932 (2022); Ali Ayoub, Andrej Stankovski, Wolfgang Kröger, and Didier Sornette, The ETH Zurich Curated Nuclear Events Database: Layout, Event Classification, and Analysis of Contributing Factors, Reliability Engineering and System Safety 213, 107,781 (2021); Precursors and startling lessons: Statistical analysis of 1250 events with safety significance from the civil nuclear sector, Reliability Engineering and System Safety 214, 107,820 (2021).

[3]The present authors would like to point out that even in critical facilities, one can leverage the knowledge of the different equipment at other facilities around the world. In the nuclear industry for example, there are some well-known suppliers providing equipment to all power plants across whole regions (e.g. most plants in the US), so one can deduce industry-wide trends in the equipment and expand the statistical pool. This is already the state of the art approach in risk assessment and component reliability studies in the nuclear industry.

equipment can improve the quality of analytical information about its condition, thus increasing the accuracy of expert recommendations or predictions.

Other respondents also expressed the view that it is difficult to set up automated risk communication systems that are fully effective in the maintenance of critical equipment. Several of the respondents had a correspondence discussion about this.

The vice president of a large electric utility insists that total automation is possible, and that subjective human feedback should be minimized. He explains that he does not believe his subordinates, and accordingly feels that the human factor should ideally be excluded from the process of collecting and transmitting information about risks, in both production and technological exploitation. Hence, critical infrastructure companies need to focus on the creation of reporting systems that can send information about production and technological exploitation directly to headquarters, bypassing the whole management hierarchy. In his experience, as soon as such a system is installed at a production site, middle managers and employees begin to send more objective information to senior management about what is happening. Obviously, people in the field understand it is impossible to hide any piece of information, because any failures will immediately be visible to headquarters. These systems also include video monitoring of the plant, which basically eliminates any deliberate safety violations by operators because they know that everything is being recorded. Many technology companies now implement fully automatic monitoring of production and technological exploitation.

However, the vice president of another electricity company disagrees with this position. His reservation is a practical one: according to his data, 75% of the electric power equipment in the country where his electricity company operates is outdated by modern standards. All modern equipment is designed from the outset with a fully-fledged automatic monitoring system installed, which detects and signals malfunctions without the involvement of maintenance personnel. But the old equipment is mainly monitored manually by specially trained personnel. Sensors can be attached to it, but they will only give data on a few key parameters—the sensor readings will not cover the full spectrum of technological exploitation needed to ensure safe operation. Therefore, in most cases, they must rely on employees who know how to operate obsolete equipment. The potential role of automated systems in monitoring the operation of equipment is very limited indeed if much of it is older. The respondent notes that human expertise still plays a large part in the electric power production process. And in the oil and gas industry—when drilling wells for example—the influence of the human factor is enormous. In this and other industries, it is impossible to fully automate all production processes, so the impact of automatic systems for monitoring critical risks will remain limited.

The head of a power plant makes the point that automating the diagnostics of technological exploitation of a critical infrastructure facility does not turn it into a brand-new facility—it is just that the managers receive more information about how it functions. According to him, everything can be automated to some degree—the question is how valuable the information obtained within the framework of such automation will be, how much this information can improve the safety of the facility. In his opinion, it is still better to invest in modernizing outdated and worn-out equipment rather than automating the monitoring systems within technological exploitation: extra information about "*dead*" hardware will have little value. In other words, it is impossible to fully automate everything—there is always a limit beyond which people must make the final decision on the operation of the equipment.

The HSE director of an electricity company believes that the first step is to understand the cost-effectiveness of automating equipment control systems. With older equipment, even fitting sensors will not always yield information that will help predict its failure. Old equipment already has faults and may be working with a significant discrepancy from the design parameters. An automated system will capture these deficiencies, but that information may be of little value to decision-makers. They already know that the equipment needs to be changed, but a company does not always have the finances to carry out a comprehensive modernization or complete replacement. By automating the monitoring of the technological operation of equipment, a company will get a huge amount of data that needs to be analyzed to inform preventive action. This can either be done by a staff of specially recruited employees or an AI system. This brings us back to the economic feasibility of automation: will all this data analysis help in predicting equipment failure and possible incidents? So far, according to the respondent, there are no effective AI solutions on the market that would show significant economic benefits from automating the monitoring process—in terms of reliably predicting future equipment operation.

The HSE manager of an oil company believes that it is dangerous to depend mainly or entirely on IT systems in the field of risk assessment and equipment monitoring. For example, imagine that a company relies exclusively on automatic systems to minimize or eliminate the human factor from the process of identifying and transmitting risks. If the sensor that collects information about the risk is out of order or there is a broken communication link somewhere, senior management may find themselves in a situation of information blackout.

The HSE manager at a utility services company cites a negative experience at his company. There was a problem on site that was simply not showing up on the automated monitoring systems, and would have been dangerous if it had gone undetected. Were it not for an employee who was doing an on-site inspection, the company would not have noticed this error, which of course meant that the systems may be overlooking other issues too, given that one significant problem was not being picked up. The shortcomings of automation at its current level are compensated by the work of highly qualified employees.

A risk management consultant specializing in oil and gas gives the following example: when the respondent worked offshore, there were many ways to detect gas leaks. However, most of the leaks were detected not by sensors and automated systems, but by people who walked around the platform and could smell the gas. The people looking for the leaks were not naive: they knew where to look for leaks, they knew where there might be trouble. Sensors may not always correctly signal a problem, and automated systems may not always correctly interpret data. Thus competent and experienced technicians add a very important layer of information interpretation, which an automated system cannot yet implement.

The chief risk officer at a national power grid company gives the following example: a good engineer who approaches a working transformer can tell by the sound whether the transformer is working well or has problems. Obviously, it is always good to have experienced specialists like this on staff who can competently analyze a situation based on incomplete information. According to this respondent, it would also be good for a critical infrastructure company to have an automated data flow, and AI to analyze the data, to complement the assessments made by the company's experienced technicians.

The executive at one electric power company considers that the focus on total automation is dangerous, as it makes critical infrastructure facilities very vulnerable in the event of unforeseen emergencies or third-party intervention.

Cyber risks should also be considered when implementing automated systems: with the growth of automation, the risk to companies from network failures and unauthorized access will grow. Cloud storage makes a critical infrastructure company vulnerable to attacks by third parties or in the event of a global emergency. Critical infrastructure can become a hostage in geopolitical instability and sanctions wars between different countries. For example, there may be a remote shutdown of equipment in a country that has come under sanctions. Many hidden remote control systems are built into modern high-tech equipment and it is very difficult to know where they are installed. High automation of modern equipment allows it to be turned off from anywhere in the world, through an automated system or through a satellite signal sent directly to the equipment. There have already been such precedents. Therefore, the development of automation and AI will need to be carefully limited if a company installs foreign software or equipment that could potentially be turned off in the event of a geopolitical confrontation. With risks like this coming into play, it is unwise to fully automate the control and monitoring of a critical infrastructure enterprise. A company should always have the backup of highly qualified employees and autonomous automated control systems, in case there is hostile interference in the operational management system. To be ready for this threat, leaders must develop the competency and loyalty of their employees, so that they can rely on them in the event of a software and hardware shutdown through external interference. In parallel with the automation of risk monitoring, it is important to develop employees, build internal channels for them to report risks to managers, and motivate them to participate in identifying and controlling risks.

The HSE head of an oilfield services company argues for the greater penetration of technology, artificial intelligence, and control systems to support people in the field of safety. One of the disadvantages of total automation is that the instrumentation would then need to be 100% reliable.[4] A high level of instrumentation redundancy—in other words, having multiple instruments monitoring the same equipment and comparing readings—will reduce the number of machine errors. Nevertheless, in the respondent's view, there are still many technological processes that cannot yet be fully automated: *"Perhaps someday we will reach the level where we no longer need people, but at the moment this is still very far away"*.

The managing director of a gas distribution company believes that a semi-automatic system of equipment operation control is the most realistic at the moment, and that a fully automated system may still face serious problems. For example, full automation increases the complexity of the entire control system. Thus in the event of failure, repair is very difficult and will probably involve a complete shutdown of critical infrastructure. This is reminiscent of the difference between old and modern cars. In older cars, most of the component systems are purely mechanical and involve little or no electronic circuitry, so they can be repaired by owners in their own garages. New cars packed with electronics are far more complex and repairing them requires a complete understanding of the interaction of all their parts, which are controlled by a central computer system whose access requires specialized proprietary equipment. Consequently, they can only be repaired by trained professionals in specialized car service centers, with the car brand's specialized hardware computer equipment. In other words, repairing an old car is a much simpler task that can potentially be done using one's own resources. This comparison can also be extended to the repair of critical infrastructure equipment.

The HSE manager of a metallurgical company puts it this way: *"The machine is made by people. The sensor is created by people. People set up the system"*. He maintains that the automation of production is necessary, because machine labor eliminates most of the risks of human injury and error when performing repetitive operations on a production line. For the next few decades at least, machines will never replace people in controlling key risks at industrial plants. But as the level of automation increases, the professionalism and competence of the people who program the machines should grow.

According to the vice president of an electric power company, the longer-term future undoubtedly belongs to equipment with an integrated automatic fault analysis system, with minimal human input. Nevertheless, over the coming decades, the influence of the human factor in industries like electric power will continue,

---

[4]The following experience of one of the authors may be illustrative in this respect, even if not directly related to critical infrastructure. It does however illustrate the issues with automation. The author bought a high-end luxury car. This car came with a fully computerized automated system to detect and report faults. But to his dismay, the author found himself constantly going to the maintenance shop as a result of faults with (I) the sensors, (II) relays between sensors and software and (III) the software. In other words, real mechanical issues were completely overshadowed by the whole automation chain that was supposed to make the car safer and more reliable. The exact opposite happened, making the owner's life miserable! Since then, this author is always trying to find a car with the minimum of automatized fault detection systems—a difficult endeavor in our world of ever more automatized "*black box*" systems.

because the existing obsolete equipment needs the supervision of experienced professionals to assess operational risks and defects. Therefore, technology companies need to simultaneously develop two systems: (I) fully automatic control over equipment where this is possible and (II) a fault and incident log maintained by the employees operating the equipment. Information from both systems should be used as an additional resource to support the traditional structure of lower and middle management reporting to a company leadership.

The vice president of a gas pipeline construction and repair company believes that, at present, priority should be given to people—who are still smarter than machines. IT solutions are just an additional element of support to people's decisions, nothing more.

The HSE senior director of an oilfield service company and the director of operations for an oil company both believe that it is impossible to fully automate the work of a critical infrastructure company: how can you possibly automate or digitize the intuition and experience of thousands of employees? If we are talking about the oil and gas industry—especially hydrocarbon production—there is too much in this industry that is out of reach, too much that cannot be measured, seen, or touched. There are too many variables that cannot be predicted in advance to be controlled by an automated system. Does this mean that partial automation is not needed—absolutely not! There have been huge positive developments in this area in recent years. Nevertheless, this is not full automation.

A lead safety manager working mainly in oil and gas believes that it is impossible at this stage to make a complete and immediate switch to robotics—although there is already a massive reduction in the workforce in the oil fields compared to the numbers in the 1970s. Has it improved production? Yes. However, humans cannot be completely excluded from the oil extraction process. 30 years ago, it was probably around 20% automation and 80% human workforce. Today, probably more like 40% automation and 60% people. However, according to this respondent, it would be impossible to have zero human input in oil production.

According to the head of HSE in a metallurgical company, for many years or even decades into the future, the human factor will remain key in diagnosing and reporting risks at industrial sites.

Prospects for artificial intelligence in the operation of critical infrastructure

The HSE manager of an industrial company which uses hazardous chemical processes highlights two advantages of the introduction of AI: (I) it minimizes the cost of finding optimal solutions and (II) information can be analyzed much more quickly for decision-making. However, he expressed skepticism about extending AI to making decisions in the operation of critical infrastructure. There are many risks if AI is allowed to independently decide on serious operation issues. Hence, the final decision must still be left to professional operators, informed by analytical information from the AI system. AI can be allowed to make secondary decisions, where the scale of any possible damage is limited. It is best used to analyze large amounts of data, creating broader analytics to inform smarter leadership decisions. AI is also helpful when modeling various scenarios for the development of a situation in the future.

The head of HSE in a mining and metallurgical company reported that AI can already be used for customized employee training. Trainees carry a gadget installed with specialized software and train their skills in the workplace. AI makes it possible to simulate different situations at a specific workplace, for a specific employee working with specific equipment, based on data collected on the operation of this equipment over many years. It can also be effective in conjunction with sensors installed on workers' overalls. From their readings, AI can analyze huge amounts of data about an employee's movements, location, breaks and working hours, and state of health at different points in time. All this is done to reduce dangerous conduct by employees and increase labor productivity. Of course, any safety violations will immediately be brought to the attention of the boss and the employee him/herself. This allows for instant correction of both deliberate violations and unwitting mistakes: generally, in the past, there has been a long gap between detecting violations and giving feedback to employees. The launch of 5G will further shorten the transmission time and connect a huge number of sensors—an example of the so-called Internet of Things.

The director of the HSE department at an electricity company also sees potential for the installation of sensors in the overalls of employees, connected to an automated system with AI elements, which would allow the system to always pinpoint their location. This would make it much easier to (I) keep them out of areas which are dangerous to enter without personal protective equipment; (II) monitor their health status—potentially lifesaving for employees who are making inspection rounds alone; and (III) monitor their activities to detect safety violations and poor performance.

The quality director of an international electricity company gives the example of a project currently being implemented in his company. The project is called "*virtual assistant*". The idea is to create a machine that can detect potentially dangerous situations on the construction sites of an electric power company. The machine can now detect whether construction workers are wearing gloves and a helmet, or whether staff are maintaining a safe distance from hazardous equipment. The technology is not yet sufficiently developed for the system to work efficiently. However, once the technological hurdle is overcome, AI will be used extensively to prevent dangerous incidents in the workplace. According to this respondent, this is a promising area which needs more investment.

The managing director of a gas distribution company gives the following example. When monitoring the condition of gas distribution equipment, they use temperature sensors normally used in fire alarm systems. These sensors have been set to a reference temperature—if the actual temperature rises above this reference value, the fire extinguisher systems are started automatically. An obvious potential problem with this system is false positives which can incur significant extra costs, so the company is currently implementing more advanced monitoring schemes to avoid this. They have begun instead to use sensors that measure the rate of temperature rise near critical infrastructure. If there is a sudden and rapid temperature rise, the fire extinguishing system is automatically activated. In this regard, solutions using artificial intelligence can monitor many different indicators, comparing the current values on a sensor with a large array of previous values to avoid false alarms and ensure that alarms go off in genuine emergencies. Vibration and noise phenomena may also need to be monitored for safety. These are not linear but logarithmic in their behavior, so linear algorithms cannot always be applied.

The HSE director of an energy company believes that, for AI to make effective predictions on the operation of equipment, there needs to be a huge array of data over a significant period of time and ideally drawn from multiple pieces of the same equipment. But there are major problems with this: (I) companies only have data for a short period in comparison with the designed lifetime of the equipment, which may be decades; (II) there is no exchange of data on the operation of similar equipment with other critical infrastructure companies, because they see each other as competitors and do not share data about their respective problems, incidents and mistakes; (III) many critical infrastructure facilities are unique, designed for specific customer conditions, locations, supply chains and regulatory constraints—so there is no comparable equipment to provide vast amounts of relevant data.

From his experience in power generation, most current IT solutions focus on collecting obvious information about the current operation of equipment, with limited simulation of possible scenarios for its future performance to predict equipment failures based on the analysis of data from many readings. The introduction of artificial intelligence is meaningless if the intention is not to get an analysis that can predict equipment failure under various operation scenarios. All this requires an array of data, powerful analysis algorithms and highly professional experts who can assess the adequacy of the analysis and correct the software in the event of logical errors. According to this respondent, 70% of operational analysis is currently carried out by people and 30% by machines. Perhaps in the next 10 years, the proportion will reach around 60%/40%. It is difficult to predict what AI solutions will be available to critical infrastructure companies in 20 years. Nevertheless, it is safe to conclude that, for many decades, people will still be the main source of information about the operational problems of technological equipment. Automation will just make their estimates more accurate. Of course, decisions on the operation of critical equipment cannot yet be transferred to AI. For the most part, AI should perform an analytical function, predicting scenarios for how a situation will develop. It can work with operators to control non-critical processes, where erroneous decisions will not lead to significant damage or put people in danger.

The principal grid studies engineer at an electric power transmission company believes that full automation is unlikely in the coming years. Translating the complex information coming from a critical infrastructure company's automated system requires experienced professionals who know exactly how to interpret the results. It is unlikely that AI will ever replace highly skilled experts.[5]

[5] In the view of the present authors, the reason is well understood in machine learning (ML) fields. ML requires the input of a great number of instances that are representative of the patterns that should be learned. However, if the data is not very representative and/or if it varies too widely, then AI will not be able to generalize from it, and ML will not work.

A consultant in nuclear safety with long experience in nuclear power plant operations believes that an automated monitoring system could create new critical risks, because glitches in AI are very hard to detect. In due course, those who created the AI system, based on their years of experience in the industry, will leave the company. Nevertheless, the next generation of people managing the system may not have the same experience of operating critical infrastructure facilities, and will not understand the failure mechanisms and consequences very well. This new generation of people will only do the part of the job they were taught, and the only people who know what's going on behind the scenes are the ones who built the AI system in the first place and have now left the company. This situation will almost certainly lead to the development of new critical risks, which the current staff do not have the experience to manage safely.

A senior HSE advisor and human factors specialist in the oil and gas industry gives the following practical example: during the active automation of the drilling process, there were more software engineers working on the drilling rigs than people operating the equipment and extracting oil. The issue was that there were so many errors in the program code that it led to constant failures, and required a huge staff of programmers in attendance. The respondent is not against automation or the introduction of AI, but he insists that all these innovations should be under human control.

Discussion: Disclosure of Critical Risks to Insurance Companies in Exchange for Reduced Premiums

As part of the in-depth interviews with heads of critical infrastructure companies around the world, the authors wanted to know their views on whether it is worth disclosing the critical risks of their businesses to insurance companies in exchange for lower insurance premiums.

In the study, 93 respondents answered this question. 57 of them (61%) reacted positively to the idea that a critical infrastructure company should disclose to the insurer everything it knows about its own critical risks, in exchange for a reduction in the insurance premium. 24 respondents (26%) expressed skepticism or were against such an exchange. 12 respondents (13%) found it difficult to answer this question.

A fair summary of the tone of some of their responses would be that "*everyone says good things to the insurance company, not bad things*". From this, the authors of the handbook would conclude that a significant number of critical infrastructure companies currently make biased and inadequate reports to insurance companies about their risks.

ARGUMENTS FOR RISK DISCLOSURE IN EXCHANGE FOR A REDUCED INSURANCE PREMIUM

The insurance company becomes an influential safety auditor

The HSE vice president of oilfield services company considers that insurers have an open-minded view of many of the problems faced by industrial companies. As such, they can become free auditors of a company's risks and of the measures taken to reduce them. If they see their role in this way, insurers can exert constant pressure on the senior management of a critical infrastructure company to solve the most serious problems it is facing.

D. Chernov et al., *Averting Disaster Before It Strikes*,
https://doi.org/10.1007/978-3-031-30772-0

The HSE head of a fertilizer mining company cites the following example from his practice. The company took out a large bank loan for the construction of a mine. One of the bank's conditions was full transparency about all the risks of the project. Essentially, the bank was concerned about environmental risks and working with third parties, and they wanted to play a very active part in the mine construction project. At first, the bank's demands seemed excessive and unnecessary to the company. However, after a few years, the company began to appreciate that having a bank on board was helping to prevent many problems. The bank had become a kind of free auditor of project risks, and a consultant on how to avoid them. They forced the company to pay attention to risks and develop measures to reduce them. For example, during the construction, most of the equipment moved through a small village. The bank forced the company to build a bypass road, to reduce the likelihood of protests and of residents turning to the regulators to complain about unbearable noise and dust from trucks. This move was very positively received by the residents, and showed them and the regulators that the company was socially responsible.

In this way, the disclosure of risks to the insurer does result in serious pressure on senior management. If it takes its interest too far, the insurance company can paralyze the work of the infrastructure company's management with hysterical demands to immediately eliminate every disclosed risk. However, one should not completely dismiss the potential benefits of meticulous insurers. They may be the only force that will motivate senior management to continuously focus on risk mitigation: there may not be a force within the company itself that will do this.

According to several respondents, having an insurance company as an additional auditor for safety matters is a win–win option that allows a critical infrastructure company to benefit from an independent external assessment of existing risks and measures to reduce them, as well as saving on insurance.

Full risk disclosure avoids very high insurance premiums

Insurance companies sometimes know that the risks of an industrial facility are very high, but cannot themselves accurately assess them. If the risk is unknown, the insurance company will lay a margin of safety for its estimates, which may be up to tens of percent of the full premium. Insurance companies set the price of an insurance premium based on their knowledge of risk: the more they know about the risk, the lower they can afford to set the premium. Take, for example, two different factories with the same equipment and output. If the insurance company knows the details of only one of the factories, then the plant about which less is known will of course be charged a higher premium, as a compensation for uncertainty—i.e. for the risks that are surmised by the insurer but remain unquantified. Insurance companies study all the possible risks—those they are aware of and those they may not be aware of. Premium pricing is based on these two criteria. Thus in order to minimize the information deficit and prevent their insurers from overestimating risks, it is beneficial for industrial companies to be open with the insurers.

ARGUMENTS AGAINST RISK DISCLOSURE IN EXCHANGE FOR REDUCING THE INSURANCE PREMIUM

If the insurance company finds out about all the risks a company is facing, then it will certainly raise the premium

There is a conflict of interest when two parties have opposing goals and motivations. The insured wants to pay the lowest possible insurance premium. Meanwhile the insurance company, and indeed the insurance industry as a whole, is focused on maximizing its profits. The insurer's task is to sell the policyholders a higher risk assessment than the policyholders themselves would make. Therefore, this proposal is disadvantageous to a critical infrastructure company: once the insurance company has seen all the risks an industrial production company faces, then even if they do not increase the premium immediately, they will do so after a while. Perhaps this will not happen in the early years, while there is still an agreement to reduce premiums in exchange for risk disclosure. However, in the long term, the insurer will have additional arguments to justify increasing the premiums.

According to the HSE manager of a production company managing a large number of hazardous chemical processes, insurance and reinsurance is a very long-term business. Indeed, insurance companies have been running for hundreds of years, adapting to many types of risks and taking the long view. Thus disclosing information about all the company's risks can have negative consequences for the company in the long term, if the insurer hikes up the premiums as soon as the discount period ends. Every year, the equipment is progressively depreciating, and if no decisions are made to reduce or control the known risks, the likelihood of a negative event will keep increasing. Thus if a "*premium amnesty*" is not in force, the premium would inevitably keep going up: insurers are primarily interested in managing their risks, and those risks are obviously increasing, especially if no effective risk reduction program is being implemented in the insured company. However, according to this respondent, it is worth noting that for the insurance industry, the best and most profitable scenario of all with its clients is zero incidents[6]; if a "*no premium hikes in return for full transparency*" deal helps bring that about, it is worth their while.

[6]The present authors need to balance this respondent's view with the reality of how insurance and reinsurance companies set their premiums, based on exchanges with and consulting for professional (re-)insurance companies. If there have been zero incidents or accidents over the previous year(s), insurance companies will be forced to significantly lower the premiums in order to attract customers. This is a well-known problem which comes up, for instance, when insuring very large projects against natural hazards. After years of no significant losses to, say, hurricanes, (re-) insurance companies feel the pain because their premiums are driven down, eroding their margin. What (re-)insurance companies need is one or two severe hurricanes, or accidents, which then justifies their high premiums for the next few years. On the other hand, obviously, they do not want too many large losses, which would then impact their capital reserves. In summary, taken to its extreme, absolute zero incidents or accidents and the convergence to zero risk is a threat to the insurance industry.

Any saving on the insurance premium is negligible compared to the risks of disclosing all risks to a third party

Risk disclosure threatens the profitability of the entire business of a critical infrastructure company: any possible saving on insurance premiums is negligible compared to the potential losses from disclosing all risks to a third party. If this information becomes public, the requirements of regulators to dramatically improve the situation to control critical risks could lead to astronomical costs, the bankruptcy of the company, and even a complete legal ban on the operation of the entire industry. If all the critical risks the company faces are highlighted to the public, then investors will lose interest in buying shares in the company for a long time.

Low trust in the ability of insurance companies to protect their customers' data

Some respondents expresses their lack of confidence in insurance companies regarding their ability to protect customer business data. Even if a critical infrastructure company has entered into a nondisclosure agreement (NDA) with the insurance company, it is not always possible to be sure that confidential information will not leave the insurance company. This is especially true in terms of control over information leakage by those employees who leave the insurance company and go to work for competitors or regulators. For example, suppose an insurance company employee is immersed in all the client company's critical infrastructure problems. Years later, he/she is lured away by the authorities to join the industry regulator's team, where he/she will be overseeing the very technology companies he/she has worked with at the insurance company… And as well as potentially incendiary information about risks, critical infrastructure companies have trade secrets that can become known to the insurance company with full disclosure of all risks.

Some insurance company representatives are not qualified to assess the risks of critical infrastructure facilities

Some respondents are very skeptical about the qualifications of auditors and insurance company representatives. Based on their experience, they conclude that many have very little understanding of their clients' businesses. Instead of experienced professional appraisers, young reps with little experience are often sent to large and complex industrial enterprises to make a comprehensive risk assessment. Unsurprisingly their conclusions are unreliable, containing many errors and misjudgments.

In some countries, insurance companies are not even particularly interested in understanding the business of companies that they insure: they will immediately reinsure their risks with global reinsurance companies in any case. The head of HSE at a mining and metallurgy company shares his negative experience with insurance companies. His company invited insurance representatives on several occasions to discuss the risks faced by their facilities. The company even invited its insurers to professional conferences—but all to no avail! The insurers were just not interested in the issues around better risk control, or in their clients' offer to help them get a better understanding of their business. The explanation for this was simple: the KPIs for insurance agents are about selling as much insurance as possible, to get a high premium for the company and a personal bonus for themselves. Since risks will be transferred to reinsurers, insurance companies have little incentive to delve into the problems of those they have insured. The respondent drily observed that the top managers of insurance companies probably have the same short-term profit KPIs as those of the industrial companies they insure. And for many leaders with these profit-driven KPIs, the consequences of mistakes in their work are less important than getting a decent annual bonus. Some representatives of insurance companies are insurance sellers, not technical experts in the operation of critical infrastructure. According to the respondent, insurance companies do not want to incur the additional costs of ordering an independent technical review of the critical infrastructure companies that they insure. Many are not even interested in whether there have been industrial accidents—they are simply involved as insurance tenders, competing with other insurance companies. Amazingly, some insurance managers do not understand what they are insuring at all.

20 years ago, a risk management consultant in the oil and gas industry had a very negative experience of trying, and failing, to get a lower premium in exchange for risk disclosure. He originally thought that the insurance industry was interested in risk–reduction. However, he discovered that he had not understood the principles of that industry. On one occasion some insurance representatives, very smart people, politely explained to him: "*The closest analogue to the insurance industry is the betting industry. We place bets. We make risky bets and we make safe bets. We are looking for a balance between risky bets and safe bets. We thrive on understanding the nuances... If we are not sure that our risks are adequately redistributed through reinsurance, then we fire the reinsurers*".[7] Insurers are focused on raising insurance premiums to make money as well as ensuring that their "*tail*" risks are under control to protect their capital reserves. When the premium goes down, so do the profits for the insurer.

The authors of the handbook would like to draw the attention of readers to the phrase "*We [the insurance industry] thrive on understanding the nuances*". In other words, a better understanding of risks allows insurers to be sure they are comfortable shouldering the insured risks, with minimal likelihood of large losses or bankruptcy. The insurance industry is driven in general to increase its knowledge of the critical risks of policyholders, but this may have costs for the critical industry. These costs should then be shared by the insurance industry.

---

[7]In fact, this analogy is incomplete. There is a large difference between betting and insurance, as follows. In betting, the lucky "*jackpot*" winner takes only a fraction of the total capital raised by all the lottery tickets sold to gamblers. Thus for the betting organization, the loss is always smaller than the total money raised and there is always a profit with zero risk of bankruptcy. By contrast, in the insurance business a large catastrophe can lead to losses exceeding by many times the total value of all premiums paid over the previous year(s)—possibly leading to a large loss or even the bankruptcy of the insurance company.

HOW TO IMPLEMENT THIS PROPOSAL ABOUT RISK TRANSPARENCY INTO THE PRACTICE OF INSURANCE COMPANIES

Extension of the term of contracts with the insurance company

Some respondents asked: "*If we disclose all our risks, what is the guarantee that, in a year, when the insurance ends, the insurer will not raise the premium?*". One of the respondents suggested the following solution: a critical infrastructure company could enter into a ten-year agreement with an insurance company that all critical risks are disclosed in exchange for reduced premiums. Over that ten-year period, the company will be able to implement a whole program to reduce disclosed risks and thereby reduce insurance costs. And with the insurer on board, the company gets a free external consultant who will persistently demand risk reductions and help find the financing for equipment modernization.

Changing the relationships between critical infrastructure companies, regulators, and insurers

Insurers need to work to earn the trust of their corporate clients. There needs to be a partnership culture between critical infrastructure companies, regulators, and insurers. For a cultural shift towards greater openness to take place, there needs to be trust between these audiences. Otherwise, a company will fear that full disclosure of risk information will be perceived negatively by the regulators and insurers, and will naturally refrain from implementing it.

Changes in legislation

The authors would recommend amending legislation on the insurance of industrial enterprises, so that there is a transparent procedure for disclosing risks and an agreed methodology for reducing premiums in proportion to the level of information transparency shown by the enterprise. It is also important that regulators can work within a framework that allows them not to penalize critical infrastructure companies for fully disclosing the risks they face. In practice, this could be achieved through a task force constituted of members, 1/3 from critical industries, 1/3 from the insurance industry and 1/3 from regulators. This task force would make propositions that would then be sent for comments to the involved parties, with a clearly designed resolution process and strict timeline.

Co-financing of risk control measures

A company and an insurer have a shared interest in reducing the likelihood of risks occurring. Thus, they can jointly finance activities to reduce accident and injury rates. Insurers could co-finance these programs from premiums received from a company: for example, the company pays the insurers the full cost of the insurance, but the insurer allocates a certain percentage to fund measures to control the company's risks. One respondent cites the example of an insurance company who co-financed a safe driving course for his company's drivers. They funded an excellent defensive driving film and other activities.

Involvement of employees at critical infrastructure sites in measures to reduce the likelihood of incidents

One of the respondents proposes an agreement with the insurance company that, if an insured event does not occur because the operators of the insured facility preventively identified the risk and averted any emergencies, the insurer will share a small part (several percent) of the premium with those operators. The employees concerned can accurately describe the identified risk, indicate what was done to reduce it, and calculate the possible damage that would have resulted from its occurrence and the insurance company's savings as a result. All operating personnel become, in effect, internal safety auditors. This will motivate others on the shop floor to identify risks themselves and reduce the likelihood of accidents. It is important that these proactive employees are rewarded publicly, with the amount made known, in order to motivate others to follow suit.

Insurance companies can help find additional funding for critical risk control measures

As part of the interaction of an insurance company with an industrial enterprise, risks can be identified that the enterprise may not have the money to reduce. An insurer could then help a company to attract cheap and long-term loans for measures to reduce the identified risks.

Insurers can bring in professional appraisers for a comprehensive assessment of a company's risks

The insurance company should engage external professional appraisers to make a high-quality risk assessment of any industrial undertaking that has agreed to disclose risks. All participants benefit from this approach, since the risk assessment and survey of mitigation measures are then carried out by (I) managers and experts from the insured company itself, (II) external professional experts and (III) representatives of the insurance company.

PRACTICAL EXAMPLES OF THE IMPLEMENTATION OF THIS IDEA

Respondents gave four examples of how this idea was successfully implemented in practice.

The HSE director of a production company managing a large number of hazardous chemical processes gives the following example from current practice: a critical infrastructure company asked insurance brokers to conduct independent audits of their manufacturing facilities, in order to get an alternative assessment of the condition of equipment. The owners of the company read the reports from these independent auditors. An independent assessment and cost estimate of its critical risks allowed the company to get additional resources for modernization from the owners, beyond the approved annual income and expenses budget. The auditors also showed the company's financiers the potential costs from the onset of risks, as part of their calculations for preventive risk mitigation. For example, the cost of restoring a production workshop after a fire is $600 million, while the modernization of the fire protection system for the workshop would cost just $30 million. Many times, after such an audit from independent experts, the company carried out urgent modernizations, and quickly corrected the identified safety issues. Once these issues were dealt with, auditors representing the insurance brokers confirmed the progress, and the brokers urged the insurance companies to lower premiums from previous years because the company was now better able to control its risks. If the company had told the insurer that all security systems were working in accordance with the requirements, and it had then turned out during an audit that the company had deliberately concealed risks, then the insurance company would have had the right to increase the premium. This motivated the company not to hide risks, but to actively discuss them with independent experts and the insurers.

The SSE manager of a gold mining company relates the following experience. The insurers of a mining company insisted that an independent global risk consultant should travel to a production site to analyze risks and suggest safety improvements. As a result of the consultant's work, any equipment classified as being a fire hazard was completely replaced with equipment made of non-combustible materials. Sprinklers were also installed in critical areas. This modernization reduced the risk of a fire at work and led to the insurance company agreeing to reduce the premium. The respondent believes that this system should be widespread: "*If you don't have accidents, you pay less*".

An HSE manager at several oil companies was involved with a company which had an ineffective and non-centralized fire extinguisher system at its production facility. When he asked what was stopping the company from installing a better system, he was told that the company was worried about the high cost involved. The respondent called the insurance company and asked, "*If we put the risks down, will the premium be lowered?*". The insurance company answered in the affirmative, because operating a facility without a proper fire suppression system is a huge risk. If the object ever catches fire, the insurance company will be forced to make a huge insurance payment. The insurance company recalculated the premium accordingly. The respondent presented their proposals to the management of the company. The difference between the annual premium currently being paid by the company and the proposed lower premium was 1/5 of the cost of installing a high-end fire suppression system. Therefore, if the company agreed to install the new system, they would recoup the installation cost over five years. Not surprisingly, the management decided to do so. In his communication with insurance companies, the HSE manager insists that insurers and policyholders are partners, that they share risks, that their relationship should be transparent. After all, if their mutually beneficial collaboration fails, they both stand to lose.

Another respondent is currently the head of HSE at a chemical company, but previously worked at an oil company which owned the world's fourth-largest oil-field. At that time, the assets of this field alone were approximately US $5–6 billion. The annual insurance premium was approximately $250 million. The respondent was asked to speak with the company's insurers to convince them to lower the premium. Simply by using the company's risk assessments, explaining in detail the results of the QRA (quantitative risk assessment) and demonstrating how much the company had done in controlling its risks, he was able to convince the insurer to reduce the premium by $25 million—10% of the figure the company had been paying.

Discussion: Impact of National Culture on Risk Information Transmission Within Critical Infrastructure Companies

Some of the respondents have worked in several countries and continents. Do they see national and/or cultural differences in the reporting and discussion of risk? All the leaders interviewed who have international work experience do indeed agree that communication about risks within organizations is significantly influenced by the peculiarities of national culture, religion, and worldview. The interviewers asked them to compile their subjective ratings of the quality of internal risk communication in the countries where they have worked. First, the respondents gave examples of countries and cultures where they felt that risk information from subordinates to superiors is significantly distorted in reports. Then they described countries where risk information is transmitted without significant distortion. They explained why they thought some countries and cultures have problems with objective feedback, while in other countries this problem does not seem to be so pronounced. This is, of course, quite a controversial issue. However, most respondents were willing to give their honest response because the interviewers guaranteed their anonymity in any research results published.

COUNTRY RATING FOR THE QUALITY OF INTERNAL RISK INFORMATION

In order not to offend readers by citing unsightly aspects of leadership behavior common in their countries and cultures, the authors will not name specific countries, but rather describe what appear from the interviews to be general characteristics of behavior in various societies and reflections on the reasons underlying them. Many respondents express the view that, in all cultures, people want to present themselves to others in the best possible light. In any society, any group of people, nobody likes to receive bad news—so nobody wants to be the bearer of bad news. The only question is how it is customary in different societies to react to it.

D. Chernov et al., *Averting Disaster Before It Strikes*,
https://doi.org/10.1007/978-3-031-30772-0

There are hierarchies in every society, but management style—the way managers manage their subordinates—differs.

	Risk information is broadcast with significant distortion	Risk information is broadcast with minor distortions
Quality of internal risk communication	Poor	Good
Political regime type	Authoritarian	Democratic
Legal regulation over the actions of government	Little or none	Significant or fully binding
Position of regulators regarding safety violations at critical infrastructure facilities	Punishment of specific employees	Preventing emergencies by punishing a company and senior management
Regulatory fines for safety violations at critical infrastructure facilities	Low	High
Power distance between managers and subordinates in most organizations	Long	Short
Model of communication between managers and subordinates inside most organizations	Monologue	Dialogue
Organizational model	Multilevel rigid hierarchy	Flattening hierarchy
Degree of influence of trade unions on senior management	Low	High
Protection of the rights of employees	Low	High
Culture of compliance with the orders of bosses	High	Low
Attitude of managers to subordinates	Disparaging	Respectful
Trust between managers and subordinates	Low	High
Relationship between subordinates	Collectivism	Individualism

All the interviewees in their own way convey the idea that the key factors affecting the quality of information sent up a company hierarchy are the power distance between managers and employees and, related to that, the traditions of authoritarian (monologue) or democratic (dialogue) governance in the country.

The main thoughts of the respondents can be summarized as follows. In some societies, the power distance between bosses and subordinates is very large and an authoritarian model of management dominates: communication is a monologue from bosses down to their subordinates. Accordingly, in these societies, the quality of transmission of risk information is low, and communication is slow. In other societies, the power differential is smaller and a democratic (dialogue) model of management dominates, so the quality of information reported about risks is high and communication is faster.

Cultures where risk information is reported with significant distortion (poor quality of internal risk communication)

As a rule, employees who communicate this way are influenced by cultural and religious norms dating back centuries, where subordinates could not disobey or openly disagree with the ruler. These societies have long been dominated by authoritarian models of government: all key decisions were made by the supreme ruler, whose will had to be carried out unquestioningly. For hundreds, if not thousands of years, social order has been dictated as a monologue from rulers to subjects. There is often no culture of the rule of law advocated by an independent judiciary—instead everything is determined by the actions and edicts of authoritarian leaders. In such societies, people respect their elders, and elders and those in power are always right and always wise. There tends to be a reverence for ritual, tradition, ancestors, and social hierarchy. To challenge the authority of the ruler in such a highly hierarchical society is to endanger the entire system. People who do so are seen as rocking the boat and threatening the stability of society.

In the past, anyone who expressed doubts about the orders of the ruler or his proposed plans and methods (with rare exceptions it was always he) were at least severely punished, and generally executed. Most subjects, seeing the consequences of disagreement with the ruler or his representatives, understood that the safest way to survive and get on in such a society was to unquestioningly obey orders from the authorities at any level. This habit of submissively accepting orders issued from above has led to the formation of enduring stereotypes in society that people should not object to the supreme power in any way. The only acceptable feedback to the ruler would be praise, flattery, gestures of loyalty or obedience and assurances that his orders are being faithfully executed. Thus for many centuries, only good news could be reported to the authorities. Bad news might be seen to imply that their orders were overly ambitious or unrealistic, and could not be fully executed. Any such suggestion would be highly dangerous if not suicidal.

This extreme and entrenched hierarchy in society still affects how corporations and critical infrastructure companies do business. For decades, the corporate system too has selected only authoritarian leaders and submissive employees. Whether he has inherited his position or shown himself the most ruthless in fighting for it, the boss determines the rules in a company (Business and industry, again reflecting wider society, have been and arguably remain overwhelmingly patriarchal). With the growth of his status, his power and control over an organization's resources grow. When his status is on the wane, his independence in making many decisions and his control of the purse strings decline with it.

If something the boss ordered goes wrong, then his subordinates had better do something fast to meet his expectations. At the same time, they had better not imply, let alone tell him, that his initial decision was a mistake. They cannot say that the decision is unworkable, or that field staff cannot implement it within the parameters defined by their superiors. Unable to discuss the situation with the boss or get the resources to deal with it, subordinates are left alone with an unsolvable problem. It is in this kind of culture that disasters occur, when deputies are afraid to argue with bosses whose ruling does not fit the circumstances. Faced with a problem that cannot be solved and a boss who does not want to hear that, the hapless subordinates are left with little choice but to hide the real situation. They tell authorities what they want to hear, not what is really happening.

In such cultures, disrespectful treatment of frontline workers is also common. At worst, the attitude of managers towards ordinary workers in these cultures could be characterized as dismissive: "*people are expendable*". If employees do not suit their superiors, or show any sign of disloyalty to their decisions, they can easily be fired, and their careers destroyed. The influence of trade unions on senior management is little more than a formality: in most cases, they cannot protect a "*disruptive*" employee in any way. In general, managers do not trust their workers and feel they must constantly monitor them. And employees do not generally trust their superiors. Thus historically, employees have tried to pull together to resist the pressure of an authoritarian and unpredictable boss. For example, they might collectively agree that no one will tell their superiors about the impossibility of solving a problem. The taboo on betraying colleagues is a normal practice in these cultures.

One of the respondents notes that some countries with this management model, according to official statistics, have the lowest injury rates in the world in certain industrial sectors. However for some unaccountable reason, major accidents with many victims regularly occur in these countries. The reason is, of course, perfectly accountable—the management of some critical infrastructure companies simply falsify the reports on many aspects of their work, in order to embellish the real situation to regulators and authorities.

In such societies, a culture of strict discipline has developed: any order from the authorities must be seen to be done. This system has its pluses and minuses.

The main disadvantage is that employees feel they cannot give objective feedback to the boss, because it goes against a national culture of silent obedience to leaders, formed over millennia and perpetuated until the present time. Moreover, they are afraid of losing their jobs: bosses can easily replace one employee with another. Some of the respondents emphasized that in these countries, people will postpone an unpleasant conversation with their superiors until the last possible moment to avoid upsetting them or pointing out previous mistakes. In addition, many employees become blinkered—they get so used to the clear execution of instructions and duties that they cannot adapt to situations that require a non-standard response. Sitting at the top of a ladder in which nobody can give honest feedback to their superiors on the next rung, it is hardly surprising if a company leadership do not adequately understand what is happening on the ground.

The upside of this culture is that managers can be sure their employees will follow their orders and instructions accurately when operating equipment. And senior management can take immediate steps to start getting objective feedback from employees—it is enough to order their obedient subordinates to report fully and honestly what is happening in the field. The CEO should simply declare that senior management want to hear good and bad news. If you explain to the workforce the importance of honest reporting, and assure them that there will be no punishments, they will very clearly comply with the new requirements. If the leadership changes the "*rules of the game*", then employees will quickly adapt. They will diligently report to their superiors because they will now see objective feedback as an important part of their work, along with productivity, diligence and efficiency: feedback will become a mandatory element of production practice. Curiously, the quality of risk communication may well be even better than in companies from countries where dialogue between employees and managers is common.

Another cultural factor which can prevent critical risks being reported to the authorities is the prevalence of a religious or spiritual stereotype, in some societies, that everything—even the onset of an emergency—happens according to fate or divine will. If this attitude is widespread in a society, many critical infrastructure workers will be fatalistic about controlling or reducing the risks they face. They may be very logical in assessing risk situations but resigned about the value of intervention. Thus according to the respondents, some highly religious employees may be less proactive in detecting and taking measures to reduce risks.

Cultures with transitional societies: subordinates distort information about risks in some situations, and send undistorted information in others (average quality of internal risk communication)

There are some societies that are moving from an authoritarian (monologue) to a democratic (dialogue) model of government. The persistence of cultural codes is such that, in most situations, employees still prefer to keep quiet about risks in dealing with their superiors. However despite the traditions of rigid hierarchy in these societies, the cult of the hero fighting for justice is also widespread: some employees may go against the decisions of their bosses and seek solutions to problems, even though they know they are risking their careers.

Fundamentally, the situation will only begin to change when the leadership take the initiative and insists that employees give honest and undistorted feedback. And this will only work if senior managers give guarantees that there will be no penalties for reporting the real situation, and that measures will be taken to resolve the problems highlighted.

Cultures where risk information is reported with minor distortions (high quality internal risk communication)

As a rule, there is more open communication about risks and problems in societies where dialogue between rulers and their subjects has been evolving for tens, and sometimes hundreds of years.[8] In some historical cases, rulers were enlightened enough to accept or even encourage feedback from their subjects, developing a more horizontal governance structure. However, the history of democratizations since 1800 suggests that in most cases, more shared power and better communication in society did not occur because farsighted elites chose it. Rather, this

[8]Ruck, D.J., Matthews, L.J., Kyritsis, T. et al. The cultural foundations of modern democracies, Nature Human Behaviour, 2020, 4, pp. 265–269, https://doi.org/10.1038/s41562-019-0769-1.

evolution was forced on them by the rise of new classes, following missteps by the elite that triggered previously latent factors.[9] But whatever the causes, dialogue between ruler and subjects in these societies was gradually structured through various systems to enable the country's inhabitants to express their opinions. In some cases, greater attention was paid to the education of the population. As people became more educated, they were able to reflect and come to more informed views on how to build the life of their community. More collective structures of administration and legislation emerged, in which any citizen could play a part. As part of this "*bargaining*" between rich and poor, enfranchising the public and establishing representative structures created mechanisms to enforce the promise of sharing future income, and thus stabilized societies.[10] Elites learned not to dismiss their subjects as stupid, uneducated, and unfit to play any part in the direction of society. Moreover, the spread in some societies of the ideology of dialogue in a broader sense was influenced by progressive religion, which postulated the idea of each person having a direct relationship with the higher power, bypassing the religious hierarchy. Gradually, such progressive societies abandoned the stereotypes of hierarchical management in many areas and came to the concept that all people are equal. In some countries, elements of the elite worked to achieve this ideal, creating quite harmonious social states where people have an active and more balanced sense of civic participation. In these societies, overconsumption and a large income gap are socially frowned upon, and many members of the elite communicate on an equal footing with ordinary people. These societies tend to follow the ancient Greek reverence for freedom, political equality, and the dignity of the person.

This model of social behavior carries over to the interaction between industrial company managers and their subordinates. Companies working in these societies attract highly educated professionals, who see themselves as essentially equal to their colleagues, regardless of their respective positions. There is an element of openness between manager and employees, and it is implied that any employee should talk to the boss if they see a problem. It would seem suspicious if they did not, because it is generally agreed that people should discuss both achievements and failures or difficulties. An employee who only ever talks about their achievements just comes across as arrogant and insincere. Thus when a subordinate handling a high-tech production site comes to senior managers to talk about a problem, they value the opinion of this highly professional employee. Managers will not always be competent in every complex technological issue, hence they should listen to the opinion of a professional who is managing critical risks on a daily basis. There are no barriers to communication—any employee is free to contact senior management. These countries follow a proverb saying "*put the fish on the table*": if you have

[9]Daniel Treisman, Democracy by mistake: how the errors of autocrats trigger transitions to freer government, American Political Science Review, 2020, 114 (3), pp. 792–810.
[10]Daron Acemoglu, James Robinson, Economic origins of dictatorship and democracy, New York: Cambridge University Press, 2006.

problem, bring it up and discuss it straight away. And this extends to working relationships, helping to avoid full-blown crises by nipping the problem in the bud.

In these countries, the relationship between the subordinate and the boss generally fits the dialogue model, and an organizational structure is often "*flattened*". Employees will not just do what is dictated from above if they do not agree with it. They will demand that management explain their decision and convince them—and they will debate the matter if they still disagree. Subordinates believe it is normal and beneficial for an organization for them to give objective feedback to managers, and ask them to be involved in solving problems. Steeped in the cultural stereotypes of a democratic society, leaders also see value in the feedback of their subordinates and are ready to calmly accept criticism and be informed by it in making decisions about the problems identified. This develops an atmosphere of dialogue between management and subordinates about the risks a company faces. Of course, nobody likes telling other people about problems. Even in open societies, information about risks is not always promptly and fully reported. Nevertheless, at the level of cultural codes and taboos, subordinates are not expected or compelled to send only good news to the boss, or to solve problems on their own without wasting his or her time.

It is common practice in these societies that people take care of their own lives, and care about the safety of others. Highly qualified professionals will require that all necessary preparatory work be carried out by their assistants according to the safety regulations. Sometimes, though, the individual agency given to ordinary employees and their high professional competence can have a negative side—they can overestimate their capabilities, take risks, and lose control of the situation. Thus when operating critical infrastructure, it makes sense to urge employees to be careful, and discourage them from making decisions on their own without prior consultation with colleagues and superiors.

Open societies are built on trust—which can significantly reduce the cost of control systems, because managers generally trust their employees. Here is an example from one of the respondents when he was training in a cross-cultural communication program as part of his professional development. The instructor of the course asked students in different countries to answer the question of how they would act in a fictional situation: *"Imagine it is late at night, and you are a passenger in a car driven by your friend. The car drives through the city at... 90 km/h, when the speed limit is 50 km/h. Accidentally, the car hits a pedestrian running across the street unexpectedly, nowhere near a pedestrian crossing. The man dies from his injuries even before the ambulance arrives. The police arrives and you are separated from your friend to be interviewed, and asked how fast the car was going. What do you tell the police? Do you (I) say 'I was asleep, I don't know anything about the speed and I didn't see the moment of the accident' or (II) tell the police that your friend was driving way over the speed limit?"*. The teacher shared his observations with the audience. He has asked this question to many people who come from a society where there is a very unequal distribution of power, a strongly collective mindset, weak rule of law and low social responsibility among individual citizens. In most cases, they choose the first option, in order not to lose their friendship by betraying their friend. He has also asked people who come from countries with a more equal distribution of power, a high level of individualism, strong rule of law and a high level of social responsibility among individual citizens. Most of these respondents choose the second option—even though their relationship with their friend would be damaged, because the friend would probably spend several years in prison. In these societies, it is considered normal that people try to reduce their own risks by reporting the unsafe behavior of others to their boss. When employees see a colleague violating safety regulations, they will immediately inform their superiors, even though this may damage or destroy working relationships. This sense that it is okay, in fact laudable, to report the bad behavior of other people allows many problems in society to be nipped in the bud.

These countries have a strong legislative tradition, and the rule of law generally applies even to the most powerful and influential members of society. In the context of critical industry too, actions and decisions at all levels are viewed through the prism of legal prosecution, and it will cost an employee more to hide risks than to report them. By communicating risks to their superiors, board controllers or regulators, ordinary employees minimize their personal risks. And if something is spelled out in an employee's instructions, there will be high legal costs to the employee if the stated requirements are not met. This also concerns, among other things, a company's requirements for employees to disclose information about risks. Employees know they can get into a lot of trouble by not following their job description to the letter because these countries have very formal legal systems with heavy fines and penalties for not following laws and regulations. Also, most workers know that if their bosses try to pressure or intimidate them into hiding risks from senior management, they can turn to powerful trade unions, regulators or in extreme cases law enforcement agencies. These bodies will side with the employee, protect them and their workplace, and force a company to compensate them for psychological damage from pressure and threats. In these countries, regulatory fines for safety violations are enormous—but regulators are focused primarily on preventing negative events. Therefore, it is beneficial for a company to identify risks and make operational decisions before a negative event occurs. Losses from dealing with the aftermath of an accident and the ensuing lawsuits are many times higher than the cost of preventive measures.

However, not everything is so smooth in these societies. Here are some examples of negative behaviors.

One interviewed executive worked for an international company, leading an HSE unit in a country in transition. He regularly sent messages about critical risks in his country to the headquarters of the company, which originated in a dialogue-oriented society; all his reports were accepted calmly by senior management, and no one was punished for revealing a difficult situation on the ground. However, while there was no indication of management displeasure or reluctance to hear negative information, no solutions or resource allocations were forthcoming from headquarters following the reports. When managers at headquarters received a problem report, the response was silence, or at best "*thank you for the information*"—and nothing else happened. Even though they knew in detail about all the problems their field staff were facing, senior managers took no action to solve them in any way. In all the other companies the respondent had worked in, there were punishments for bad news. In this company, which came from a country where dialogue had been instilled for hundreds of years, it was absolutely safe to bring bad news to the leadership... but there was no hope that they would do anything about it.

If a country has very strong trade unions that vigilantly protect the rights of employees, then some employees may exploit this and "*play the system*" in their own selfish interests. For example, they can inform senior management about risks and immediately declare that they refuse to go to work. During forced downtime, they are paid full wages. However, subsequent analysis of the reported risk can show that it did not pose a direct and immediate threat to employees, and that work could have continued in parallel with measures to control the risk. Or again, they can take sick leave on some fictitious and unverifiable ground—stress or anxiety after talking with colleagues, putting their back out, or whatever—in order to receive free wages while they are off. Such actions skew the statistics on injuries at work, and increase information noise, confusing the picture about a company's operational risks.

The HSE director of an oilfield services company reports that, in "*conversational*" cultures, people believe they have the right to question authorities and challenge their orders. Excessive ambition, narcissistic individualism, arrogant overestimation of one's own skills and abilities, even a sense of one's own indestructibility—these can lead to the "*cowboy effect*", where people want to do everything their own way and seem to think that normal procedures and rules do not apply to them. Most dangerously, these maverick rebels can sometimes overestimate their ability to control risks, which can cause incidents and accidents at work. In short, if you want people to do exactly what you say, go to a country with a monologue model of communication between leaders and subordinates; if you want people to be free to give objective feedback, take risks and challenge current procedures, then go to a country where dialogue is promoted.

The head of risk management in a renewable energy company working in several countries with a more open communication culture maintains that, in his experience, this does not always lead to a "*cowboy effect*" in the workplace. On the contrary, in some "*overly cautious*" countries, people are very prudent about security issues, pessimistic in their outlook and unwilling to take risks. This attitude sometimes leads people to exaggerate the magnitude of risks, or the likelihood of them occurring. Consequently, in these countries people are more willing to reveal bad news and risks but are afraid to take risks. In this regard, the vice president of HSE for a global oilfield services company shared some of his observations. The "*overly cautious*" countries are the best in the world for the free transmission of risk information from employees to managers, due to their very flat organizational structures and the widespread culture of dialogue. The distance between managers and competent employees is very short. Employees understand that the more they say about risks the more reliable the system will become, because managers will make the right decisions in time based on the best possible information. But this can create a problem for the production plan. Leaders from "*overly cautious*" countries cannot take risks—thus in the respondent's experience, if they are put in charge of oil production units, they tend to set very modest production indicators. It is great to work with people from these countries in the field of safety, but you should not expect high production performance from them.

Risk information is not reported because employees lack a basic understanding of the risks they are running (off the scale)

Some of the respondents cite examples of countries where the level of literacy and education of ordinary workers is so low that many lack even a basic understanding of how complex technological systems work. Uneducated people simply do not understand the risks.

One respondent gives the following example. The director of one mine, in a country thousands of kilometers away from headquarters, estimates that 65% of the ordinary workers of the mine cannot even read their name. Most of the workforce are young people with little education and no work experience. If the manager explains in detail what they need to do, they will do it without question. These employees do not have the experience to compare what the manager expects there with work practice at other mines. They take what the manager says as "*the gospel truth*", beyond doubt. Therefore, it makes little sense to ask whether they are hiding or reporting risks. Most of them simply cannot provide any relevant feedback to the manager: they simply do not understand the risks and complexities of the technological systems at the mine. In this regard, another respondent had worked in a company where they introduced a risk assessment scale among miners who had no education at all. Instead of text and slogans, the mining company use crocodile images on the mining equipment. One crocodile is low risk, two crocodiles are medium, three crocodiles are high. Ordinary miners are advised not to approach equipment with more than one crocodile. Only specialists with training and work experience have approval to work on this equipment.

The high mortality rate in these countries, especially among newborns, makes many people fatalists. Death is a phenomenon that people encounter every day in the circle of their relatives and friends. Coupled with the extreme poverty of the population and the fierce competition for jobs among men, this makes many employees "*fearless*" when working in hazardous industries. To complete a task or meet a target, to earn at least some money, they will take unreasonable risks and violate the simplest safety rules. The main thing for many employees is to bring money home and feed the family; it does not matter whether they die or get injured. In the public perception, such employees are heroes, risking their lives to ensure the survival of their loved ones. With this combination of a low educational level, a fatalistic resignation about death or injury and the urgent priority of earnings, ground level workers simply do not report problems at work. In most cases, they do not even think about improving conditions or making the operation more efficient. If the problem is going to affect their earnings, they will go straight to the supervisor looking for a solution. For example, if equipment grinds to a halt and the operators will not be paid for down time, they will complain. However, if it is more or less working, they will not pay attention to any deviations.

CROSS-INFLUENCE OF DIFFERENT NATIONAL CULTURES WORKING AT THE SAME INDUSTRIAL SITE

People with different national cultures can influence each other, forming a new company culture. Here is a real-world case study from the vice president of HSE for a global oilfield services company. An ordinary employee died at one of the fields. This was a mine, run by two managers from the middle tier: one from a country dominated by dialogue and one from a country dominated by monologue. They had worked together for 15 years. Both agreed not to tell senior management at headquarters about the death. The company had a corporate culture that promoted calm discussion of any problems that arise. However, these two leaders, despite their national differences in culture, had worked together so well over the years that they had formed their local culture of hiding risks at this field.

A manager responsible for the operation and maintenance of turbomachinery in an electric power company operating across several countries has also noticed this effect. He says that when international companies operate in countries with an authoritarian type of political system, they hire specialists from all over the world as well as residents to work in their offices. This mixture of employees creates a unique corporate culture where people do not work entirely in the manner of the host country, because they are used to working according to the values of an international company and their own cultural background. Company leaders from the host country, who are used to monologue, may behave according to the principles of dialogue communication, and leaders who are used to dialogue in their own country may follow the monologue model common in this country. This exchange informs the behavior of managers across the spectrum of national cultures, giving them a broader scope in their decision-making.

MOVING SAFETY MANAGEMENT MODELS TO DIFFERENT COUNTRIES

There are many types of civilizations and nation states—the latter are a more recent development, since the state sovereignty principle rooted in the Peace of Westphalia in 1648—with very different characteristics. The scale and reach of their values is also different.

The head of HSE in a multinational chemical company says that senior managers at their headquarters have the illusion that the whole world operates according to the same laws as the parent country. Wherever a company has subsidiaries, they expect business to be conducted in the same way and assume that this approach should suit the workforce at all their sites. This is certainly not the case, and the reality of work in other countries soon corrects these prescriptive ideas and expectations.

In fact, it is often impossible to transfer successful technologies from one country to another. Each culture has its own characteristics of behavior that leave a significant imprint on how people behave in risky situations. Leaders need to understand these differences well, and take them into account when managing people in different countries. Therefore, it is also worth adapting solutions that are effective at one site in one country when implementing them at another site elsewhere. You cannot come into a country and impose cultural patterns that completely contradict the national culture there.

Managers need to have a good understanding of cultural differences to maximize the benefits and minimize the downsides. A large industrial company should have sociologists, psychologists, and political scientists on its staff so that management messages are correctly accented, and new ideas or initiatives are correctly "*planted*" into the heads of employees in different countries.

Here is an example from the oil industry. When employees were trained to improve their driving skills, those from countries with a dialogue culture lost in the driving skills competition to their colleagues from countries where monologue dominates. The former took risks which they were not asked or expected to take during the driving tests, while the latter did not. Not surprisingly, people used to more prescriptive and authoritarian leadership followed the instructions they were given, and those used to greater individual liberty took the initiative beyond what was required of them, taking unnecessary risks when driving.

In one country, it is normal for a passenger to drink beer in a car, in another country other motorists may well call the police and report them. In one country, a driver who calls the police for something like this would be called a "*snitch*" or a *"busybody"*. In another, it will be assumed that the caller is worried that the driver with the drinking passenger will succumb to the temptation to have a drink too and cause an accident: such vigilant care will not be considered officious but will meet general approval. After all, the caller is thinking about the safety of the whole society, and such proactive behavior can prevent critical developments in many situations. Both attitudes are embedded in the cultural code of their respective countries.

One of the executives interviewed believes that the level of safety culture that exists more widely in a society affects the level at industrial facilities there. A company may try to create a very high safety culture at its sites but, after leaving the gate of the plant, an employee finds himself in a society where violation of safety rules is the norm. For example, violation of traffic rules when driving: it may be generally agreed that it is safest to drive at the speed of the flow, rather than sticking to the speed limit. Thus, employees at enterprises that create advanced labor protection systems are forced to violate driving safety rules in their daily life to comply with the usual behavior of motorists in their country. In a similar way, the public safety culture can negatively influence employee behavior within an enterprise. The opposite can also happen: if the wider society pays increased attention to life safety issues, then within companies working in such societies, employees will be more receptive to advanced solutions in the field of occupational health and safety.

One interviewee is the HSE head of an oilfield services company, with more than 40 years of industry experience in many countries. He also believes that in a global company employing people from a wide range of cultures, a company's culture may not be able to override the norms of social behavior previously laid down for particular individuals, or the environment in the countries where a company operates.

However, another senior manager in the HSE field disagrees with this. In his opinion, it is necessary to draw a clear dividing line between a society and the industrial enterprises working within it. The point is that competent management of safety issues can neutralize the negative national characteristics of a society in matters of safety. He gives an example of this from his practice. In his youth, when he was serving in the army as a sergeant and deputy platoon commander, he oversaw people of different nationalities, psychological types, and so on. And if the commander cannot cope with bringing such a motley crew of subordinates to a common denominator, it is the commander's fault, not the soldiers'. Returning to industrial safety issues, he maintains that everything depends on the lead set by owners and shareholders—any national factor is secondary. But some leaders who cannot get their disparate employees to work as a coherent team tend to blame not themselves, but the conflicting national characteristics of the employees.

The respondent cites a multinational chemical company that operates plants all over the world, in countries at all levels of development. In order to direct the entire workforce towards common safety goals, the company tries to consider country specificities, and to find ways of influencing the corporate culture to neutralize any negative impact of national and regional culture. As a result, the safety culture prevalent among employees at their sites can be significantly different from the stereotypes and principles prevalent in the populations outside. The respondent gave another example of several coal mines in a country with a strong tradition of dialogue between managers and employees. In these mines, the rank-and-file miners are mainly immigrants from transitional or authoritarian societies. The mines have a fantastic level of safety. In this instance, people with a very different cultural background are coming to an enterprise where the principle of dialogue in communication is widespread, and accepting the safety principles and rules established there—despite their own national characteristics, which might normally influence their attitudes towards safety.

On the contrary, even if an industrial plant is in a country where public safety is taken very seriously, an aggressive management strategy to maximize profit and performance can create an extremely unfavorable safety culture. Examples like these indicate that employees and managers, with different national stereotypes in the field of safety, are adjusting to the corporate rules and requirements of owners in the country where they work. Thus, this respondent maintains that any national stereotypes that would tend to encourage unsafe behavior can be overridden by the competent work of leaders to build a safety culture that is distinct from national characteristics. Accordingly, the respondent advises the leaders of critical infrastructure companies to immerse themselves in the specificities of local cultures to understand how they can influence local staff in a way that works with their culture. Injury and accident levels do not need to differ from country to country. The external environment and values of the society where an industrial plant is located can and do have a strong impact on safety there—but only in enterprises where owners and senior management have not engaged in safety, and done the work to cultivate a strong safety culture of their own.

Several interviewees share the observation that, if a company's senior management are actively involved in maintaining very high safety standards, employees recruited from different countries and cultures will gradually learn to behave in accordance with those standards. Gradually, the corporate culture of a company "*wears down*" the cultural characteristics of individuals: 80% of employees' actions begin to be determined by the norms of the system, and national cultural habits make up only 20%. Respondents have worked or visited enterprises with an amazing safety record in countries where the national mindset does not strongly prioritize public or occupational safety. The main reason for the success of these enterprises is the safety leadership of the senior management, and the persistence of their efforts over a long period of time. Thus, managers should not focus on the perceived national and cultural characteristics of employees—they should set a high bar in the field of safety, and purposefully move towards this goal for many years. If they do, they will achieve a positive result. It all depends on the desire of the owners and the readiness of senior management.

Printed in the USA
CPSIA information can be obtained
at www.ICGtesting.com
LVHW010747161123
764114LV00004B/7